A Text Book of ENGINEERING MATHEMATICS

VOLUME - II

Dr. Rajesh Pandey
M.Sc., Ph.D.
Assistant Professor/Reader
Department of Mathematics
Sherwood College of Engineering,
Research and Technology Lucknow
Faizabad Road, Barabanki (U.P.)

Lucknow

Published by

Khushnuma Complex Basement
7, Meerabai Marg (Behind Jawahar Bhawan)
Lucknow 226 001 U.P. (INDIA)
Tel. : 91-522-2209542, 2209543, 2209544, 2209545
Fax : 0522-4045308
E-Mail : word-press@hotmail.com

First Edition 2010

Price: Rs. 275/-

ISBN 978-93-80257-12-9

Composed & Designed at :
Panacea Computers
3rd Floor, Agrawal Sabha Bhawan
Subhash Mohal, Sadar Cantt. Lucknow-226 002
Tel. : 0522-2483312, 9335927082, 9452295008
E-mail : prasgupt@rediffmail.com

Printed at:
Salasar Imaging Systems
C-7/5, Lawrence Road Industrial Area
Delhi - 110 035
Tel. : 011-27185653, 9810064311

Dedicated
to
LORD
SHRI HARI VISHNU

SYLLABUS

U.P. Technical University, Lucknow

[Revised Syllabus: w.e.f. the Session 2009-2010]

Unit - I : Differential Equations

Linear differential equations of nth order with constant coefficients, Complementary functions and particular integrals, Simultaneous linear differential equations, Solution of second order differential equation by changing dependent and independent variables, Method of variation of parameters, Applications to engineering problems (without derivation).

Unit – II : Series Solutions and Special Functions

Series solution of ordinary differential equations of 2nd order with variable coefficients (Frobenius Method), Bessel and Legendre equations and their solutions, Properties of Bessel functions and Legendre polynomials.

Unit – III : Laplace Transform

Laplace transform, Existence theorem, Laplace transform of derivatives and Integrals, Inverse Laplace transform, Unit step function, Dirac delta function, Laplace transform of periodic functions, Convolution theorem, Application to solve simple linear and simultaneous differential equations.

Unit – IV : Fourier Series and Partial Differential Equations

Periodic functions, Trigonometric series, Fourier series and period 2π, Eulers formulae, Functions having arbitrary period, Change of interval, Even and odd functions, Half range sine and cosine series, Harmonic analysis.

Solution of first order Lagrange's linear partial differential equations, Linear partial differential equations with constant coefficients of 2nd order and their classifications – parabolic, elliptic and hyperbolic with illustrative examples.

Unit – V : Applications of Partial Differential Equations

Method of separation of variables for solving partial differential equations, Wave equation up to two dimensions, Laplace equation in two-dimensions, Heat conduction equations up to two-dimensions, Equations of transmission lines.

Preface

Mathematics has proved itself as a fundamental discipline for all branches of science, Engineering and Technology with ever-spreading applications in all spheres. Thus, the need for conceptualizing the framework followed by ratioal applications becomes imperative. This book has been designed to serve as a text for students of engineering which covers the syllabus of second semester mathematics for the engineering students of various universities, with special reference to U.P. Technical University Lucknow (U.P). The book can be used as a reference book for degree, Honours and postgraduate students of all Indian Universities and also for I.A.S. and P.C.S competitions.

The author is indebted to the faculty members and other friends for the advice and their help in preparing the manuscript of the book. I am extremely thankful to Prof. Vijay Kumar Kapoor, Retd. Professor and Head of Department of Applied Science and Humanities, Kamla Nehru Institute of Technology (Autonomus Institute of Govt. of U.P.) Sultanpur U.P. for suggestion & opinion, encourragement time to time.

I am very greatful to Dr. G.C Chaubey for valuable and useful suggestions.

I offer my deepest and most sincere gratitude to my parents and my wife Smt. Arti Pandey, daughter Sakshi Pandey and son Tejasvi Kumar Pandey, who have been constant sources of inspiration and encouragement throughout the entire period during this endure.

I take this opportunity to express my deep sense of gratitude to Mr. Suneel Gomber, Director, Word Press (Publishing Division) and the other staff of Word Press for publishing this book in short time.

Any suggestions for improving the contents will be appreciated.

Dr. Rajesh Pandey
Email: rajesh.pandy69@gmail.com

Basic Results and Concepts

I. GENERAL INFORMATION

1. Greek Letters Used

α alpha	θ theta	κ kappa	τ tau
β beta	ϕ phi	μ mu	χ chi
γ gamma	ψ psi	ν nu	ω omega
δ delta	ξ xi	π pi	Γ cap. gamma
ε epsilon	η eta	ρ rho	Δ cap. delta
i iota	ζ zeta	σ sigma	Σ cap. sigma
	λ lambda		

2. Some Notations

$\in$ belongs to	$\cup$ union	$\notin$ doesnot belong to
$\cap$ intersection	$\Rightarrow$ implies	/ such that
$\Leftrightarrow$ implies and implied by		

3. Unit Prefixes Used

Multiples and Submultiples	Prefixes	Symbols
10^3	kilo	k
10^2	hecto	h
10	deca	da
10^{-1}	deci*	d
10^{-2}	centi*	c
10^{-3}	milli	m
10^{-6}	micro	μ

* The prefixes 'deci' and 'centi' are only used with the metre, e.g., Centimeter is a recognized unit of length but Centigram is not a recognized unit of mass.

4. Useful Data

$e = 2.7183$	$1/e = 0.3679$	$\log_e 2 = 0.6931$	$\log_e 3 = 1.0986$
$\pi = 3.1416$	$1/\pi = 0.3183$	$\log_e 10 = 2.3026$	$\log_{10} e = 0.4343$
$\sqrt{2} = 1.4142$	$\sqrt{3} = 1.732$	1 rad. $= 57^0 17' 45''$	$1^0 = 0.0174$ rad.

5. Systems of Units

Quantity	F.P.S. System	C.G.S. System	M.K.S. System
Length	foot (ft)	centimetre (cm)	metre (m)
Mass	pound (lb)	gram (gm)	kilogram (kg)
Time	second (sec)	second (sec)	second (sec)
Force	lb. wt.	dyne	newton (nt)

6. Conversion Factors

1 ft. = 30.48 cm = 0.3048 m	1m = 100 cm = 3.2804 ft.
1 ft^2 = 0.0929 m^2	1 acre = 4840 yd^2 = 4046.77 m^2
1ft^3 = 0.0283 m^3	1 m^3 = 35.32 ft^3
1 m/sec = 3.2804 ft/sec.	1 mile /h = 1.609 km/h.

II. ALGEBRA

1. Quadratic Equation : $ax^2 + bx + c = 0$ has roots

$$\alpha = \frac{-b + \sqrt{(b^2 - 4ac)}}{2a}, \ \beta = \frac{-b - \sqrt{(b^2 - 4ac)}}{2a}$$

$$\alpha + \beta = -\frac{b}{a}, \quad \alpha\beta = \frac{c}{a}.$$

Roots are equal if $b^2 - 4ac = 0$

Roots are real and distinct if $b^2 - 4ac > 0$

Roots are imaginary if $b^2 - 4ac < 0$

2. Progressions

(i) Numbers a, a + d, a + 2d. ..... are said to be in Arithmetic Progression (A.P.)

Its nth term $T_n = a + \overline{n-1}\,d$ and sum $S_n = \frac{n}{2}(2a + \overline{n-1}\,d)$

(ii) Numbers a, ar, ar^2, ... are said to be in Geometric Progression (G.P.)

Its nth term $T_n = ar^{n-1}$ and sum $S_n = \frac{a(1-r^n)}{1-r}$, $S_\infty = \frac{a}{1-r}$ $(r < 1)$

(iii) Numbers 1/a, 1/(a + d), 1/(a + 2d),.... are said to be in Harmonic Progression (H.P.) (i.e., a sequence is said to be in H.P. if its reciprocals are in A.P. Its nth term $T_n = 1/(a + \overline{n-1}\,d)$.)

(iv) If a and b be two numbers then their

Arithmetic mean = $\frac{1}{2}(a + b)$, Geometric mean = $\sqrt{ab}$, Harmonic mean = 2ab/(a + b)

(v) Natural numbers are 1,2,3 ...,n.

$$\Sigma n = \frac{n(n+1)}{2}, \ \Sigma n^2 = \frac{n(n+1)(2n+1)}{6}, \ \Sigma n^3 = \left\{\frac{n(n+1)}{2}\right\}^2$$

(vi) Stirling's approximation. When n is large $n! \sim \sqrt{2\pi n}\ .\ n^n\ e^{-n}$.

3. Permutations and Combinations

$$^nP_r = \frac{n!}{(n-r)!};\ ^nC_r = \frac{n!}{r!\,(n-r)!} = \frac{^nP_r}{r!}$$

$$^nC_{n-r} = {^nC_r},\ ^nC_0 = 1 = {^nC_n}$$

4. Binomial Theorem

(i) When n is a positive integer

$(1+x)^n = 1 + {^nC_1}x + {^nC_2}x^2 + {^nC_3}x^3 + ... + {^nC_n}x^n$.

(ii) When n is a negative integer or a fraction

$$(1+x)^n = 1 + nx + \frac{n(n-1)}{1.2}x^2 + \frac{n(n-1)(n-2)}{1.2.3}x^3 + ...\infty.$$

5. Indices

(i) $a^m . a^n = a^{m+n}$

(ii) $(a^m)^n = a^{mn}$

(iii) $a^{-n} = 1/a^n$

(iv) $n\sqrt{a}$ (i.e., nth root of a) $= a^{1/n}$.

6. Logarithms

(i) Natural logarithm log x has base e and is inverse of e^x.

Common logarithm $\log_{10}x = M \log x$ where $M = \log_{10}e = 0.4343$.

(ii) $\log_a 1 = 0$; $\log_a 0 = -\infty (a > 1)$; $\log_a a = 1$.

(iii) $\log(mn) = \log m + \log n$; $\log(m/n) = \log m - \log n$; $\log(m^n) = n \log m$.

III. GEOMETRY

1. Coordinates of a point : Cartesian (x ,y) and polar (r , θ).

Then $x = r\cos\theta$, $y = r\sin\theta$

or $r = \sqrt{(x^2+y^2)}$, $\theta = \tan^{-1}\left(\frac{y}{x}\right)$.

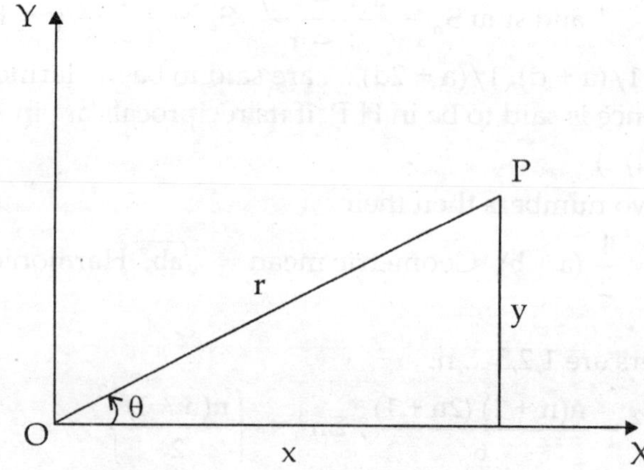

x

Distance between two points

(x_1, y_1) and $(x_2, y_2) = \sqrt{[(x_2 - x_1)^2 + (y_2 - y_1)^2]}$

Points of division of the line joining (x_1, y_1) and (x_2, y_2) in the ration $m_1 : m_2$ is

$$\left(\frac{m_1x_2 + m_2x_1}{m_1 + m_2}, \frac{m_1y_2 + m_2y_1}{m_1 + m_2}\right)$$

In a triangle having vertices (x_1, y_1), (x_2, y_2) and (x_3, y_3)

(i) area $= \frac{1}{2}\begin{vmatrix} x_1 & y_1 & 1 \\ x_2 & y_2 & 1 \\ x_3 & y_3 & 1 \end{vmatrix}$.

(ii) Centroid (point of intersection of medians) is

$$\left(\frac{x_1 + x_2 + x_3}{3}, \frac{y_1 + y_2 + y_3}{3}\right)$$

(iii) Incentre (point of intersection of the internal bisectors of the angles) is

$$\left(\frac{ax_1 + bx_2 + cx_3}{a + b + c}, \frac{ay_1 + by_2 + cy_3}{a + b + c}\right)$$

where a, b, c are the lengths of the sides of the triangle.

(iv) Circumcentre is the point of intersection of the right bisectors of the sides of the triangle.

(v) Orthocentre is the point of intersection of the perpendiculars drawn from the vertices to the opposite sides of the triangle.

2. Straight Line

(i) Slope of the line joining the points (x_1, y_1) and $(x_2, y_2) = \frac{y_2 - y_1}{x_2 - x_1}$

Slope of the line $ax + by + c = 0$ is $-\frac{a}{b}$ i.e., $-\frac{\text{coeff, of } x}{\text{coeff, of } y}$

(ii) Equation of a line:

(a) having slope m and cutting an intercept c on y – axis is $y = mx + c$.

(b) cutting intercepts a and b from the axes is $\frac{x}{a} + \frac{y}{b} = 1$.

(c) passing through (x_1, y_1) and having slope m is $y - y_1 = m(x - x_1)$

(d) Passing through (x_1, y_2) and making an $\angle\theta$ with the x – axis is

$$\frac{x - x_1}{\cos\theta} = \frac{y - y_1}{\sin\theta} = r$$

(e) through the point of intersection of the lines $a_1x + b_1y + c_1 = 0$ and $a_2x + b_2y + c_2 = 0$ is $a_1x + b_1y + c_1 + k(a_2x + b_2y + c_2) = 0$

(iii) Angle between two lines having slopes m_1 and m_2 is $\tan^{-1}\frac{m_1 - m_2}{1 - m_1m_2}$

Two lines are parallel if $m_1 = m_2$

Two lines are perpendicular if $m_1m_2 = -1$

Any line parallel to the line $ax + by + c = 0$ is $ax + by + k = 0$

Any line perpendicular to $ax + by + c = 0$ is $bx - ay + k = 0$

(iv) Length of the perpendicular from $(x_1, y_1)$of the line $ax + by + c = 0$. is $\frac{ax_1 + by_1 + c}{\sqrt{(a^2 + b^2)}}$.

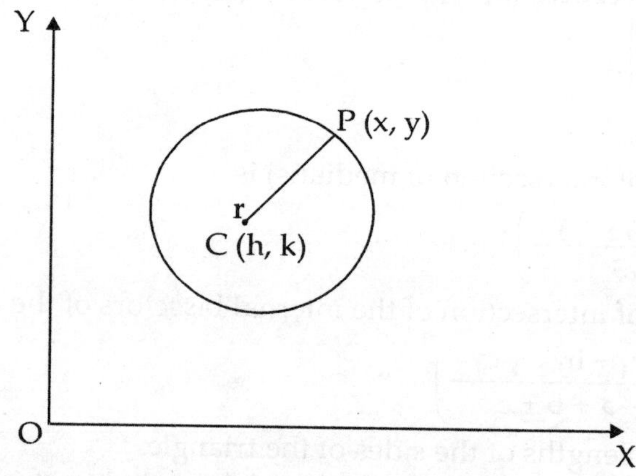

3. Circle

(i) Equation of the circle having centre (h, k) and radius r is $(x - h)^2 + (y - k)^2 = r^2$

(ii) Equation $x^2 + y^2 + 2gx + 2fy + c = 0$ represents a circle having centre (–g, –f) and radius = $\sqrt{(g^2 + f^2 - c)}$.

(iii) Equation of the tangent at the point (x_1, y_1) to the circle $x^2 + y^2 = a^2$ is $xx_1 + yy_1 = a^2$.

(iv) Condition for the line $y = mx + c$ to touch the circle $x^2 + y^2 = a^2$ is $c = a\sqrt{(1 + m^2)}$.

(v) Length of the tangent from the point (x_1, y_1) to the circle $x^2 + y^2 + 2gx + 2fy + c = 0$ is $\sqrt{(x_1^2 - y_1^2 + 2gx_1 + 2fy_1 + c)}$.

4. Parabola

(i) Standard equation of the parabola is $y^2 = 4ax$.

Its parametric equations are $x = at^2$, $y = 2at$.

Latus – rectum LL' = 4a, Focus is S (a,0)

Directrix ZM is $x + a = 0$.

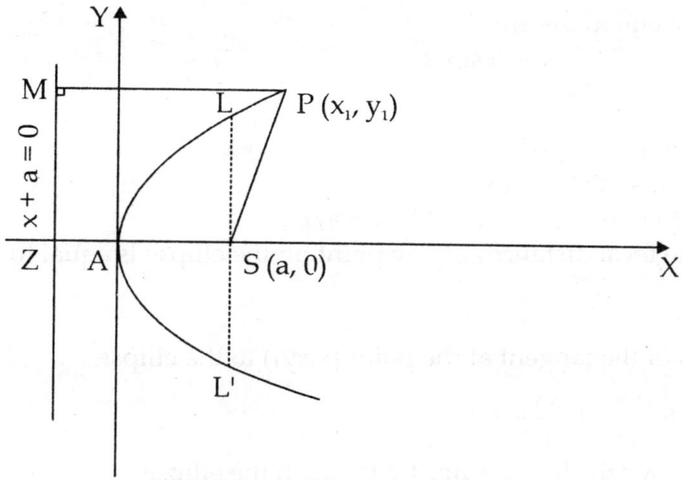

(ii) Focal distance of any point P (x_1, y_1) on the parabola
$y^2 = 4ax$ is $SP = x_1 + a$
(iii) Equation of the tangent at (x_1, y_1) to the parabola
$y^2 = 4ax$ is $yy_1 = 2a(x + x_1)$
(iv) Condition for the line $y = mx + c$ to touch the parabola
$y^2 = 4ax$ is $c = a/m$.
(v) Equation of the normal to the parabola $y^2 = 4ax$ in terms of its slope m is
$y = mx - 2am - am^3$.

5. Ellipse

(i) Standard equation of the ellipse is

$$\frac{x^2}{a^2} + \frac{y^2}{b^2} = 1.$$

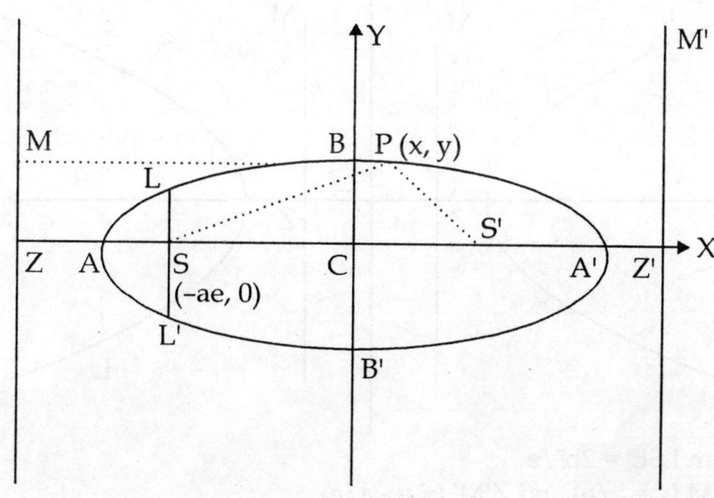

Its parametric equations are

$x = a \cos \theta, \qquad y = b \sin \theta.$

Eccentricity $e = \sqrt{(1 - b^2 / a^2)}$.

Latus - rectum LSL' = $2b^2/a$.

Foci S (— ae, 0) and S' (ae, 0)

Directrices ZM (x = - a/e) and Z'M' (x = a/e.)

(ii) Sum of the focal distances of any point on the ellipse is equal to the major axis i.e.,

SP + S'P = 2a.

(iii) Equation of the tangent at the point (x_1, y_1) to the ellipse

$\frac{x^2}{a^2} + \frac{y^2}{b^2} = 1$ is $\frac{xx_1}{a^2} + \frac{yy_1}{b^2} = 1$.

(iv) Condition for the line y = mx + c to touch the ellipse

$\frac{x^2}{a^2} + \frac{y^2}{b^2} = 1$ is $c = \sqrt{(a^2m^2 + b^2)}$.

6. Hyperbola

(i) Standard equation of the hyperbola is

$$\frac{x^2}{a^2} - \frac{y^2}{b^2} = 1.$$

Its parametric equations are

$x = a \sec \theta, y = b \tan \theta.$

Eccentricity $e = \sqrt{(1 + b^2 / a^2)}$,

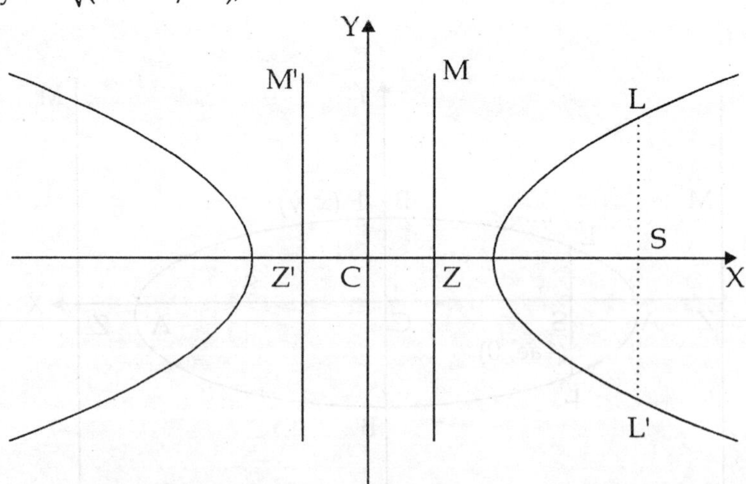

Latus - rectum LSL' = $2b^2/a$.

Directrices ZM (x = a/e) and Z'M' (x = - a/e).

(ii) Equation of the tangent at the point (x_1, y_1) to the hyperbola

$\frac{x^2}{a^2} - \frac{y^2}{b^2} = 1$ is $\frac{xx_1}{a^2} - \frac{yy_1}{b^2} = 1$.

(iii) Condition for the line y = mx + c to touch the hyperbola

$\frac{x^2}{a^2} - \frac{y^2}{b^2} = 1$ is $c = \sqrt{(a^2m^2 - b^2)}$

(iv) Asymptotes of the hyperbola $\frac{x^2}{a^2} - \frac{y^2}{b^2} = 1$ are $\frac{x}{a} + \frac{y}{b} = 0$ and $\frac{x}{a} - \frac{y}{b}$ 0.

(v) Equation of the rectangular hyperbola with asymptotes as axes is $xy = c^2$. Its parametric equations are x = ct, y = c/t.

7. Nature of the a Conic

The equation $ax^2 + 2hxy + by^2 + 2gx + 2fy + c = 0$ represents

(i) a pair of lines, if $\begin{vmatrix} a & h & g \\ h & b & f \\ g & f & c \end{vmatrix}$ $(= \Delta) = 0$

(ii) a circle, if a = b, h = 0, $\Delta \neq 0$

(iii) a parabola, if $ab - h^2 = 0$,c $\Delta \neq 0$

(iv) an ellipse, if $ab - h^2 > 0$, $\Delta \neq 0$

(v) a hyperbola, if $ab - h^2 < 0$, $\Delta \neq 0$

and a rectangular hyperbola if in addition, a + b = 0.

8. Volumes and Surface Areas

Solid	**Volume**	**Curved Surface Area**	**Total Surface Area**
Cube (side a)	a^3	$4a^2$	$6a^2$
Cuboid (length l, breadth b, height h)	lbh	$2(l + b)h$	$2(lb + bh + hl)$
Sphere (radius r)	$\frac{4}{3}\pi r^3$	–	$4\pi r^2$
Cylinder (base radius r, height h)	$\pi r^2 h$	$2\pi rh$	$2\pi r(r + h)$
Cone	$\frac{1}{3}\pi r^2 h$	πrl	$\pi r(r + l)$
where slant height l is given by $l = \sqrt{(r^2 + h^2)}$.			

IV. TRIGONOMETRY

1.

$\theta^o = 0$	0	30	45	60	90	180	270	360
$\sin\theta$	0	½	$1/\sqrt{2}$	$\frac{\sqrt{3}}{2}$	1	0	-1	0
$\cos\theta$	1	$\frac{\sqrt{3}}{2}$	$1/\sqrt{2}$	$1/2$	0	-1	0	1
$\tan\theta$	0	$1/\sqrt{3}$	1	$\sqrt{3}$	∞	0	$-\infty$	0

2. Any t-ratio of $(n.\ 90^0 \pm \theta) = \pm$ same ratio of θ, when n is even.
$= \pm$ co – ratio of θ, when n is odd.
The sign + or — is to be decided from the quadrant in which $n.\ 90^0 \pm \theta$ lies.

e.g., $\sin 570^0 = \sin(6 \times 90^0 + 30^0) = -\sin 30^0 = -\frac{1}{2}$;

$\tan 315^0 = \tan(3 \times 90^0 + 45^0) = -\cot 45^0 = -1.$

3. $\sin(A \pm B) = \sin A \cos B \pm \cos A \sin B$

$\cos(A \mp B) = \cos A \cos B \pm \sin A \sin B$

$\sin 2A = 2\sin A \cos A = 2\tan A/(1 + \tan^2 A)$

$\cos 2A = \cos^2 A - \sin^2 A = 1 - 2\sin^2 A = 2\cos^2 A - 1 = \frac{1 - \tan^2 A}{1 + \tan^2 A}.$

4. $\tan(A \pm B) = \frac{\tan A \pm \tan B}{1 \mp \tan A \tan B}$; $\tan 2A = \frac{2\tan A}{1 - \tan^2 A}.$

5. $\sin A \cos B = \frac{1}{2}[\sin(A + B) + \sin(A - B)]$

$\cos A \sin B = \frac{1}{2}[\sin(A + B) - \sin(A - B)]$

$\text{coa } A \cos B = \frac{1}{2}[\cos(A + B) + \cos(A - B)]$

$\sin A \sin B = \frac{1}{2}[\cos(A - B) - \cos(A + B)].$

6. $\sin C + \sin D = 2\sin\frac{C + D}{2}\cos\frac{C - D}{2}$

$\sin C - \sin D = 2\cos\frac{C + D}{2}\sin\frac{C - D}{2}$

$\cos C + \cos D = 2\cos\frac{C + D}{2}\cos\frac{C - D}{2}$

$\cos C - \cos D = -2\sin\frac{C + D}{2}\sin\frac{C - D}{2}$

7. $a\sin x + b\cos x = r\sin(x + \theta)$

$a \cos x + b \sin x = r \cos (x - \theta)$

where $a = r \cos, \theta, b = r \sin\theta$ so that $r = \sqrt{(a^2 + b^2)},\ \theta = \tan^{-1}\left(\frac{b}{a}\right)$

8. In any ΔABC:

(i) $a/\sin A = b/\sin B = c/\sin C$ (sine formula)

(ii) $\cos A = \dfrac{b^2 + c^2 - a^2}{2bc}$. (cosine formula)

(iii) $a = b \cos C + c \cos B$ (Projection formula)

(iv) Area of $\Delta ABC = \frac{1}{2} bc \sin A = \sqrt{s(s-a)(s-b)(s-c)}$ where $s = \frac{1}{2}(a + b + c)$.

9. Series

(i) Exponential Series: $e^x = 1 + \dfrac{x}{1!} + \dfrac{x^2}{2!} + \dfrac{x^3}{3!} + \ldots\ldots\infty$

(ii) sin x, cos x, sin hx, cos hx series

$$\sin x = x - \frac{x^3}{3!} + \frac{x^5}{5!} - \ldots\ldots\infty,$$

$$\cos x = 1 - \frac{x^2}{2!} + \frac{x^4}{4!} - \ldots\ldots\infty$$

$$\sin h x = x + \frac{x^3}{3!} + \frac{x^5}{5!} + \ldots\ldots\infty, \quad \cos h x = 1 + \frac{x^2}{2!} + \frac{x^4}{4!} + \ldots\ldots\infty$$

(iii) Log series

$$\log (1 + x) = x - \frac{x^2}{2} + \frac{x^3}{3} - \ldots\ldots\infty, \quad \log (1 - x) = -\left(x + \frac{x^2}{2} + \frac{x^3}{3} + \ldots\infty\right)$$

(iv) Gregory series

$$\tan^{-1} x = x - \frac{x^3}{3} + \frac{x^5}{5} - \ldots\ldots\infty, \tan h^{-1} x = \frac{1}{2} \log \frac{1 + x}{1 - x} = x + \frac{x^3}{3} + \frac{x^5}{5} + \ldots\infty.$$

10. (i) Complex number : $z = x + iy = r (\cos \theta + i \sin \theta) = re^{i\theta}$

(ii) Euler's theorem: $\cos \theta + i \sin \theta = e^{i\theta}$

(iii) Demoivre's theorem: $(\cos \theta + i\sin \theta)^n = \cos n\theta + i \sin n \theta$.

11. (i) Hyperbolic functions: $\sin h x = \dfrac{e^x - e^{-x}}{2}$; $\cos h x = \dfrac{e^x + e^{-x}}{2}$;

$$\tan h x = \frac{\sin h x}{\cosh x};\ \cot h x = \frac{\cos h x}{\sin h x};\ \sec h x = \frac{1}{\cos h x};\ \mathrm{cosec}\, h x = \frac{1}{\sin hx}$$

(ii) Relations between hyperbolic and trigonometric functions:

$\sin ix = i \sin h x$; $\cos h x = \cos h x$; $\tan ix = i \tan h x$.

(iii) Inverse hyperbolic functions;

$\sin h^{-1}x = \log[x + \sqrt{x^2 + 1}\,]$; $\cosh^{-1}x = \log[x + \sqrt{x^2 - 1}]$; $\tan h^{-1} x = \frac{1}{2} \log \frac{1+x}{1-x}$.

V. CALCULUS

1. Standard limits:

(i) $\underset{x \to a}{\text{Lt}} \frac{x^n - a^n}{x - a} = na^{n-1}$, n any rational number

(ii) $\underset{x \to 0}{\text{Lt}} \frac{\sin x}{x} = 1$

(iii) $\underset{x \to 0}{\text{Lt}} (1 + x)^{1/x} = e$

(iv) $\underset{x \to \infty}{\text{Lt}} x^{1/x} = 1$

(v) $\underset{x \to 0}{\text{Lt}} \frac{a^x - 1}{x} = \log_e a$.

2. Differentiation

(i) $\frac{d}{dx}(uv) = u\frac{dv}{dx} + v\frac{du}{dx}$

$\frac{d}{dx}\left(\frac{u}{v}\right) = \frac{v\,du/dx - u\,dv/dx}{v^2}$

$\frac{du}{dx} = \frac{du}{dy} \cdot \frac{dy}{dx}$ (chain Rule)

$\frac{d}{dx}(ax + b)^n = n\,(ax + b)^{n-1} \cdot a$

(ii) $\frac{d}{dx}(e^x) = e^x$

$\frac{d}{dx}(a^x) = a^x \log_e a$

$\frac{d}{dx}(\log_e x) = 1/x$

$\frac{d}{dx}(\log_a x) = \frac{1}{x \log a}$.

(iii) $\frac{d}{dx}(\sin x) = \cos x$

$\frac{d}{dx}(\cos x) = -\sin x$

$\frac{d}{dx}(\tan x) = \sec^2 x$

$\frac{d}{dx}(\cot x) = -\text{cosec}^2 x$

$\frac{d}{dx}(\sec x) = \sec x \tan x$

$\frac{d}{dx}(\text{cosec}\, x) = -\text{cosec}\, x \cot x$.

(iv) $\frac{d}{dx}(\sin^{-1}x) = \frac{1}{\sqrt{(1 - x^2)}}$

$\frac{d}{dx}(\cos^{-1}x) = \frac{-1}{\sqrt{(1 - x^2)}}$

$\frac{d}{dx}(\tan^{-1}x) = \frac{1}{1 + x^2}$

$\frac{d}{dx}(\cot^{-1}x) = \frac{-1}{1 + x^2}$

$\frac{d}{dx}(\sec^{-1}x) = \frac{1}{x\sqrt{(x^2 - 1)}}$

$\frac{d}{dx}(\text{cosec}^{-1}x) = \frac{-1}{x\sqrt{(x^2 - 1)}}$.

(v) $\frac{d}{dx}(\sin h\, x) = \cos h\, x$

$\frac{d}{dx}(\cos h\, x) = \sin h\, x$

$\frac{d}{dx}(\tan h\, x) = \text{sech}^2 x$

$\frac{d}{dx}(\cot h\, x) = -\text{cosec}\, h^2 x$.

(vi) $D^n (ax + b)^m = m\,(m - 1)\,(m - 2) \ldots\ldots (m - n + 1)\,(ax + b)^{m-n} \cdot a^n$

$D^n \log(ax+b) = (-1)^{n-1}(n-1)!\, a^n/(ax+b)^n$

$D^n(e^{mx}) = m^n e^{mx}$ $\qquad$ $D^n(a^{mx}) = m^n(\log a)^n . a^{mx}$

$$D^n \begin{bmatrix} \sin(ax+b) \\ \cos(bx+c) \end{bmatrix} = (a^2+b^2)^{n/2} e^{ax} \begin{bmatrix} \sin(bx+c+n\tan^{-1} b/a) \\ \cos(bx+c+n\tan^{-1} b/a) \end{bmatrix}.$$

(vii) Leibnitz theorem: $(uv)_n$

$= u_n + {}^nC_1 u_{n-1}v_1 + {}^nC_2 u_{n-2}v_2 + \ldots + {}^nC_r u_{n-r}v_r + \ldots + {}^nC_n v_n.$

3. Integration

(i) $\int x^n\, dx = \dfrac{x^{n+1}}{n+1}$ $(n \neq -1)$ $\qquad$ $\int \dfrac{1}{x}\, dx = \log_e x$

$\int e^x\, dx = e^x$ $\qquad$ $\int a^x\, dx = a^x/\log_e a$

(ii) $\int \sin x\, dx = -\cos x$ $\qquad$ $\int \cos x\, dx = \sin x$

$\int \tan x\, dx = -\log \cos x$ $\qquad$ $\int \cot x\, dx = \log \sin x$

$$\int \sec x\, dx = \log(\sec x + \tan x) = \log \tan\left(\frac{\pi}{4} + \frac{x}{2}\right)$$

$$\int \operatorname{cosec} x\, dx = \log(\operatorname{cosec} x - \cot x) = \log \tan\left(\frac{x}{2}\right)$$

$\int \sec^2 x\, dx = \tan x$ $\qquad$ $\int \operatorname{cosec}^2 x\, dx = -\cot x.$

(iii) $\int \dfrac{dx}{a^2+x^2} = \dfrac{1}{a}\tan^{-1}\dfrac{x}{a}$ $\qquad$ $\int \dfrac{dx}{\sqrt{(a^2-x^2)}} = \sin^{-1}\dfrac{x}{a}$

$\int \dfrac{dx}{a^2-x^2} = \dfrac{1}{2a}\log\dfrac{a+x}{a-x}$ $\qquad$ $\int \dfrac{dx}{\sqrt{(a^2+x^2)}} = \sin h^{-1}\dfrac{x}{a}$

$\int \dfrac{dx}{x^2-a^2} = \dfrac{1}{2a}\log\dfrac{x-a}{x+a}$ $\qquad$ $\int \dfrac{dx}{\sqrt{(x^2-a^2)}} = \cos h^{-1}\dfrac{x}{a}.$

(iv) $\int \sqrt{(a^2-x^2)}\, dx = \dfrac{x\sqrt{(a^2-x^2)}}{2} + \dfrac{a^2}{2}\sin^{-1}\dfrac{x}{a}$

$$\int \sqrt{(a^2+x^2)}\, dx = \frac{x\sqrt{(a^2+x^2)}}{2} + \frac{a^2}{2}\sinh^{-1}\frac{x}{a}. = \frac{x}{2}\sqrt{(a^2+x^2)} + \frac{a^2}{2}\log\frac{x+\sqrt{(a^2+x^2)}}{a}$$

$$\int \sqrt{(x^2-a^2)}\, dx = \frac{x\sqrt{(x^2-a^2)}}{2} + \frac{a^2}{2}\cosh^{-1}\frac{x}{a}. = \frac{x}{2}\sqrt{(x^2-a^2)} - \frac{a^2}{2}\log\frac{x+\sqrt{(x^2-a^2)}}{a}$$

(v) $\int e^{ax}\sin bx\, dx = \dfrac{e^{ax}}{a^2+b^2}(a\sin bx - b\cos bx)$

$$\int e^{ax} \cos bx\, dx = \frac{e^{ax}}{a^2 + b^2}(a \cos bx + b \sin bx)$$

(vi) $\int \sin h\, x\, dx = \cos h\, x$ $\qquad$ $\int \cos h\, x\, dx = \sin h\, x$

$\int \tan h\, x\, dx = \log \cos h\, x$ $\qquad$ $\int \cot h\, x\, dx = \log \sin h\, x$

$\int \sec h^2\, x\, dx = \tan h\, x$ $\qquad$ $\int \operatorname{cosech}^2 x\, dx = -\ \cot h\, x.$

(vii) $\int_0^{\pi/2} \sin^n x\, dx = \int_0^{\pi/2} \cos^n x\, dx$

$$= \frac{(n-1)(n-3)(n-5)\ldots.}{n(n-2)(n-4)} \times \left(\frac{\pi}{2}, \text{ only if n is even}\right)$$

$$\int_0^{\pi/2} \sin^m x \cos^n x\, dx = \frac{(m-1)(m-3)\ldots. \times (n-1)(n-3)\ldots.}{(m+n)(m+n-2)(m+n-4)\ldots.}$$

$$\times \left(\frac{\pi}{2}, \text{ only if both m and n are even}\right)$$

(viii) $\int_0^a f(x)\, dx = \int_0^a f(a-x)\, dx$

$\int_{-a}^a f(x)\, dx = 2 \int_0^a f(x)\, dx$, if f (x) is an even function.

= 0, if f(x) is an odd function.

$\int_0^{2a} f(x)\, dx = 2 \int_0^a f(x)\, dx$, if $f(2a - x) = f(x)$

= 0, if $f(2a - x) = -f(x)$.

VI. Coordinate systems

	Polar coordinates (r, θ)	Cylindrical coordinates (ρ, ϕ, z)	Spherical polar coordinates (r, θ, ϕ)
Coordinate transformations	$x = r \cos\theta$ $y = r \sin\theta$	$x = \rho \cos\phi$ $y = \rho \sin\phi$ $z = z$	$x = r \sin\theta \cos\phi$ $y = r \sin\theta \sin\phi$ $z = r \cos\theta$
Jacobian	$\frac{\partial(x, y)}{\partial(r, \theta)} = r$	$\frac{\partial(x, y, z)}{\partial(\rho, \phi, z)} = \rho$	$\frac{\partial(x, y, z)}{\partial(r, \theta, \phi)} = r^2 \sin\theta$
(Arc – length)^2	$(ds)^2 = (dr)^2 + r^2 (d\theta)^2$ $dx\, dy = r d\theta\, dr$	$(ds)^2 = (d\rho)^2 + \rho^2 (d\phi)^2 + (dz)^2$	$(ds)^2 = (dr)^2 + r^2 (d\theta)^2 + (r \sin\theta)^2 (d\phi)^2$
Volume- element		$dV = \rho\, d\rho\, d\phi\, dz$	$dV = r^2 \sin\theta\, dr\, d\theta\, d\phi$

Contents

UNIT - I
Differential Equations

Chapter 1

Basic Concepts of Differential Equations

INTRODUCTION

Differential equations are of fundamental importance in engineering mathematics because many physical laws and relations appear mathematically in the form of such equations. The mathematical formulation of many problems in science, engineering, Economics, sociology, physiology, Biology, Finance and management, give rise to differential equations. For example, the problem of motion of a satellite, the flow of current in an electric circuit, the growth of a population, the changes in price of commodities, decay of radioactive substance, cooling of a body etc. lead to differential equations. Each of the above problems are characterised by some laws which involve the rate of change of one or more quantities, with respect to the other quantities. The laws characterising these problems when expressed mathematically, become equations involving derivatives and such equations are called differential equations.

DEFINITION

Any relation between known functions and an unknown function is called a differential equation if it involves the differential coefficient (or coefficients) of the unknown function.

It is usual to denote the unknown function by y. Finding the unknown function is called solving or integrating the differential equation. The solution or integral of the differential equation is also called its primitive, because the differential equation can be regarded as a relation derived from it.

Equations such as

(i) $$(x^2 - y^2)\frac{dy}{dx} = xy \qquad (1)$$

(ii) $$\rho = \frac{\left\{1 + \left(\frac{dy}{dx}\right)^2\right\}^{3/2}}{\frac{d^2y}{dx^2}} \qquad (2)$$

(iii) $$x^3 \frac{d^3y}{dx^3} + 2x^2 \frac{d^2y}{dx^2} + 3x \frac{dy}{dx} - 3y = x^2 + x \qquad (3)$$

(iv) $$x\frac{\partial u}{\partial x} + y \frac{\partial u}{\partial y} + z \frac{\partial u}{\partial z} = xyz \qquad (4)$$

(v) $$\frac{\partial^2 u}{\partial x^2} = K \frac{\partial^2 u}{\partial y^2} \qquad (5)$$

(vi) $$y = 2x\frac{dy}{dx} + y^2 \left(\frac{dy}{dx}\right)^3 \qquad (6)$$

Which involve differential coefficients are called the differential equations.

Differential equations which involve only one independent variable are called ordinary differential equations. Equations (i) (ii), (iii) and (vi) are of this type.

Differential equations which involve two or more independent variables are called partial differential equations. Equations (iv) and (v) are of this type

The order of a differential Equation. The order of a differential equation is the order of the highest derivative involving in the equation.

The Degree of a differential Equation. The degree of a differential equation is the degree of the highest order derivative involving in the equation, When the equation is free from radicals and fractional powers.

For example– The differential equations

$$\frac{dy}{dx} + xy = a \qquad (1)$$

$$\frac{d^2y}{dx^2} + x\frac{dy}{dx} = 2 \qquad (2)$$

$$\left(\frac{d^3y}{dx^3}\right)^4 - 6x^2 \left(\frac{dy}{dx}\right)^2 + e^x = \sin xy \qquad (3)$$

(U.P.T.U. 2009)

$$\frac{d^2y}{dx^2} = \left[1 + \left(\frac{dy}{dx}\right)^5\right]^{1/3} \qquad (4)$$

The equation (1) is of the first order and first degree. Equation (2) is of second order and first degree. Equation (3) is of third order and fourth degree. Equation (4) is of second order and third degree

Formation of A Differential Equation

Example 1. Find the differential equation of the family of circles of radius r whose centre lies on the x axis. (I.A.S. 1993, 95, 96)

Solution. The equation of the circle with radius r and centre on x axis is

$$(x-a)^2 + y^2 = r^2 \qquad (1)$$

Differentiating (1) with respect to x, we get

$$2(x-a) + 2y\frac{dy}{dx} = 0$$

Eliminating 'a' between (1) and (2), we get

$$\left(-y\frac{dy}{dx}\right)^2 + y^2 = r^2$$

or $$y^2\left[\left(\frac{dy}{dx}\right)^2 + 1\right] = r^2$$

Which is the required differential equation.

Example 2. Find the differential equation of the family of parabolas with foci at the origin and axis along the x-axis. (I.A.S. 1994)

Solution. The equation of the parabolas with foci at the origin and axis along the x axis is given by

$$\sqrt{x^2 + y^2} = \frac{x + 2a}{\sqrt{1^2}}$$

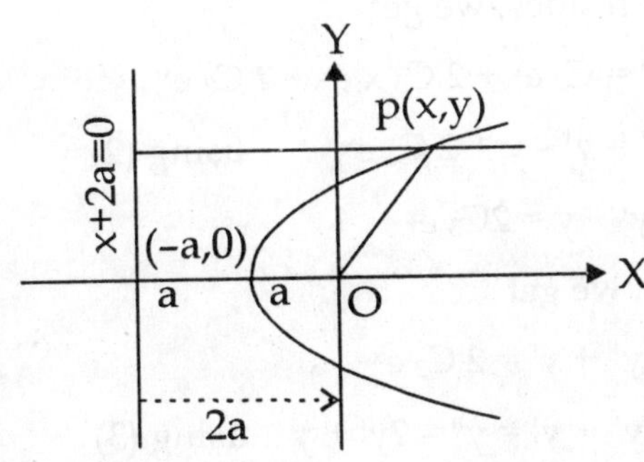

or $$x^2 + y^2 = x^2 + 4ax + 4a^2$$

or $$y^2 = 4a\,(x + a) \qquad (1)$$

Differentiating with respect to x, we get

$$2y\frac{dy}{dx} = 4a$$

$$\Rightarrow y\frac{dy}{dx} = 2a \qquad (2)$$

Eliminating a between (1) and (2), we get

$$y^2 = 2y\frac{dy}{dx}\left[x + \frac{1}{2}y\frac{dy}{dx}\right]$$

or $$y\left(\frac{dy}{dx}\right)^2 + 2x\frac{dy}{dx} - y = 0$$

which is the required differential equation.

Example 3. Determine the differential equation whose set of independent solution is $\{e^x, xe^x, x^2 e^x\}$

(U.P.T.U. 2002)

Solution. Here we have

$$y = C_1e^x + C_2\,x\,e^x + C_3\,x^2\,e^x \qquad (1)$$

Differentiating both sides of (1) w.r.t "x" we get

$$y' = C_1e^x + C_2\,x\,e^x + C_2\,e^x + C_3\,x^2\,e^x + C_3\,2x\,e^x$$

$$\Rightarrow \quad y' = y + C_2\,e^x + 2\,C_3\,x\,e^x$$

$$\Rightarrow \quad y' - y = C_2\,e^x + 2\,C_3x\,e^x \qquad (2)$$

Again differentiating both sides, we get

$$y'' - y' = C_2\,e^x + 2\,C_3\,x\,e^x + 2\,C_3\,e^x$$

$$\Rightarrow \quad y'' - y' = y' - y + 2\,C_3\,e^x \quad \text{using (2)}$$

$$\Rightarrow \quad y'' - 2y' + y = 2C_3\,e^x \qquad (3)$$

Again differentiating (3), we get

$$y''' - 2y'' + y' = 2\,C_3\,e^x$$

$$\Rightarrow \quad y''' - 2y'' + y' = y'' - 2y' + y \quad \text{using (3)}$$

$\Rightarrow$ $$y''' - 3y'' + 3y' - y = 0$$

or $$\frac{d^3y}{dx^3} - 3\frac{d^2y}{dx^2} + 3\frac{dy}{dx} - y = 0$$

is the required differential equation

EXERCISE

1. Form the differential equation of simple harmonic motion given by $x = A\cos(nt + \alpha)$

Ans. $\frac{d^2x}{dt^2} + n^2x = 0$

2. Obtains the differential equation of all circles of radius a and centre (h, k) and hence prove that the radius of curvature of a circle at any point is constant.

Ans. $\left[1 + \left(\frac{dy}{dx}\right)^2\right]^3 = a^2\left(d^2y/dx^2\right)^2$

3. Show that $v = \frac{A}{r} + B$ is a solution of $\frac{d^2v}{dr^2} + \frac{2}{r}\frac{dv}{dr} = 0$

4. Show that $Ax^2 + By^2 = 1$ is the solution of $x\left\{y\frac{d^2y}{dx^2} + \left(\frac{dy}{dx}\right)^2\right\} = y\frac{dy}{dx}$

5. By eliminating the constants a and b obtain differential equation of which $xy = ae^x + be^{-x} + x^2$ is a constant.

Objective Type of Questions

Each question possesses four alternative answers, but only one answer is correct tick mark the correct one.

1. Degree and order of the differential equation $\sqrt{2\left(\frac{dy}{dx}\right)^3 + 4} = \left(\frac{d^2y}{dx^2}\right)^{3/2}$ are respectively.

(a) order 2, degree 3 (b) order 1, degree 3

(c) order 3, degree 2 (d) order 3, degree 1

Ans. (a)

2. The degree of the differential equation $\left[y + x\left(\frac{d^2y}{dx^2}\right)^2\right]^{1/4} = \frac{d^3y}{dx^3}$ is given by

(a) 2 (b) 3 (c) 4 (d) 1

(R.A.S. 1993)

Ans. (c)

3. The order of the differential equation $\left[1 + \left(\frac{d^3y}{dx^3}\right)^2\right]^{4/3} = \frac{d^2y}{dx^2}$ is given by

(a) 1 (b) 2 (c) 3 (d) 4

(R.A.S. 1993)

Ans. (c)

4. If $x = A\cos(mt - \alpha)$ then the differential equation satisfying this relation is

(a) $\frac{dx}{dt} = 1-x^2$

(b) $\frac{d^2x}{dt^2} = -\alpha^2 x$

(c) $\frac{d^2x}{dt^2} = -m^2x$

(d) $\frac{dx}{dt} = -m^2x$

(I.A.S. 1993)

5. The equation $y\frac{dy}{dx} = x$ represents a family of

(a) Circles

(b) hyperbola

(c) parabolas

(d) ellipses

(U.P.P.C.S. 1995)

Ans. (b)

Chapter 2

Differential Equations of First Order and First Degree

INTRODUCTION

An equation of the form $F\left(x, y, \frac{dy}{dx}\right) = 0$ in which x is the independent variable and $\frac{dy}{dx}$ appears with first degree is called a first order and first degree differential equation. It can also be written in the form $\frac{dy}{dx} = f(x, y)$ or in the form Mdx + Ndy = 0, where M and N are functions of x and y. Generally, it is difficult to solve the first order differential equations and in some cases they may not possess any solution. There are certain standard types of first order, first degree equations. In this chapter we shall discuss the methods of solving them..

VARIABLES SEPARABLE

If the equation is of the form $f_1(x)\,dx = f_2(y)\,dy$ then, its solution, by integration is $\int f_1(x)dx = \int f_2(y)\,dy + C$

where C is an arbitrary constant.

Example 1. Solve $\frac{dy}{dx} = e^{x-y} + x^2 e^{-y}$

Solution. The given equation is

$$\frac{dy}{dx} = e^{x-y} + x^2 e^{-y}$$

or
$$e^y\,dy = (e^x + x^2)\,dx$$

Integrating, $e^y = e^x + \frac{x^3}{3} + c$, where C is an arbitrary constant is the required solution.

Example 2. Solve $\left(y - x\frac{dy}{dx}\right) = a\left(y^2 + \frac{dy}{dx}\right)$

Solution. The given equation is

$$y - x\frac{dy}{dx} = ay^2 + a\frac{dy}{dx}$$

or $$(a + x)\frac{dy}{dx} = (y - ay^2)$$

or $$\frac{dy}{y - ay^2} = \frac{dx}{a + x}$$

or $$\frac{dy}{y - ay^2} = \frac{dx}{a + x}$$

or $$\frac{dy}{y(1 - ay)} = \frac{dx}{x + a}$$

or $$\left[\frac{a}{1 - ay} + \frac{1}{y}\right]dy = \frac{dx}{x + a},$$ **resolving into partial fractions.**

Integrating, $[-\log(1 - ay) + \log y] = \log(x + a) + \log C$

where C is an arbitrary constant

or $$\log\left(\frac{y}{1 - ay}\right) = \log\{C(x + a)\}$$

or $$\frac{y}{1 - ay} = C(x + a)$$

or $y = C(x + a)(1 - ay)$ is the required solution.

Example 2. Solve $(x + y)^2\frac{dy}{dx} = a^2$

Solution. Let $x + y = v$

then from (1) on differentiating with respect to x, we have

$$1 + \frac{dy}{dx} = \frac{dv}{dx}$$

or $$\frac{dy}{dx} = \left(\frac{dv}{dx} - 1\right)$$

Substituting these values from (1) and (2) in the given equation, we get

$$v^2 \left[\frac{dv}{dx} - 1\right] = a^2$$

or $$v^2 \frac{dv}{dx} = a^2 + v^2$$

or $$\frac{v^2}{a^2 + v^2} dv = dx$$

or $$\frac{a^2 + v^2 - a^2}{a^2 + v^2} dv = dx$$

or $$\left[1 - \frac{a}{a^2 + v^2}\right] dv = dx$$

Integrating,

$$v - a^2 \frac{1}{a} \tan^{-1}\left(\frac{v}{a}\right) = x + c$$

where c is an arbitrary constant

or $$v - a \tan^{-1} (v/a) = x + c$$

or $$(x + y) - a \tan \{(x + y)/a\} = x + c, \text{ from (1)}$$

or $$y - a \tan^{-1} \{(x + y)/a\} = C$$

Method of Solving Homogeneous Differential Equation

It is a differential equation of the form

$$\frac{dy}{dx} = \frac{\phi (x, y)}{\psi (x, y)}, \tag{1}$$

where $\phi (x, y)$ and $\psi (x, y)$ are homogeneous functions of x and y at the same degree, n (say)

Such equation can be solved by putting $y = vx$

where $$\frac{dy}{dx} = v + x \frac{dy}{dx} \tag{2}$$

Now the given differential equation is

$$\frac{dy}{dx} = \frac{\phi (x, y)}{\psi (x, y)} = \frac{x^n \phi (y/x)}{x^n \psi (y/x)}$$

or $$\frac{dy}{dx} = \frac{\phi\,(y/x)}{\psi\,(y/x)} = f\left(\frac{y}{x}\right), \text{ say} \qquad (3)$$

Substituting the values of y and $\frac{dy}{dx}$ from, (1) and (2) in (3),

we have $$v + x\frac{dv}{dx} = f(v)$$

or $$x\frac{dv}{dx} = f(v) - v$$

or $$\frac{dv}{f(v) - v} = \frac{dx}{x}$$

The variables has now been separated and its solution is

$\int \frac{dv}{f(v) - v} = \log x + c$ where c is an arbitrary constant.

After integration v should be replaced by (y/x) to get the required solution.

Example 4. Solve $x\,dy - y\,dx = \sqrt{(x^2 + y^2)}\,dx$

Solution. The given equation can be rewritten as

$$x\frac{dy}{dx} - y = \sqrt{(x^2 + y^2)}$$

or $$x\left(v + x\frac{dv}{dx}\right) - vx = \sqrt{(x^2 + v^2x^2}, \text{ putting } y = vx$$

or $$x\frac{dv}{dx} = \sqrt{(1 + v^2)}$$

or $$\frac{dv}{\sqrt{(1 + v^2)}} = \frac{dx}{x}$$

Integrating, $\log\left\{v + \sqrt{(v^2 + 1)}\right\} = \log x + \log c,$ where c is an arbitrary constant

or $$\left\{v + \sqrt{(v^2 + 1)}\right\} = Cx$$

or $$\left[y + \sqrt{(y^2 + x^2)}\right]/x = Cx \qquad \because v = y/x$$

or $$y + \sqrt{(y^2 + x^2)} = Cx^2$$

Example 5. Solve $(1 + e^{x/y})\,dx + e^{x/y}\,[1 - (x/y)]\,dy = 0$ (U.P.P.C.S. 1999)

Solution. The given equation may be rewritten as

$$e^{x/y}\left(1 - \frac{x}{y}\right) + (1 + e^{x/y})\frac{dx}{dy} = 0$$

or $$e^{v}(1 - v) + (1 + e^{v})\left(v + y\frac{dv}{dy}\right) = 0$$

$$\text{putting } x = vy \text{ or } \frac{dx}{dy} = v + y\frac{dv}{dy}$$

or $$e^{v} - ve^{v} + v + ve^{v} + (1 + e^{v})\,y\frac{dv}{dy} = 0$$

or $$(v + e^{v}) + (1 + e^{v})\,y\frac{dv}{dy} = 0$$

or $$\frac{(1 + e^{v})}{v + e^{v}}\,dv + \frac{dy}{y} = 0$$

Integrating, $\log(v + e^{v}) + \log y = \log C$, when C is an arbitrary constant

or $$\log\{(v + e^{v})\,y\} = \log C$$

or $$(v + e^{v})\,y = C$$

or $$\left[\frac{x}{y} + e^{x/y}\right]y = C, \text{ putting } v = x/y$$

or $$(x + y\,e^{x/y}) = C$$

Equations Reducible to Homogeneous Form.

Consider the equation $$\frac{dy}{dx} = \frac{ax + by + C}{a'x + b'y + C'} \qquad (1)$$

where $a:b \neq a':b'$

If C and C' are both zero, the equation is homogenous and can be solved by the method of homogeneous equation. If C and C' are not both zero, we change the variables so that constant terms are no longer present, by the substitutions $x = X + h$ and $y = Y + k$ Where h and k are constants yet to be chosen. (2)

From (2) dx = dX and dy = dY

and so (1) reduces to $\frac{dY}{dX} = \frac{a\,(X + h) + b\,(Y + k) + C}{a'\,(X + h) + b'\,(Y + k) + C'}$

or
$$\frac{dY}{dX} = \frac{aX + bY + (ah + bk + c)}{a'\,X + b'\,Y + (a'\,h + b'\,k + C')} \qquad (3)$$

Now, choose h and k such that

$$ah + bk + c = 0 \text{ and } a'h + b'k + c' = 0 \qquad (4)$$

Then (3) reduces to $\frac{dY}{dX} = \frac{aX + bY}{a'X + b'Y}$, which is homogeneous and can be solved by the substitution Y = vX. Replacing X and Y in the solution so obtained by x-h and y-k [from (3)] respectively

We can get the required solution is terms of x and y, the original variables.

A Special Case When a:b = a':b'

In this case we cannot solve the equations given by (4) above and the differential equation is of the form

$$\frac{dy}{dx} = \frac{ax + by + C}{kax + kby + C'} \qquad (5)$$

In this case the differential equation is solve by putting

$$v = ax + by \qquad (6)$$

Differentiating both sides of (6) with respect to x, we get

$$\frac{dv}{dx} = a + b\,\frac{dy}{dx} \text{ or } \frac{dy}{dx} = \frac{1}{b}\left(\frac{dv}{dx} - a\right)$$

∴ The equation (5) reduces to

$$\frac{1}{b}\left(\frac{dv}{dx} - a\right) = \frac{v+C}{kv + C},\ \text{or } \frac{dv}{dx} = a + \frac{b(v+C)}{kv+C'}$$

The variables, are now separable, and we can determine v in terms of x. Replacing v by ax + by in this solution, we can obtain the final solution.

Example 6. Solve (2x + y – 3) dy = (x + 2y – 3) dx

Solution. The given differential equation is $\frac{dy}{dx} = \frac{x + 2y - 3}{2x + y - 3}$ (1)

putting $x = X + h$ and $y = Y + k$ (2)

the equation (1) reduce to $\dfrac{dY}{dX} = \dfrac{(X + h) + 2(Y + k) - 3}{2(X + h) + (Y + k) - 3}$

or $$\frac{dY}{dX} = \frac{X + 2Y + (h + 2K - 3)}{2X + Y + (2h + K - 3)} \quad (3)$$

Choose h and k such that

$$h + 2k - 3 = 0 \text{ and } 2h + k - 3 = 0 \quad (4)$$

Solving the equations (4) we have $h = 1 = k$

$\therefore$ From (2) $x = X + 1$ and $y = Y + 1$

or $X = x - 1$ and $Y = y - 1$ (5)

Also (3) reduce to $\dfrac{dY}{dX} = \dfrac{X + 2Y}{2X + Y}$

putting $Y = vX$, $v + X\dfrac{dv}{dX} = \dfrac{X + 2vX}{2X + vX}$

or $$v + X\frac{dv}{dX} = \frac{1 + 2v}{2 + v}$$

or $$X\frac{dv}{dX} = \frac{1 + 2v}{2 + v} - v = \frac{1 + 2v - 2v - v^2}{2 + v} = \frac{1 - v^2}{2 + v}$$

or $$\frac{2 + v}{1 - v^2}\,dv = \frac{dX}{X}$$

or $$\left[\frac{1}{2}\,\frac{1}{1 + v} + \frac{3}{2}\,\frac{1}{1 - v}\right] dv = \frac{dX}{X}$$

Integrating, $\dfrac{1}{2}\log(1 + v) - \dfrac{3}{2}\log(1 - v) = \log X + \log C$, where C is an arbitrary constant

or $$\log\left[\frac{1 + v}{(1 - v)^3}\right] = 2\log(CX) = \log(CX)^2$$

or $$\frac{1 + v}{(1 - v)^3} = (CX)^2$$

or $$\frac{1 + Y/k}{(1 - Y/k)^3} = C^2X^2$$

or $$\frac{X + Y}{(X - Y)^3} = C^2$$

or $$\frac{(x - 1) + (y - 1)}{\{(x - 1) - (y - 1)\}^3} = C^2 \text{ from (5)}$$

or $$\frac{x + y - 2}{(x - y)^3} = C^2$$

Example 7. Solve $(2x + 2y + 3)\, dy - (x + y + 1)\, dx = 0$

Solution. The given equation is $\frac{dy}{dx} = \frac{x + y + 1}{2x + 2y + 3}$ (1)

put $x + y + 1 = v$

Differentiating

$$1 + \frac{dy}{dx} = \frac{dv}{dx} \text{ or } \frac{dy}{dx} = \frac{dv}{dx} - 1$$

Therefore (1) reduce to $\frac{dv}{dx} - 1 = \frac{v}{2v + 1}$

or $$\frac{dv}{dx} = \frac{3v + 1}{2v + 1}$$

or $$\frac{2v + 1}{3v + 1}\, dv = dx$$

or $$\left(\frac{2}{3} + \frac{1}{3\,(3v + 1)}\right) dv = dx$$

Integrating, $\frac{2}{3} v + \frac{1}{9} \log (3v + 1) = x + c$, where C is an arbitrary constant

or $6v + \log (3v + 1) = 9x + C_1$, where $C_1 = 9C$

or $6\,(x + y + 1) + \log (3x + 3y + 4) = 9x + C_1$

or $6y - 3x + \log (3x + 3y + 4) = C_2$ where $C_2 = C_1 + 6$

Linear Differential Equations

A differential equation of the form $\frac{dy}{dx} + Py = Q$

where P and Q are constants or functions of x alone (and not of y) is called a linear differential equation of the first order in y

its integrating factor $= e^{\int P\,dx}$

Multiplying both sides of (1) by this integrating factor (I.F.) and then integrating we get

$y.\, e^{\int Pdx} = C + \int Q.\, e^{\int Pdx}\, dx$, where C is an arbitrary constant, is the complete solution of (1)

Example 8. Solve $\frac{dy}{dx} + 2y \tan x = \sin x$, given that $y = 0$ when $x = \pi/3$.

(U.P.P.C.S. 2003)

Solution. Here $P = 2 \tan x$ and $Q = \sin x$

$\therefore$ Integrating factor $= e^{\int Pdx} = e^{\int 2\tan x\, dx}$

$= e^{2\log \sec x}$

$= e^{\log (\sec x)^2} = \sec^2 x$

Multiplying the given equation by $\sec^2 x$, we get

$$\sec^2 x \left(\frac{dy}{dx} + 2y \tan x\right) = \sin x \sec^2 x$$

or $\frac{d}{dx}(y \sec^2 x) = \sec x \tan x$

Integrating both sides with respect to x, we get

$y \sec^2 x = C + \int \sec x \tan x\, dx$, where C is an arbitrary constant

or $$y \sec^2 x = C + \sec x \qquad (1)$$

it is given that when $x = \pi/3$, $y = 0$

$\therefore$ from (1) $0 \times \sec^2 \frac{\pi}{3} = C + \sec \frac{\pi}{3}$

or $\quad 0 = C + 2 \quad \therefore \sec \frac{\pi}{3} = 2$

∴ from (1) the required solution is

$$y \sec^2 x = -2 + \sec x$$

or $$y = -2 \cos^2 x + \cos x$$

Example 9. Solve $(1 + y^2)\, dx + (x - e^{\tan^{-1} y})\, dy = 0$

(Bihar P.C.S. 2002, I.A.S. 2006)

Solution. The given equation can be written as

$$\frac{dx}{dy} + \frac{x}{1+y^2} = \frac{e^{\tan^{-1} y}}{1+y^2}$$

Therefore the integrating factor $= e^{\int \frac{1}{1+y^2} dy}$

$$= e^{\tan^{-1} y}$$

Multiplying both sides of (1) by the integrating factor and integrating, we have

$$x.\, e^{\tan^{-1} y} = C + \int \frac{e^{\tan^{-1} y}}{1+y^2} \times e^{\tan^{-1} y}\, dy$$

where C is an arbitrary costant

or $$x\, e^{\tan^{-1} y} = C + \int e^{2t}\, dt, \text{ where } t = \tan^{-1} y$$

$$= C + \frac{1}{2} e^{2t}$$

or $$x.\, e^{\tan^{-1} y} = C + \frac{1}{2} e^{2 \tan^{-1} y}$$

Example 10. Solve $\frac{dy}{dx} + y \cos x = \frac{1}{2} \sin 2x$ (I.A.S. 2004)

Solution. Here $P = \cos x$ and $Q = \frac{1}{2} \sin 2x = \sin x \cos x$

∴ Integrating factor $= e^{\int P dx} = e^{\int \cos x\, dx} = e^{\sin x}$

Multiplying the given equation by the integrating factor $e^{\sin x}$ and integrating with respect to x, we get

$$y.\, e^{\sin x} = C + \int e^{\sin x} \sin x \cos x\, dx,$$ where C is an arbitrary constant.

or $$y.\, e^{\sin x} = C + \int e^{t}\, t\, dt, \qquad \text{where } t = \sin x$$

$$= C + t.\, e^{t} - e^{t}$$

$$= C + e^{\sin x} (\sin x - 1)$$

or $$y.\, e^{\sin x} = C + e^{\sin x} (\sin x - 1)$$

Equations Reducible to the Linear Form.

The equation

$$\frac{dy}{dx} + Py = Qy^n \qquad (1)$$

Where P and Q are constants or functions of x alone and n is a constant other than zero or unity is called the extended form of linear equation or Bernoulli's Equation.

This type of equation can be reduced to the linear form on dividing by y^n and putting $\frac{1}{y^{n-1}}$ equal to v

Dividing (1) by y^n, we get $$\frac{1}{y^n}\frac{dy}{dx} + P.\frac{1}{y^{n-1}} = Q \qquad (2)$$

Put $\frac{1}{y^{n-1}} = v$ or $y^{-n+1} = v$

Differentiating both sides with respect to x, get

$$(-n+1)\, y^{-n} \frac{dy}{dx} = \frac{dv}{dx}$$

or $$\frac{1}{y^n}\frac{dy}{dx} = \frac{1}{1-n}\frac{dv}{dx}$$

Making these substitutions in (2), we have

$$\frac{1}{(1-n)}\frac{dv}{dx} + Pv = Q$$

or $$\frac{dv}{dx} + P\,(1-n)\, v = Q\,(1-n)$$

which is linear in v and can be solve by method of linear differential equation.

Example 11. Solve $\frac{dy}{dx} + x \sin 2y = x^3 \cos^2 y$ (U.P.P.C.S. 1994)

Solution. Dividing both sides of the given equation by $\cos^2 y$

we get

$$\sec^2 y \frac{dy}{dx} + x\,(2 \tan y) = x^3$$

putting $\tan y = v$ or $\sec^2 y \frac{dy}{dx} = \frac{dv}{dx}$

the above equation reduces to $\frac{dv}{dx} + 2xv = x^3$ (1)

which is a linear equation. whose integrating factor

$$= e^{\int 2x\,dx} = e^{x^2}$$

Multiplying both sides of (1) by the integrating factor and integrating, we have

$$v.\, e^{x^2} = C + \int x^3 e^{x^2} dx, \text{ where C is an arbitrary constant}$$

or $$v.\, e^{x^2} = C + \frac{1}{2} \int x^2 e^{x^2}\, 2x\,dx$$

$$= C + \frac{1}{2} \int t.\, e^t\, dt, \text{ where } t = x^2$$

$$= C + \frac{1}{2} \left(t\, e^t - \int e^t\, dt\right)$$

or $$e^{x^2} \tan y = C + \frac{1}{2} e^t\, (t - 1) \qquad \because v = \tan y$$

or $$e^{x^2} \tan y = C + \frac{1}{2} e^{x^2} (x^2 - 1)$$

or $$2 \tan y = 2\, Ce^{-x^2} + (x^2 - 1)$$

Example 2. Solve $\frac{dz}{dx} + \frac{z}{x}\log z = \frac{z}{x^2}(\log z)^2$

(I.A.S. 2001, U.P.P.C.S., 1999)

Solution. Dividing both sides of the given equation by z (log z)², we get

$$\frac{1}{z(\log z)^2}\frac{dz}{dx} + \frac{1}{(\log z)}\frac{1}{x} = \frac{1}{x^2}$$

putting $\frac{1}{\log z} = v$ or $-\frac{1}{z(\log z)^2}\frac{dz}{dx} = \frac{dv}{dx}$, the above equation reduces to

$$-\frac{dv}{dx} + \frac{v}{x} = \frac{1}{x^2}$$

or $\frac{dv}{dx} - \frac{1}{x}v = -\frac{1}{x^2}$, which is linear equation in v and whose integrating factor

$$= e^{-\int \frac{1}{x} dx} = e^{-\log x} = e^{\log x^{-1}}$$

$$= \frac{1}{x}$$

Proceeding is the usual way, the solution of above equation is

$$v\frac{1}{x} = C - \int \frac{1}{x^2}\frac{1}{x}\,dx$$

or $$\frac{1}{x(\log z)} = C + \frac{1}{2x^2}$$

Example 13. Solve y sin 2x dx – (1 + y² + cos²x) dy = 0

(I.A.S. 1996)

Solution. The given equation can be rewritten as

$$y \sin 2x \frac{dx}{dy} - \cos^2 x = 1 + y^2$$

or $$\sin 2x \frac{dx}{dy} - \frac{1}{y}\cos^2 x = \frac{1 + y^2}{y} \quad (1)$$

putting $-\cos^2 x = v$ or $-2\cos x(-\sin x)\,dx = dv$

or $$\sin 2x\,dx = dv$$

the equation (1) reduce to

$$\frac{dv}{dy} + \frac{1}{y} v = \frac{1 + y^2}{y} \qquad (2)$$

Which is linear in v with y as independent variable

Its integrating factor $= e^{\int \frac{1}{y} dy} = e^{\log y} = y$

Multiplying both sides of (2) by y and integrating, we get

$$vy = C + \int (1 + y^2)\, dy, \text{ where C is constant of integration}$$

or $$-y \cos^2 x = C + y + \frac{1}{3} y^3 \qquad \because v = -\cos^2 x$$

or $$y \cos^2 x + C + y + \frac{1}{3} y^3 = 0$$

Example 14. Solve x (dy/dx) + y = y^2 log x

(U.P.P.C.S. 1995)

Solution. The given equation can be written as

$$\frac{1}{y^2} \frac{dy}{dx} + \frac{1}{x} \frac{1}{y} = \frac{1}{x} \log x \qquad (1)$$

putting $-\frac{1}{y} = v$ or $\frac{1}{y^2} \frac{dy}{dx} = \frac{dv}{dx}$ in (1), we get

$$\frac{dv}{dx} - \frac{1}{x} v = \frac{1}{x} \log x \qquad (2)$$

which is in the standand form of the linear equation and integrating factor $e^{-\int \frac{1}{x} dx = e^{-\log x}} = \frac{1}{x}$

Multiplying both sides of (2) by the integrating factor and integrating, we get

$$v. \frac{1}{x} = C + \int \frac{1}{x^2} \log x\, dx \text{ where C is an arbitrary constant}$$

or $$\frac{v}{x} = C + \int t e^{-t}\, dt, \text{ putting } t = \log x$$

or $$\frac{v}{x} = C + \left[-t\, e^{-t} + \int e^{-t}\, dt\right]$$

$$= C - (t+1)\, e^{-t} = C - (1 + \log x)\, e^{-\log x}$$

or $$-\frac{1}{xy} = C - (1 + \log x)\left(\frac{1}{x}\right) \qquad \because v = -\frac{1}{y}$$

or $$1 = (1 + \log x)\, y - C\, xy$$

Example 15. Solve $x\,(dy/dx) + y \log y = x\, ye^{x}$

(I.A.S. 2003, M.P.P.C.S. 1996)

Solution. Dividing both sides of the given equation by y, we get

$$\frac{x}{y}\frac{dy}{dx} + (\log y) = x\, e^{x}$$

or $$\frac{1}{y}\frac{dy}{dx} + \frac{1}{x}(\log y) = e^{x}$$

putting $v = \log y$

or $$\frac{dv}{dx} = \frac{1}{y}\frac{dy}{dx}$$, the above equation reduce to

$$\frac{dv}{dx} + \frac{1}{x}\, v = e^{x} \qquad (1)$$

which is a linear equation in v

Its integrating factor $= e^{\int \frac{1}{x} dx} = e^{\log x} = x$

Multiplying both sides of (1) by the integrating factor x and integrating, we have

$v.\, x = C + \int x\, e^{x}\, dx$, where C is an arbitrary constant

or $$v.\, x = C + x\, e^{x} - \int 1.\, e^{x}\, dx$$

or $$(\log y)\, x = C + x\, e^{x} - e^{x}$$

Exact Differential Equations

A differential equation which can be obtained by direct differentiation of some function of x and y is called exact differential equation, consider the equation

$$Mdx + Ndy = 0 \text{ is exact}$$

$$\text{if } \frac{\partial M}{\partial y} = \frac{\partial N}{\partial x}$$

where M and N are the functions of x and y.

Solution of a exact differential equation is

$$\underset{\substack{\text{Regarding}\\ \text{y as a}\\ \text{constant}}}{\int M\,dx} + \underset{\substack{\text{only those}\\ \text{terms of N}\\ \text{not containing x}}}{\int N\,dy} = C$$

Example 16. Solve

$$\left\{y\left(1+\frac{1}{x}\right)+\cos y\right\}dx + (x + \log x - x \sin y)\,dy = 0$$

(I.A.S. 1993)

Solution. Here $M = y\left(1+\frac{1}{x}\right)\cos y$ and $N = x + \log x - x \sin y$

$$\therefore \frac{\partial M}{\partial y} = \left(1+\frac{1}{x}\right) - \sin y \text{ and } \frac{\partial N}{\partial x} = 1 + \frac{1}{x} - \sin y$$

$$\Rightarrow \frac{\partial M}{\partial y} = \frac{\partial N}{\partial x}, \text{ hence the given equation is exact}$$

Regarding y as constant, $\int M dx = \int\left\{y\left(1+\frac{1}{x}\right)+\cos y\right\}dx$

$$= y\int\left(1+\frac{1}{x}\right)dx + \cos y\int dx$$

$$= y\,(x + \log x) + (\cos y)\,x \qquad (1)$$

Also no new term is obtained by integrating N with respect to y.

$\therefore$ From (1), the required solution is

$y\,(x + \log x) + x \cos x = C$, where C is an arbitrary constant

Example 17. Verify that the equation

$(x^4 - 2xy^2 + y^4)\,dx - (2x^2y - 4xy^3 + \sin y)\,dy = 0$ is exact and solve it

(U.P.P.C.S. 1996)

Solution. Here $M = x^4 - 2xy^2 + y^4$, $\quad N = -2x^2 y + 4xy^3 - \sin y$

$$\therefore \frac{\partial M}{\partial y} = -4xy + 4y^3 \text{ and } \frac{\partial N}{\partial x} = -4xy + 4y^3$$

$$\therefore \frac{\partial M}{\partial y} = \frac{\partial N}{\partial x}, \text{ hence the given equation is exact}$$

Regarding y as constant, $\int M dx = \int (x^4 - 2xy^2 + y^4)\, dx$

$$= \frac{x^5}{5} - x^2y^2 + y^4x \qquad (1)$$

Integrating N with respect to y, we get

$$\int (-2x^2y + 4\,xy^3 - \sin y)\, dy = -x^2y^2 + xy^4 + \cos y$$

omitting from this the term $-x^2y^2$ and xy^4 which are already occuring in (1) we get cos y

∴ From (1) and (2) the required solution is

$$\frac{x^5}{5} - x^2y^2 + y^2x + \cos y = C$$

where C is an arbitrary constant

Rules for finding integrating factor

A differential equation of the type Mdx + Ndy = 0 which is not exact can be made exact by multiplying the equation by some function of x and y, which is called an integrating factor.

A few methods (without proof) are given below for finding the integrating factor in certain cases.

Method I. Integrating factor found by inspection.

In the case of some differential equation the integrating factor can be found by inspection. A few exact differentials are given below which would help students (if they commit these to memory) in finding the integrating factors.

(a) $d\left(\frac{y}{x}\right) = \frac{x\,dy - y\,dx}{x^2}$ (b) $d\left(\frac{x}{y}\right) = \frac{y\,dx - x\,dy}{y^2}$

(c) $d\,(xy) = x\,dy + y\,dx$ (d) $d\left(\frac{x^2}{y}\right) = \frac{2yx\,dx - x^2\,dy}{y^2}$

(e) $d\left(\frac{y^2}{x}\right) = \frac{2xy\,dy - y^2\,dx}{x^2}$ (f) $d\left(\frac{y^2}{x^2}\right) = \frac{2x^2y\,dy - 2xy^2\,dx}{x^4}$

(g) $d\left(\tan^{-1}\frac{x}{y}\right) = \frac{y\,dx - x\,dy}{x^2 + y^2}$ (h) $d\left(\tan^{-1}\frac{y}{x}\right) = \frac{x\,dy - y\,dx}{x^2 + y^2}$

(i) $d\left(\frac{1}{2}\log(x^2 + y^2)\right) = \frac{x\,dx + y\,dy}{x^2 + y^2}$ (j) $d\left(-\frac{1}{xy}\right) = \frac{x\,dy + y\,dx}{x^2y^2}$

(k) $d\log\left(\frac{x}{y}\right) = \frac{y\,dx - x\,dy}{xy}$ (l) $d\log\left(\frac{y}{x}\right) = \frac{x\,dy - y\,dx}{xy}$

(m) $d\left(\frac{e^x}{y}\right) = \frac{ye^x\,dx - e^x\,dy}{y^2}$

Example 18. Solve $(1 + xy)\,y\,dx + (1 - xy)\,x\,dy = 0$

(I.A.S. 1992, M.P.P.C.S. 1974)

Solution. The given eqution can be written as

$$(y\,dx + x\,dy) + (xy^2\,dx - x^2y\,dy) = 0$$

or $$d(yx) + xy^2\,dx - x^2y\,dy = 0$$

Dividing both sides of this equation by x^2y^2, we get

$$\frac{d(yx)}{x^2y^2} + \frac{1}{x}\,dx - \frac{1}{y}\,dy = 0$$

or $$\frac{1}{z^2}\,dz + \frac{1}{x}\,dx - \frac{1}{y}\,dy = 0, \quad \text{where } z = xy$$

Integrating, $-\frac{1}{z} + \log x - \log y = C$, where C is an arbitrary constant

or $$-\frac{1}{xy} + \log x - \log y = C, \text{ putting } z = xy$$

or $$\log(x/y) = C + \frac{1}{xy}$$

Example 19. Solve $(xy^2 + 2x^2y^3)\,dx + (x^2y - x^3y^2)\,dy = 0$

(U.P.P.C.S. 1993)

Solution. The given equation can be rewritten as

$$xy^2 (1 + 2xy)\, dx + x^2y (1 - xy)\, dy = 0$$

or $$y (1+ 2xy)\, dx + x (1 - xy)\, dy = 0$$

or $$(y\, dx + x\, dy) + 2xy^2\, dx - x^2y\, dy = 0$$

or $$d (xy) + 2xy^2\, dx - x^2y\, dy = 0$$

Dividing both sides of this equation by x^2y^2, we get

$$\frac{d(xy)}{x^2 y^2} + \frac{2}{x} dx - \frac{1}{y} dy = 0$$

or $$\left(\frac{1}{z^2}\right) dz + (z/x)\, dx - (1/y)\, dy = 0, \text{ where } z = xy$$

Integrating, $-\frac{1}{z} + 2 \log x - \log y = C$, where C is an arbitrary constant

or $$-\left(1/_{xy}\right) + \log\left(x^2/_{y}\right) = C, \text{ putting } z = xy$$

or $$\log\left(x^2/_{y}\right) = C + \left(1/_{xy}\right)$$

Example 20. Solve $y \sin 2x\, dx = (1 + y^2 + \cos^2 x)\, dy$

(I.A.S. 1996)

Solution. The given equation can be written as

$$2y \sin x \cos x\, dx - \cos^2 x\, dy = (1 + y^2)\, dy$$

or $$-d\,[y \cos^2 x] = (1 + y^2)\, dy$$

Integrating, $-y \cos^2 x = y + \frac{y^3}{3} + C$, where C is an arbitrary constant.

Method II. In the differential equation $Mdx + Ndy = 0$, If $M = y f_1(xy)$ and $N = x f_2(xy)$, then $\frac{1}{Mx - Ny}$ is an integrating factor, Provided $Mx - Ny \neq 0$

Example 21. Solve

$(xy \sin xy + \cos xy)\, y\, dx + (xy \sin xy - \cos xy)\, x dy = 0$

(Bihar P.C.S. 2007, U.P.P.C.S. 1990, 94)

Solution. Integrating factor

$$= \frac{1}{(xy \sin xy + \cos xy)\, xy - (xy \sin xy - \cos xy)\, xy}$$

$$= \frac{1}{2yx \cos xy}$$

Multiplying both sides of the given equation by this integrating factor, we get

$$\frac{1}{2}\left(\tan xy + \frac{1}{xy}\right) y\, dx + \frac{1}{2}\left(\tan\, xy - \frac{1}{xy}\right) x\, dy = 0$$

or $$(\tan xy)(y\, dx + x\, dy) + \frac{1}{x}\, dx - \frac{1}{y}\, dy = 0$$

or $$(\tan xy)\, d(xy) + \frac{1}{x}\, dx - \frac{1}{y}\, dy = 0$$

or $$\tan z\, dz + \frac{1}{x}\, dx - \frac{1}{y}\, dy = 0, \text{ where } z = xy$$

Integrating term by term, we get

$\log(\sec z) + \log x - \log y = \log C$, where C is an arbitrary constant

or $$\log\left\{\frac{x \sec z}{y}\right\} = \log C$$

or $$\frac{x}{y}(\sec z) = C$$

or $$x \sec(xy) = Cy$$

Method III. In the differential equation $Mdx + Ndy = 0$ if $\frac{1}{N}\left(\frac{\partial M}{\partial y} - \frac{\partial N}{\partial x}\right)$ is a function of x alone, say f(x), then the integrating factor is $e^{\int f(x)\, dx}$

Example 22. solve $(x^2 + y^2 + 1)\, dx - 2xy\, dy = 0$

(U.P.P.C.S. 1988, 82)

Solution. Here $M = x^2 + y^2 + 1$ and $N = -2xy$

$$\therefore \frac{\partial M}{\partial y} = 2y \quad \text{and} \quad \frac{\partial M}{\partial x} = -2y$$

Therefore, $\frac{1}{N}\left(\frac{\partial M}{\partial y}-\frac{\partial N}{\partial x}\right)=\frac{1}{-2xy}(2y+2y)=-\frac{2}{x}$, which is a function of x alone. Hence method III is applicable

Here $f(x) = -2/x$

$\therefore$ Integrating factor $= e^{\int f(x)\,dx} = e^{-\int \frac{2}{x}dx}$

$$= e^{-2\log x} = 1/x^2$$

Multiplying the given equation by this integrating factor $1/x^2$, we get

$$\frac{1}{x^2}\left(x^2+y^2+1\right)dx-\frac{1}{x^2}(2xy)\,dy=0$$

or $$\left[1+\frac{y^2}{x^2}+\frac{1}{x^2}\right]dx-2\frac{y}{x}dy=0$$

or $$\left[1+\frac{1}{x^2}\right]dx+\left[\frac{y^2}{x^2}dx-2\frac{y}{x}dy\right]$$ which is an exact

or $$\left[1+\frac{1}{x^2}\right]dx+d\left(-y^2/x\right)=0 \text{ using method I}$$

Integrating term by term, we get

$$x-\frac{1}{x}+\left(-y^2/x\right)=C, \text{ where C is constant of integration}$$

or $$x^2-1-y^2=Cx \text{ is the required solution.}$$

Method IV. In the equation $Mdx + Ndy = 0$

If $\frac{1}{M}\left(\frac{\partial N}{\partial x}-\frac{\partial M}{\partial y}\right)$ is a function of y alone, say f(y),

then the integrating factor is $e^{\int f(y)\,dy}$

Example 23. Solve $(xy^3 + y)\,dx + 2\,(x^2y^2 + x + y^4)\,dy = 0$

Solution. Here $M = xy^3 + y$ and $N = 2x^2y^2 + 2x + 2y^4$

$$\therefore \frac{\partial M}{\partial y}=2xy^2+1, \text{ and } \frac{\partial N}{\partial x}=4xy^2+2$$

$$\therefore \frac{1}{M}\left(\frac{\partial N}{\partial x} - \frac{\partial M}{\partial y}\right) = \frac{1}{(xy^3 + y)}\left\{\left(4xy^2 + 2\right) - \left(3xy^2 + 1\right)\right\}$$

$$= \frac{1}{y\,(xy^2 + 1)}(xy^2 + 1) = \frac{1}{y},$$ which is a function of y alone and equal to f(y) say.

Then integrating factor $= e^{\int f(y)\,dy} = e^{\int \frac{1}{y}\,dy}$

$$= e^{\log y} = y$$

Multiplying the given equation by integrating factor y we get

$(xy^4 + y^2)\,dx + (2x^2 y^3 + 2xy + 2y^5)\,dy = 0$

which is an exact differential equation and solving by the method of exact, we have

$$3x^2y^4 + 6xy^2 + 2y^6 = 6\,C$$ is the required solution.

Example 24. Solve $(3x^2 y^4 + 2xy)\,dx + (2x^3y^3 - x^2)\,dy = 0$

(U.P.P.C.S. 2001)

Solution. Here $M = 3x^2 y^4 + 2xy$, $N = 2x^3 y^3 - x^2$

and $\frac{\partial M}{\partial y} = 12\,x^2y^3 + 2x$, $\quad \frac{\partial N}{\partial x} = 6x^2y^3 - 2x$

As $\frac{\partial M}{\partial y} \neq \frac{\partial N}{\partial x}$, so the equation is not exact in this form. Thus, we have to find the integrating factor by trial. In the present case, we see that

$$\frac{1}{M}\left(\frac{\partial N}{\partial x} - \frac{\partial M}{\partial y}\right) = \frac{6x^2y^3 - 2x - 12x^2y^3 - 2x}{3x^2y^4 + 2xy} = -\frac{2}{y}$$, is the function of y alone

The integrating factor is $e^{-\int \frac{2}{y}\,dy} = \frac{1}{y^2}$

Thus, the differential equation becomes

$(3x^2y^2 + 2x/y)\,dx + (2x^3\,y - x^2/y^2)\,dy = 0$

Which is an exact, as $\frac{\partial M}{\partial y} = 6x^2y - 2x/y^2 = \frac{\partial N}{\partial x}$

Its solution is

$$\int (3x^2 y^2 + 2x/y)\, dx = 0$$

or $$x^3 y^2 + x^2/y = C$$

Where C is an arbitrary constant.

Method V. If the equation Mdx + N dy = 0 is homogeneous then $\dfrac{1}{Mx + Ny}$ is an integrating factor, Provided $Mx + Ny \neq 0$

Example 25. Solve $x^2y\, dx - (x^3 + y^3)\, dy = 0$

Solution. Here $M = x^2y$ and $N = -x^3 - y^3$

$\therefore\ Mx + Ny = x^3y - x^3y - y^4 = -y^4 \neq 0$

$\therefore$ Integrating factor $= \dfrac{1}{Mx + Ny} = -\dfrac{1}{y^4} \neq 0$

Multiplying the given equation by this integrating factor $-1/y^4$, we get

$$-\frac{x^2}{y^3}\, dx + \left(\frac{x^3}{y^4} + \frac{1}{y}\right) dy = 0$$

In this form of the equation, $M = \dfrac{-x^2}{y^3}$ and $N = \dfrac{x^3}{y^4} + \dfrac{1}{y}$

$$\therefore\ \frac{\partial M}{\partial y} = \frac{3x^2}{y^4} \text{ and } \frac{\partial N}{\partial x} = \frac{3x^2}{y^4}$$

$$\Rightarrow\ \frac{\partial M}{\partial y} = \frac{\partial N}{\partial x}$$

Hence this equation is an exact

Solving, we get

$$x^3 = 3y^3 (\log y - C)$$

Method VI. If the equation be of the form $x^a y^b (my\, dx + nx\, dy) + x^c y^d (py\, dx + qx\, dy) = 0$

where a, b, c, d, m, n, p and q are constants, then the integrating factor is $x^h y^k$, where h and k can be obtained by applying the condition that after multiplication by $x^h y^k$ the equation is exact.

Example 26. Solve $(y^2 + 2x^2y)\,dx + (2x^3 - xy)\,dy = 0$

or $\qquad y(y\,dx - x\,dy) + 2x^2(y\,dx + x\,dy) = 0$ **(U.P.P.C.S. 2000)**

Solution. The given equation can be rewritten as

$$y(y + 2x^2)\,dx + x(2x^2 - y)\,dy = 0$$

which is of the form as given in method VI

Let $x^h y^k$ be an integrating factor

Multiplying the given equation by $x^h y^k$, we get $(x^h y^{k+2} + 2x^{h+2} y^{k+1})\,dx + (2x^{h+3} y^k - x^{h+1} y^{k+1})\,dy = 0$

Here $M = x^h y^{k+2} + 2x^{h+2} y^{k+1}$ and $N = 2x^{h+3} y^k - x^{h+1} y^{k+1}$

$$\therefore \frac{\partial M}{\partial y} = (k+2)\,x^h y^{k+1} + 2(k+1)\,x^{h+2} y^k \qquad (1)$$

and
$$\frac{\partial N}{\partial x} = 2(h+3)\,x^{h+2} y^k - (h+1)\,x^h y^{k+1} \qquad (2)$$

If the equation (A) be exact we must have $\dfrac{\partial M}{\partial y} = \dfrac{\partial N}{\partial x}$

Or $\quad (k+2)\,x^h y^{k+1} + 2(k+1)\,x^{h+2} y^k = -(h+1)\,x^h y^{k+1} + 2(h+3)\,x^{h+2} y^k$ from (1) and (2)

Equating coefficients of $x^h y^{k+1}$ and $x^{h+2} y^k$ on both sides we get $k + 2 = -(h+1)$ and $2(k+1) = 2(h+3)$

hence solving we get

$$h = -\frac{5}{2},\ k = -\frac{1}{2}$$

$\therefore$ The integrating factor $= x^h y^k = x^{-5/2} y^{-1/2}$

Multiplying the given by $x^{-5/2} y^{-1/2}$, we get

$(x^{-5/2} y^{3/2} + 2x^{-1/2} y^{1/2})\,dx + (2x^{1/2} y^{-1/2} - x^{-3/2} y^{1/2})\,dy = 0$

In this form, we have $\dfrac{\partial M}{\partial y} = \dfrac{\partial N}{\partial x}$, the equation is exact

$\therefore$ Regarding y as constant

$$\int M\,dx = \int (x^{-\frac{5}{2}} y^{3/2} + 2x^{-1/2} y^{1/2})\,dx$$

$$= \frac{2}{3} x^{-3/2} y^{3/2} + 4x^{1/2} y^{1/2}$$

Also no new term is obtained by integrating N with respect to y, Hence the required solution is

$$- \frac{2}{3} x^{-3/2} y^{3/2} + 4x^{1/2} y^{1/2} = C$$

Where C is an arbitrary constant

EXERCISE

Solve the following differential equations

1. $(x^2 - 2x + 2y^2)\,dx + 2xy\,(1 + \log x^2)\,dy = 0$

(I.A.S. 1991)

Ans. $x^2 - 4x + 4y^2 \log x + 2y^2 = C$

2. $x^2 \frac{dy}{dx} = x^2 + xy + y^2$

(U.P.P.C.S. 1998)

Ans. $x = C\,e^{\tan^{-1}\left(\frac{y}{x}\right)}$

3. $(3y + 2x + 4)\,dx - (4x + 6y + 5)\,dy = 0$ **(U.P.P.C.S. 2001)**

Ans. $14\,(2x + 3y) - 9 \log (14x + 21y + 22) = 49x + C$

4. $xy - \frac{dy}{dx} = y^3 e^{-x^2}$

(I.A.S. 1998)

Ans. $e^x = y^2 (x + C)$

5. Solve the initial value problem $\frac{dy}{dx} = \frac{x}{x^2 y + y^3}$, $y(0) = 0$

(I.A.S. 1997)

Ans. $(x^2 + y^2 + 2) = 2e^{y^2/2}$

6. Show that the equation $(4x + 3y + 1)\,dx + (3x + 2y + 1)\,dy = 0$ represents a family of hyperbolas having as asymptotes the line $x + y = 0$, $2x + y + 1 = 0$

(I.A.S. 1998)

7. The equations of motion of a particle are given by $\frac{dx}{dt} + wy = 0$, $\frac{dy}{dt} - wx = 0$, Find the path of the particle and so that it is a circle.

(U.P.T.U. 2009)

Hint. $y(t) = C_1 \cos wt + C_2 \sin wt$

$x(t) = C_2 \cos wt - C_1 \sin wt$

so $x^2 + y^2 = C_1^2 + C_2^2 = R^2$

Objective Type of Questions

Choose a correct answer from the four answers given in each of the following questions:

1. The differential equation $y\,dx + x\,dy = 0$ represents a family of

(U.P.P.C.S. 1999)

(a) Circles (b) Ellipses

(c) Cycloids (d) Rectangular hyperbolas

Ans. (d)

2. The solution of $(xy^2 + 1)\,dx + (x^2y + 1)\,dy = 0$ is

(U.P.P.C.S. 1999)

(a) $x^2y^2 + 2x^2 + 2y^2 = C$ (b) $x^2y^2 + x^2 + y^2 = C$

(c) $x^2y^2 + x + y = C$ (d) $x^2y^2 + 2x + 2y = C$

Ans. (d)

3. The differential equation for the family of all tangents to the parabola $y^2 = 2x$

(U.P.P.C.S. 1999)

(a) $2x\,(y')^2 + 1 = 2yy'$ (b) $2xy + 1 = 2yy'$

(c) $2x^2y' + 1 = 2yy'$ (d) $2\,(y')^2 + x = 2yy'$

Ans. (a)

4. The general solution of the differential equation $(1 + x)\,y\,dx + (1 - y)\,x\,dy = 0$ is

(U.P.P.C.S. 2000)

(a) $xy = C\,e^{x-y}$ (b) $x + y = C\,e^{xy}$

(c) $xy = C\,e^{y-x}$ (d) $x - y = C\,e^{xy}$

Ans. (c)

5. An integrating factor of the differential equation $(1 + x^2)\dfrac{dy}{dx} + 2xy = \cos x$ is

(U.P.P.C.S. 2000)

(a) $1 + x^2$ (b) $\dfrac{1}{1+x^2}$

(c) $\log(1 + x^2)$ (d) $-\log(1 + x^2)$

Ans. (a)

6. If $\dfrac{dy}{dx} = e^{-2y}$, $y = 0$ when $x = 5$, then the value of x for $y = 3$ is

(U.P.P.C.S. 2000)

(a) $(e^6 + 9)/2$ (b) e^5

(c) $\log_e 6$ (d) $e^6 + 1$

Ans. (a)

7. The solution of $(x - 1)\,dy = y\,dx$, $y(0) = -5$ is

(a) $y = 5(x - 1)$ (b) $y = -5(x - 1)$

(c) $y = 5(x + 1)$ (d) $y = -5(x + 1)$

Ans. (a)

8. The differential equation $x\,dx - y\,dy = 0$ represents a family of

(U.P.P.C.S. 2001)

(a) Circles (b) Ellipse

(c) rectangular hyperbolas (d) Cycloids

Ans. (c)

9. The solution of the differential equation $(x + 2y)\,dy - (2x - y)\,dx = 0$ is

(U.P.P.C.S. 2001)

(a) $x^2 + y^2 - 2xy = C$ (b) $xy + y^2 + x^2 = C$

(c) $x^2 + 4xy + y^2 = C$ (d) $xy + y^2 - x^2 = C$

Ans. (d)

10. In the differential equation $\dfrac{x\,dy}{dx} + my = e^{-x}$, if the integrating factor is $\dfrac{1}{x^2}$, then the value of m is

(a) 2 (b) -2

(c) 1 (d) -1

Ans. (b)

11. The solution of the variable separable equation $(x^2 + 1)(y^2 - 1)\,dx + xy\,dy = 0$ is

(I.A.S. 1988)

(a) $y^2 - 1 = x^2 + 1 + C$ (b) $\log(y^2 - 1) = C \log(x^2 + 1)$

(c) $y^2 = 1 + C\dfrac{e^{-x^2}}{x^2}$ (d) none of these

Ans. (c)

12. The solution of the differential equation $\dfrac{dy}{dx} = e^{x-y} + x^2 e^{-y}$ is

(M.P.P.C.S. 1991, R.A.S. 1993, U.P.P.C.S. 1995)

(a) $e^y = e^x + \dfrac{1}{3}x^3 + C$ (b) $e^y = e^{-x} + \dfrac{1}{3}x^3 + C$

(c) $e^y = e^x + x^3 + C$ (d) $e^{-y} = \dfrac{1}{3}x^3 + e^x + C$

Ans. (a)

13. If $\dfrac{1}{M}\left(\dfrac{\partial N}{\partial x} - \dfrac{\partial M}{\partial y}\right) = f(y)$ a function of y alone, then the integrating factor of $Mdx + Ndy = 0$ is

(M.P.P.C.S. 1991, 93)

(a) $e^{-\int f(y)\,dy}$ (b) $e^{\int f(y)\,dy}$

(c) $f(y)\int e^{f(y)}dy$ (d) $\int e^{f(y)} f(y)\,dy$

Ans. (b)

14. The solution of the differential equation $(x + y)^2 \dfrac{dy}{dx} = a^2$ is given by

(I.A.S. 1994)

(a) $y + x = a \tan\left(\dfrac{y - c}{a}\right)$ (b) $(y - x) = a \tan(y - c)$

(c) $y - x = \tan\left(\dfrac{y - c}{a}\right)$ (d) $a(y - x) = \tan\left(\dfrac{y - c}{a}\right)$

Ans. (a)

15. $Pdx + x \sin y\, dy = 0$ is exact, then P can be

(MP.P.P.C.S. 1994)

(a) $\sin y + \cos y$ (b) $-\sin y$

(c) $x^2 - \cos y$ (d) $\cos y$

Ans. (c)

16. The solution of the differential equation $(x - y^2)\, dx + 2xy\, dy = 0$ is

(I.A.S. 1993)

(a) $ye^{y^2/x} = A$ (b) $x\, e^{y^2/x} = A$

(c) $y\, e^{x/y^2} = A$ (d) $x\, e^{x/y^2} = A$

17. The solution of the equation $\dfrac{dy}{dx} + 2xy = 2xy^2$ is

(I.A.S. 1994)

(a) $y = \dfrac{cx}{1 + e^{x^2}}$ (b) $y = \dfrac{1}{1 - c\,e^{x}}$

(c) $y = \dfrac{1}{1 + c\,e^{x^2}}$ (d) $y = \dfrac{cx}{1 + e^{x^2}}$

Ans. (c)

18. The homogeneous differential equation

$$M(x, y)\, dx + N(x, y)\, dy = 0$$

can be reduced to a differential equation in which the variables are separable, by the substitution-

(I.A.S. 1996)

(a) $y = vx$ (b) $xy = v$

(c) $x + y = v$ (d) $x - y = v$

Ans. (a)

19. The solution of the differential equation

$$\frac{dy}{dx} + \frac{y}{x} = x^2$$

Under the condition that y = 1, when x = 1 is

(I.A.S. 1996)

(a) $4xy = x^3 + 3$ (b) $4xy = y^4 + 3$

(c) $4xy = x^4 + 3$ (d) $4xy = y^3 + 3$

Ans. (c)

20. The necessary condition to exact the differential equation Mdx + Ndy = 0 will be

(MP.P.C.S. 1993, R.A.S. 1995)

(a) $\frac{\partial M}{\partial y} = \frac{\partial N}{\partial x}$ (b) $\frac{\partial M}{\partial x} = \frac{\partial N}{\partial y}$

(c) $\frac{\partial^2 M}{\partial x^2} = \frac{\partial^2 N}{\partial y^2}$ (d) $\frac{\partial^2 M}{\partial y^2} = \frac{\partial^2 N}{\partial x^2}$

21. If I_1, I_2 are integrating factors of the equations $xy' + 2y = 1$ and $xy' - 2y = 1$ then

(M.P.P.C.S. 1994)

(a) $I_1 = -I_2$ (b) $I_1 I_2 = x$

(c) $I_1 = x^2 I_2$ (d) $I_1 I_2 = 1$

Ans. (d)

22. The family of conic represented by the solution of the differential equation $(4x + 3y + 1)\,dx + (3x + 2y + 1)\,dy = 0$ is

(U.P.P.C.S. 1994)

(a) Circles (b) Parabolas

(c) hyperbolas (d) ellipse

Ans. (c)

Chapter 3

Linear Differential Equations with Constant Coefficients and Applications

INTRODUCTION

A differential equation is of the form

$$\frac{d^n y}{dx^n} + a_1 \frac{d^{n-1} y}{dx^{n-1}} + a_2 \frac{d^{n-2} y}{dx^{n-2}} + \ldots\ldots\ldots + a_n y = Q \qquad (1)$$

where $a_1, a_2, \ldots\ldots. a_n$ are constants and Q is a function of x only, is called a linear differential equation of nth order. Such equations are most important in the study of electro-mechanical vibrations and other engineering problems.

The operator $\frac{d}{dx}$ is denoted by D.

$\therefore D^n y + a_1 D^{n-1} y + \ldots\ldots\ldots\ldots + a_n y = Q$

or $\qquad f(D)\, y = Q$

where $f(D) = D^n + a_1 D^{n-1} + \ldots\ldots\ldots\ldots + a_n$

Solution of the Differential Equation

If the given equation is

$$\frac{d^n y}{dx^n} + a_1 \frac{d^{n-1} y}{dx^{n-1}} + a_2 \frac{d^{n-2} y}{dx^{n-2}} + \ldots\ldots\ldots + a_n y = 0 \qquad (1)$$

or $\qquad (D^n + a_1 D^{n-1} + a_2 D^{n-2} + \ldots\ldots\ldots + a_n)\, y = 0 \qquad (2)$

Let $y = e^{mx}$

$\Rightarrow \qquad D^r y = m^r e^{mx}, 1 \leq r \leq n$

$\therefore$ Then from equation (2)

$m^n + a_1 m^{n-1} + a_2 m^{n-2} + \ldots\ldots\ldots + a_n) e^{mx} = 0$

$y = e^{mx}$ is a solution of (1), if

$m^n + a_1 m^{n-1} + a_2 m^{n-2} + ... + a_n = 0$

This equation is called the auxiliary equation.

Case 1. When Auxiliary Equation has Distinct and real Roots

Let m_1, m_2, ... m_n are distinct roots of the auxiliary equation, then the general solution of (1) is $y = C_1 e^{m_1 x} + C_2 e^{m_2 x} + ... + C_n e^{m_n x}$

where $C_1, C_2, ... C_n$ are arbitrary constants

Illustration. Solve the differential equation

$$\frac{d^2y}{dx^2} + 3\frac{dy}{dx} - 54\,y = 0$$

Solution. The given equation is

$$(D^2 + 3D - 54)\,y = 0$$

Here auxiliary equation is

$$m^2 + 3m - 54 = 0$$

or $$(m + 9)(m - 6) = 0$$

$\Rightarrow$ $$m = 6, -9$$

Hence the general solution of the given differential equation is $y = C_1 e^{6x} + C_2 e^{-9x}$

Case II. When Auxiliary Equation has real and some equal roots.

If the auxiliary equation has two roots equal, say $m_1 = m_2$ and others are distinct say $m_3, m_4, ... m_n$. In this Case the general solution of the equation (1) is

$y = (C_1 + C_2 x)\, e^{m_1 x} + C_3 e^{m_3 x} + ... + C_n e^{m_n x}$

where $C_1, C_2, C_3 .... C_n$ are arbitrary constants

Illustration. Solve the differential equation

$$(D^4 - D^3 - 9D^2 - 11D - 4)\,y = 0$$

Solution. The auxiliary equation of the give equation is

$m^4 - m^3 - 9m^2 - 11m - 4 = 0$

or $$(m + 1)^3 (m - 4) = 0$$

$\Rightarrow$ $$m = -1, -1, -1, 4$$

Hence, the required solution is

$$y = (C_1 + C_2 x + C_3 x^2)\, e^{-x} + C_4 e^{4x}$$

or $$y = (C_1 x^2 + C_2 x + C_3) e^{-x} + C_4 e^{4x}$$

Case III. When the auxiliary equation has imaginary roots

If there are one pair of imaginary roots say $m_1 = \alpha + i\beta$, $m_2 = \alpha - i\beta$ i.e. $\alpha \pm i\beta$ say then the required solution is

$e^{x\,(\text{Real part})}$ [C_1 {cos (imaginary part) x} + C_2 {sin (imaginary part) x}]

i.e. $$e^{\alpha x} [C_1 \cos \beta x + C_2 \sin \beta x]$$

or $$y = C_1 e^{\alpha x} \cos (\beta x + C_2)$$

Illustration. Solve $(D^2 - 2D + 5)\, y = 0$

Solution. Here the auxiliary equation is

$$m^2 - 2m + 5 = 0$$

or $$m = \frac{1}{2} \left[2 \pm \sqrt{(4 - 20)}\right] = 1 \pm 2i$$

∴ The required solution is

$y = e^x (C_1 \cos 2x + C_2 \sin 2x)$

Particular Integral (P.I.)

when the equation is

$$D^n + a_1 D^{n-1} + \ldots\ldots\ldots + a_n)\, y = Q$$

or $$f(D)\, y = Q$$

The general solution of $f(D)\, y = Q$ is equal to the sum of the general solution of $f(D)\, y = 0$ called complementary function (C.F.) and any particular integral of the equation $f(D)\, y = Q$

∴ General solution = C.F. + P.I.

A particular integral of the differential equation

$f(D)\, y = Q$ is given by $\frac{1}{f(D)} Q$

Methods of finding Particular integral

(A)

Case I. P.I., when Q is of the form of e^{ax}, where a is any constant and $f(a) \neq 0$

we know that $\quad D(e^{ax}) = a\, e^{ax}$

$$D^2 (e^{ax}) = a^2 e^{ax}$$

$$D^3 (e^{ax}) = a^3 e^{ax}$$

In general $D^n (e^{ax}) = a^n e^{ax}$

$$\therefore f(D)(e^{ax}) = f(a) e^{ax}$$

or $\frac{1}{f(D)} f(D) e^{ax} = \frac{1}{f(D)} f(a) e^{ax}$

or $e^{ax} = f(a) \frac{1}{f(D)} e^{ax}$ $\because$ f(a) is constant

or $\frac{1}{f(a)} e^{ax} = \frac{1}{f(D)} e^{ax}$

Hence P.I. = $\frac{1}{f(D)} e^{ax} = \frac{1}{f(a)} e^{ax}$, if $f(a) \neq 0$

Case II. P.I., when Q is of the form of e^{ax}, and f(a) = 0

Then $\frac{1}{f(D)} (e^{ax}) = e^{ax} \frac{1}{f(D+a)} 1$, which shows that if e^{ax} is brought to the left from the right of $\frac{1}{f(D)}$, then D should be replaced by (D + a)

Another method for Exceptional Case

If f(a) = 0, then

P.I. = $\frac{1}{f(D)} e^{ax} = x \frac{1}{f'(D)} e^{ax}$

$$= x \frac{e^{ax}}{f'(a)}, \text{ if } f'(a) \neq 0$$

If f' (a) = 0, then

P.I. = $\frac{1}{f(D)} e^{ax}$

$$= x^2 \frac{e^{ax}}{f''(a)}, \text{ if } f''(a) \neq 0$$

Example 1. Solve $(D^2 - 2D + 5) y = e^{-x}$

Solution. Here auxiliary equation is $m^2 - 2m + 5 = 0$, whose roots are

$$m = -1 \pm 2i$$

$\therefore$ C.F. $= e^{-x} [C_1 \cos 2x + C_2 \sin 2x]$, where C_1 and C_2 are arbitrary constants

and $$\text{P.I.} = \frac{1}{D^2 - 2D + 5} e^{-x}$$

$$= \frac{1}{(-1)^2 - 2(-1) + 5} e^{-x} \qquad \because \text{ here } a = -1$$

$$= \frac{1}{8} e^{-x}$$

$\therefore$ The required solution is y = C.F. + P.I.

i.e. $y = e^{-x} (C_1 \cos 2x + C_2 \sin 2x) + \frac{1}{8} e^{-x}$

Example 2. Solve $(D - 1)^2 (D^2 + 1)^2 y = e^x$

Solution. Here the auxiliary equation is

$$(m - 1)^2 (m^2 + 1)^2 = 0$$

or $$m = 1, 1, \pm i, \pm i$$

$\therefore$ C.F. $= (C_1 + C_2 x) e^x + (C_3 x + C_4) \cos x + (C_5 x + C_6) \sin x$

where c's are arbitrary constant

and $$\text{P.I.} = \frac{1}{(D - 1)^2 (D^2 + 1)^2} e^x$$

$$= \frac{1}{(D - 1)^2} \frac{1}{(1^2 + 1)^2} e^x$$

$$= \frac{1}{(D - 1)^2} \frac{1}{(2)^2} e^x$$

$$= \frac{1}{(D - 1)^2} \frac{1}{4} e^x$$

$$= e^x \frac{1}{(D + 1 - 1)^2} \frac{1}{4}$$

$$= e^x \frac{1}{D^2}\frac{1}{4} = \frac{1}{4} e^x \frac{1}{D^2}(1) = \frac{1}{4} e^x \frac{x^2}{2} = \frac{1}{8} x^2 e^x$$

∴ The required solution is y = C. F + P.I.

or $$y = (C_1 x + C_2) e^x + (C_3 x + C_4) \cos x + (C_5 x + C_6) \sin x + \frac{1}{8} x^2 e^x$$

Example 3. Solve $(D + 2)(D - 1)^3 y = e^x$

Solution. Here the auxiliary equation is

$$(m + 2)(m - 1)^3 = 0$$

or $$m = -2 \text{ and } m = 1 \text{ (thrice)}$$

Therefore C. F = $C_1 e^{-2x} + (C_2 x^2 + C_3 x + C_4) e^x$, where C_1, C_2 and C_3 are constants and

$$\text{P.I.} = \frac{1}{(D + 2)(D - 1)^3} e^x$$

$$= \frac{1}{(1 + 2)(D - 1)^3} e^x$$

$$= \frac{1}{3}\frac{1}{(D - 1)^3} e^x = \frac{1}{3} e^x \frac{1}{\{(D + 1) - 1\}} 1$$

$$= \frac{1}{3} e^x \frac{1}{D^3} 1$$

$$= \frac{1}{3} e^x \left(\frac{1}{6} x^3\right) = \frac{1}{18} x^3 e^x$$

∴ the required solution is y = C.F + P.I

or $$y = C_1 e^{-2x} + (C_2 x^2 + C_3 x + C_4) e^x + \frac{1}{18} x^3 e^x$$

(B) (i) P.I. when Q is of the form sin ax or cos ax and f $(-a^2) \neq 0$

$$\frac{1}{f(D^2)} \sin ax = \frac{1}{f(-a^2)} \sin ax, \text{ if } f(-a^2) \neq 0$$

$$\& \frac{1}{f(D^2)} \cos ax = \frac{1}{f(-a^2)} \cos ax, \text{ if } f(-a^2) \neq 0$$

(ii) P.I. when Q is of the form sin ax or cos ax and f $(-a^2) = 0$

Let $f(D) = D^2 + a^2$ and $Q = \sin ax$

Then $\quad \text{P.I.} = \dfrac{1}{D^2 + a^2} \sin ax$

$= \dfrac{1}{D^2 + a^2}$ (Imaginary part of e^{iax})

$= \text{Imaginary part of } \dfrac{1}{D^2 + a^2} e^{iax}$

$= \text{I.P. of } e^{iax} \dfrac{1}{\{(D + ia)^2 + a^2\}} 1$

$= \text{I.P. of } e^{iax} \dfrac{1}{(D^2 + 2iaD + i^2a^2 + a^2)} 1$

$= \text{I.P. of } e^{iax} \dfrac{1}{2iaD\left[1 + \dfrac{D}{2ia}\right]} 1$

$= \text{I.P. of } \dfrac{e^{iax}}{2ia} \dfrac{1}{D} \left(1 - \dfrac{D}{2ia} + \dots\dots\dots\right) 1$

$= \text{I.P. of } \dfrac{e^{iax}}{2ia} \dfrac{1}{D} 1 = \text{I.P. of } \dfrac{e^{iax}}{2ia} x$

$= \text{I.P. of } \dfrac{e^{iax}\, xi}{2i^2\, a}$

$= \text{I.P. of } \dfrac{1}{2} \left(-\dfrac{x}{a}\right) i (\cos ax + i \sin ax)$

$= \text{I.P. of } \dfrac{1}{2} \left(-\dfrac{x}{a}\right) (i \cos ax - \sin ax)$

$= -\dfrac{1}{2} \dfrac{x}{a} \cos ax$

$\therefore \dfrac{1}{D^2 + a^2} \sin ax = -\dfrac{x}{2a} \cos ax$

Now if $f(D) = D^2 + a^2$ and $Q = \cos ax$

Then $\text{P.I.} = \dfrac{1}{D^2 + a^2} \cos ax = \dfrac{1}{D^2 + a^2}$ (Real part of e^{iax})

$$= \text{Real part (or R.P.) of } \frac{1}{D^2 + a^2} e^{iax}$$

$$= \text{R.P. of } -\frac{1}{2}\left(\frac{x}{a}\right) (i \cos ax - \sin ax)$$

$$= \frac{x}{2a} \sin ax$$

$$\therefore \frac{1}{D^2 + a^2} \cos ax = \frac{x}{2a} \sin ax$$

Example 4. Solve $(D^2 + D + 1)\, y = \sin 2x$

Solution. Here the auxiliary equation is $m^2 + m + 1 = 0$

which gives $m = -\frac{1}{2} \pm i\left(\frac{1}{2}\sqrt{3}\right)$

$$\therefore \text{C.F.} = e^{-x/2}\left\{C_1 \cos\left(\frac{1}{2} x \sqrt{3}\right) + C_2 \sin\left(\frac{1}{2} x\sqrt{3}\right)\right\}$$

where C_1 and C_2 are arbitrary constants

$$\text{and P.I.} = \frac{1}{D^2 + D + 1} \sin 2x$$

$$= \frac{1}{(-2)^2 + D + 1} \sin 2x \text{ replacing } D^2 \text{ by } -2^2$$

$$= \frac{1}{D - 3} \sin 2x$$

$$= \frac{1}{(D - 3)(D + 3)} (D + 3) \sin 2x$$

$$= \frac{1}{D^2 - 9} (D + 3) \sin 2x = \frac{1}{-2^2 - 9} (D + 3) \sin 2x$$

$$= -\frac{1}{13} (D + 3) \sin 2x = \frac{-1}{13} [D (\sin 2x) + 3 \sin 2x]$$

$$= -\frac{1}{13} [2 \cos 2x + 3 \sin 2x]$$ Since D means differentiation with respect to x

$\therefore$ The complete solution is

$$y = e^{-x/2}\left[C_1 \cos\left(\frac{1}{2}x\sqrt{3}\right) + C_2 \sin\left(\frac{1}{2}x\sqrt{3}\right)\right] - \frac{1}{13}(2\cos 2x + 3\sin 2x)$$

Example 5. Solve $(D^2 - 5D + 6)\,y = \sin 3x$

Solution. Its auxiliary equation is $m^2 - 5m + 6 = 0$ which gives

$m = 2, 3$

$\therefore$ C.F. $= C_1 e^{2x} + C_2 e^{3x}$, where C_1 and C_2 are arbitrary constants

and
$$\text{P.I.} = \frac{1}{D^2 - 5D + 6}\sin 3x$$

$$= \frac{1}{-3^2 - 5D + 6}\sin 3x, \text{ replacing } D^2 \text{ by } -3^2$$

$$= \frac{1}{-(5D+3)}\sin 3x = \frac{-1}{(5D+3)(5D-3)}(5D-3)\sin 3x$$

$$= \frac{-1}{(25D^2 - 9)}(5D-3)\sin 3x$$

$$= \frac{-1}{\{25(-3^2) - 9\}}(5D-3)\sin 3x$$

$$= \frac{1}{234}[5D(\sin 3x) - 3\sin 3x]$$

$$= \frac{1}{234}[5 \times 3\cos 3x - 3\sin 3x]$$

$$= \frac{1}{78}(5\cos 3x - \sin 3x)$$

Hence the required solution is

$$y = C_1 e^{2x} + C_2 e^{3x} + \frac{1}{78}(5\cos 3x - \sin 3x)$$

Example 6. Solve $(D^3 + D^2 - D - 1)\,y = \cos 2x$

Solution. Its auxiliary equation is $m^3 + m^2 - m - 1 = 0$

or $(m^2 - 1)(m + 1) = 0$

or $m = 1, -1, -1$

$\therefore$ C.F. $= C_1 e^{x} + (C_2 x + C_3)\,e^{-x}$, where C_1, C_2 and C_3 are arbitrary constants and $\quad \text{P.I.} = \dfrac{1}{D^3 + D^2 - D - 1} \cos 2x$

$$= \frac{1}{D(D^2) + D^2 - D - 1} \cos 2x$$

$$= \frac{1}{D(-2^2) + (-2^2) - D - 1} \cos 2x = -\frac{1}{5}\,\frac{1}{D+1} \cos 2x$$

$$= -\frac{1}{5}\,\frac{D-1}{(D-1)(D+1)} \cos 2x$$

$$= -\frac{1}{5}\,\frac{1}{D^2 - 1}(D-1)\cos 2x = -\frac{1}{5}\,\frac{1}{(-2^2-1)}[D(\cos 2x) - \cos 2x]$$

$$= \frac{1}{25}(-2 \sin 2x - \cos 2x)$$

Therefore the complete solution is

$$y = \text{C.F.} + \text{P.I.}$$

or $\quad y = C_1 e^x + (C_2 x + C_3) e^{-x} - \dfrac{1}{25}(2 \sin 2x + \cos 2x)$

Example 7. Solve $(D^2 - 4D + 4)\, y = \sin 2x$, given that $y = 1/8$ and $Dy = 4$ when $x = 0$. Find also the value of y when $x = \pi/4$. $\left(\text{Here } D \equiv \dfrac{d}{dx}\right)$

Solution. Its auxiliary equation is $m^2 - 4m + 4 = 0$ or $m = 2, 2$

$\therefore$ C.F. $= (C_1 x + C_2) e^{2x}$, where C_1 and C_2 are arbitrary constants

and $\text{P.I.} = \dfrac{1}{D^2 - 4D + 4} \sin 2x = \dfrac{1}{(-2^2 - 4D + 4)} \sin 2x$

$$= -\frac{1}{4}\,\frac{1}{D} \sin 2x = -\frac{1}{4}\int \sin 2x \, dx = \frac{1}{8} \cos 2x$$

$\therefore$ The solution of the given equation is

$$y = \text{C.F.} + \text{P.I.} = (C_1 x + C_2) e^{2x} + \frac{1}{8} \cos 2x \qquad (1)$$

$\therefore \quad Dy = C_1 e^{2x} + 2(C_1 x + C_2) e^{2x} - \dfrac{1}{4} \sin 2x$

or $\quad Dy = (2C_1 x + 2C_2 + C_1) e^{2x} - \frac{1}{4} \sin 2x$

According to the problem $y = \frac{1}{8}$ and $Dy = 4$ when

$x = 0$, so from (1) and (2) we get

$$\frac{1}{8} = C_2 e^0 + \frac{1}{8} \cos o = C_2 + \frac{1}{8} \text{ or } C_2 = 0$$

And $4 = (2C_2 + 4) e^o = \frac{1}{4} \sin o = C_1 \because C_2 = 0$

or $\quad C_1 = 4$

Substituting these values of C_1, C_2 in (1) we get

$$y = 4xe^{2x} + \frac{1}{8} \cos 2x$$

when $x = \pi/4$, $y = 4\left(\frac{\pi}{4}\right) e^{\pi/2} + \frac{1}{8} \cos(\pi/2) = \pi e^{\pi/2}$

Example 8. Solve $(D^3 + a^2 D) y = \sin ax$

Solution. Here the auxiliary equation is

$$m^3 + a^2 m = 0 \text{ or } m = 0, \pm ai$$

$\therefore$ C.F. $= C_1 + C_2 \cos ax + C_3 \sin ax$, where C_1, C_2 and C_3 are arbitrary constants and

$$\text{P.I.} = \frac{1}{D^3 + a^2 D} \sin ax = \frac{1}{D^3 + a^2 D} (\text{Imaginary part of } e^{iax})$$

$$= \text{I.P. of } \frac{1}{D^3 + a^2 D} e^{iax}$$

$$= \text{I.P. of } e^{iax} \frac{1}{(D + ia)^3 + a^2 (D + ia)} 1$$

$$= \text{I.P. of } e^{iax} \frac{1}{D^3 + 3\,ia\,D^2 - 2a^2 D} 1$$

$$= \text{I.P. of } e^{iax} \frac{1}{-2a^2 D \left[1 - \frac{3i}{2a} D - \frac{1}{2a^2} D^2\right]}$$

$$= \text{I.P. of } \frac{e^{iax}}{-2a^2} \frac{1}{D} \left(1 - \frac{3i}{2a} D - \frac{1}{2a^2} D^2\right)^{-1} 1$$

$$= \text{I.P. of } \frac{e^{iax}}{-2a^2} \frac{1}{D} \left(1 - \frac{3i}{2a} D - \frac{1}{2\,a^2} D^2\right)^{-1} 1$$

$$= \text{I.P. of } \frac{e^{iax}}{-2a^2} \frac{1}{D} \left(1 + \frac{3i}{2a} D + \ldots\ldots\ldots\right) (1)$$

$$= \text{I.P. of } \frac{e^{iax}}{-2a^2} \frac{1}{D} (1) = \text{I.P. of } \frac{e^{iax}}{-2a^2} (x)$$

$$= -\left(\frac{1}{2a^2}\right) x \sin ax$$

$\therefore$ The required solution is y = C.F + P.I

or $$y = C_1 + C_2 \cos ax + C_3 \sin ax - \frac{x}{2a^2} \sin ax$$

Example 9. Solve $\frac{d^4y}{dx^4} - m^4y = \sin mx$

(I.A.S. 1991)

Solution. We have $\frac{d^4y}{dx^4} - m^4y = \sin mx$

or $$(D^4 - m^4)\, y = \sin mx, \; D = \frac{d}{dx}$$

Here the auxiliary equation is

$$M^4 - m^4 = 0$$

$\Rightarrow$ $$(M^2 - m^2)(M^2 + m^2) = 0$$

$\therefore$ M = –m, m, ± mi

Therefore, C_1 F = $C_1 e^{mx} + C_2 e^{-mx} + C_3 \cos mx + C_4 \sin mx$

where C_1, C_2, C_3 and C_4 are arbitrary constants

$$\text{P.I.} = \frac{1}{D^4 - m^2} \sin mx = \frac{1}{(D^2 - m^2)(D^2 + m^2)} \sin mx$$

$$= \frac{1}{(D^2+m^2)}\left\{\frac{1}{D^2-m^2}\sin mx\right\} = \frac{1}{D^2+m^2}\left\{\frac{1}{-m^2-m^2}\sin mx\right\} \text{ replacing } D^2 \text{ by } -m^2$$

$$= \frac{1}{D^2+m^2}\left(-\frac{1}{2m^2}\sin mx\right) = -\frac{1}{2m^2}\left(\frac{1}{D^2+m^2}\sin mx\right)$$

$$= -\frac{1}{2m^2}\left(-\frac{x}{2m}\cos mx\right) = \frac{x}{4m^3}\cos mx$$

$$\because \frac{1}{D^2+a^2}\sin ax = -\frac{x}{2a}\cos ax, \text{ if } f(-a^2) = 0$$

Hence the required solution is y = C.F + P.I

or $y = C_1 e^{mx} + C_2 e^{-mx} + C_3 \cos mx + C_4 \sin mx + \frac{x}{4m^3}\cos mx$

Example 10. Solve $\frac{d^3y}{dx^3} - 3\frac{d^2y}{dx^2} + 4\frac{dy}{dx} - 2y = e^x + \cos x$

(I.A.S. 1999, U.P.T.U. 2001, 2006)

Solution. The given can be rewritten as

$(D^3 - 3D^2 + 4D - 2)\, y = e^x + \cos x,\ D \equiv \frac{d}{dx}$

Here the auxiliary equation is

$$m^3 - 3m^2 + 4m - 2 = 0$$

$$\Rightarrow \quad (m-1)(m^2 - 2m + 2) = 0$$

i.e. $m = 1, 1 \pm i$

$\therefore$ C.F. $= C_1 e^x + e^x (C_2 \cos x + C_3 \sin x)$, where C_1, C_2 and C_3 are arbitrary constants.

$$\& \text{ P.I.} = \frac{1}{D^3 - 3D^2 + 4D - 2}(e^x + \cos x)$$

$$= \frac{1}{(D-1)(D^2-2D+2)} e^x + \frac{1}{D^3 - 3D^2 + 4D - 2} \cos x$$

$$= \frac{1}{(D-1)} \left\{ \frac{1}{1-2+2} e^x \right\} + \frac{1}{(-1)^2 D - 3(-1^2) + 4D - 2} \cos x$$

$$= \frac{1}{D-1} e^x + \frac{1}{3D+1} \cos x$$

$$= e^x \frac{1}{(D+1)-1} 1 + \frac{(3D-1)}{9D^2-1} \cos x$$

$$= e^x \frac{1}{D}.1 + \frac{3D-1}{9(-1^2)-1} \cos x$$

$$= x e^x - \frac{1}{10} (3D-1) \cos x$$

$$= x e^x - \frac{1}{10} (3D \cos x - \cos x)$$

$$= x e^x - \frac{1}{10} (-3 \sin x - \cos x)$$

$$= x e^x + \frac{1}{10} (3 \sin x + \cos x)$$

Hence, the required solution is y = C.F + P.I

$$y = C_1 e^x + e^x (C_2 \cos x + C_3 \sin x) + x e^x + \frac{1}{10} (3 \sin x + \cos x)$$

(C) To find P.I. when Q is of the form x^m

In this case $P.I = \frac{1}{f(D)} x^m$, where m is a positive integer

To evaluate this we take common the lowest degree from f(D), so that the remaining factor reduces to the form [1 + F (D)] or [1 – F (D)]. Now take this factor in the numerator with a negative index and expand it by Binomial theorem in powers of D upto the term D^m, (Since other higher derivatives of x^m will be zero) and operate upon x^m. The following examples will illustrate the method.

Example 11. Solve $\frac{d^2y}{dx^2} + \frac{dy}{dx} - 6y = x$

Solution. The given equation can be written as

$$(D^2 + D - 6) y = x$$

Its auxiliary equation is $m^2 + m - 6 = 0$ or $m = 2, -3$

$\therefore$ C.F. = $C_1 e^{2x} + C_2 e^{-3x}$, where C_1 and C_2 are arbitrary constants and $$\text{P.I.} = \frac{1}{D^2 + D - 6}\, x = \frac{1}{-6\left(1 - \frac{1}{6} D - \frac{1}{6} D^2\right)}\, x$$

$$= -\frac{1}{6}\left[1 - \frac{1}{6}(D + D^2)\right]^{-1} x$$

$$= -\frac{1}{6}\left[1 + \frac{1}{6}(D + D^2) + \ldots\ldots\ldots\ldots\right] x$$

$$= -\frac{1}{6}\left(x + \frac{1}{6}\right) = \frac{-1}{36}(6x + 1)$$

∴ The required solution is y = C.F + P.I

$$y = C_1 e^{2x} + C_2 e^{-3x} - \frac{1}{36}(6x + 1)$$

Example 12. Solve $(D^3 - D^2 - 6D)\, y = x^2 + 1$ Where $D \equiv \frac{d}{dx}$

(Bihar P.C.S. 1993)

Solution. Here the auxiliary equation is $m^3 - m^2 - 6m = 0$

or $$m(m^2 - m - 6) = 0$$

or $$m(m - 3)(m + 2) = 0$$

or $$m = 0, 3, -2$$

∴ $C_1 F = C_1 e^{0x} + C_2 e^{3x} + C_3 e^{-2x}$

or C.F. $= C_1 + C_2 e^{3x} + C_3 e^{-2x}$, where C_1, C_2 and C_3 are arbitrary constants.

and $$\text{P.I.} = \frac{1}{D^2 - D^2 - 6D}(x^2 + 1)$$

$$= \frac{1}{-6D\left(1 + \frac{1}{6} D - \frac{1}{6} D^2\right)}(x^2 + 1)$$

$$= -\frac{1}{6D}\left[1 - \left(\frac{D}{6} - \frac{D^2}{6}\right) + \left(\frac{D}{6} - \frac{D^2}{6}\right)^2 \ldots\ldots\ldots\right](x^2 + 1)$$

$$= -\frac{1}{6D}\left[1 + x^2 - \frac{1}{6} D(x^2) + \frac{7}{36} D^2(x^2)\right] \qquad \because D(1) = 0$$

$$= -\frac{1}{6D}\left(1 + x^2 - \frac{1}{3}x + \frac{7}{18}\right)$$

$$= -\frac{1}{6}\int\left(1 + x^2 - \frac{x}{3} + \frac{7}{18}\right)dx \qquad \because D \equiv \frac{d}{dx}$$

$$= -\frac{1}{6}\left[x + \frac{x^3}{3} - \frac{1}{6}x^2 + \frac{7}{18}x\right] = -\frac{1}{6}\left(\frac{25}{18}x + \frac{x^3}{3} - \frac{x^2}{6}\right)$$

$\therefore$ The required solution is y = C.F. + P.I.

or
$$y = C_1 + C_2 e^{3x} + C_3 e^{-2x} - \frac{25}{108}x - \frac{x^3}{18} + \frac{x^2}{36}$$

Example 13. Solve $\frac{d^2y}{dx^2} - 2\frac{dy}{dx} + 2y = x + e^x \cos x$

(U.P.T.U. 2002)

Solution. Given differential equation is

$$(D^2 - 2D + 2)\, y = x + e^x \cos x$$

Here the auxiliary equation is

$$m^2 - 2m + 2 = 0 \Rightarrow m = 1 \pm i$$

$\therefore$ C.F = $e^x (C_1 \cos x + C_2 \sin x)$, where C_1 and C_2 are arbitrary constants,

and P.I. $= \frac{1}{D^2 - 2D + 2}(x + e^x \cos x)$

$$= \frac{1}{D^2 - 2D + 2}x + \frac{1}{D^2 - 2D + 2}(e^x \cos x)$$

$$= \frac{1}{2\left[1 - D + \frac{D^2}{2}\right]}x + e^x \frac{1}{\left\{(D+1)^2 - 2(D+1) + 2\right\}}\cos x$$

$$= \frac{1}{2}\left(1 - D + \frac{D^2}{2}\right)^{-1}(x) + e^x \frac{1}{D^2 + 1}\cos x \quad \text{Here } f(-a^2) = 0$$

$$= \frac{1}{2}\left(1 + D - \frac{D^2}{2} + \ldots\ldots\ldots\right)(x) + e^x \frac{x}{2}\sin x$$

$$= \frac{1}{2}\left\{x + D(x) - \frac{1}{2}D^2(x)\right\} + \frac{xe^x}{2}\sin x$$

$$= \frac{1}{2}(x + 1) + \frac{x\,e^x}{2}\sin x$$

$\therefore$ The required solutions y = C.F + P.I

or $y = e^x (C_1 \cos x + C_2 \sin x) + \frac{1}{2}(x+1) + \frac{x e^x}{2} \sin x$

Example 15. Find the complete solution of

$$\frac{d^2y}{dx^2} - 3\frac{dy}{dx} + 2y = x e^{3x} + \sin 2x$$ **(U.P.T.U. 2003)**

Solution. The given differential equation is

$$(D^2 - 3D + 2)\, y = x e^{3x} + \sin 2x$$

Here the auxilliary equation is

$$m^2 - 3m + 2 = 0 \Rightarrow (m-1)(m-2) = 0 \Rightarrow m = 1, 2$$

$\therefore$ C.F $= C_1 e^x + C_2 e^{2x}$, where C_1 and C_2 being arbitrary constants

$$\& \quad \text{P.I.} = \frac{1}{D^2 - 3D + 2}(x e^{3x} + \sin 2x)$$

$$= \frac{1}{D^2 - 3D + 2}(x e^{3x}) + \frac{1}{D^2 - 3D + 2}(\sin 2x)$$

$$= e^{3x} \frac{1}{(D+3)^2 - 3(D+3) + 2}(x) + \frac{1}{-2^2 - 3D + 2}(\sin 2x)$$

$$= e^{3x} \frac{1}{D^2 + 3D + 2}(x) + \frac{1}{-3D - 2}\sin 2x$$

$$= e^{3x} \frac{1}{2} \frac{1}{\left(1 + \frac{3}{2}D + \frac{D^2}{2}\right)}(x) - \frac{1}{(3D+2)}(\sin 2x)$$

$$= \frac{1}{2} e^{3x} \left(1 + \frac{3D}{2} + \frac{D^2}{2}\right)^{-1}(x) - \frac{3D - 2}{9D^2 - 4}(\sin 2x)$$

$$= \frac{1}{2} e^{3x} \left[1 - \frac{3D}{2} - \frac{D^2}{2} \ldots\ldots\ldots\right](x) - \frac{(3D-2)}{9(-2^2) - 4}(\sin 2x) \text{ replacing } D^2 \text{ by} -2^2$$

$$= \frac{1}{2} e^{3x} \left(x - \frac{3}{2} \right) + \frac{1}{40} (3D \sin 2x - 2 \sin 2x)$$

$$= \frac{1}{2} e^{3x} \left(x - \frac{3}{2} \right) + \frac{1}{40} (6 \cos 2x - 2 \sin 2x)$$

$$= e^{3x} \left(\frac{x}{2} - \frac{3}{4} \right) + \frac{1}{20} (3 \cos 2x - \sin 2x)$$

Hence the required solution is y = C.F + P.I

or $y = C_1 e^x + C_2 e^{2x} + e^{3x} \left(\frac{x}{2} - \frac{3}{4} \right) + \frac{1}{20} (3 \cos 2x - \sin 2x)$

Example 16. A body executes damped forced vibrations given by the equation.

$$\frac{d^2x}{dt^2} + 2k \frac{dx}{dt} + b^2x = e^{-kt} \sin wt$$

Solve the equation for both the cases, when $w^2 \neq b^2 - k^2$ and $w^2 = b^2 - k^2$

[U.P.T.U. 2001, 03, 04 (C.O) 2005]

Solution. We have $\frac{d^2x}{dt^2} + 2k \frac{dx}{dt} + b^2x = e^{-kt} \sin wt$

or $(D^2 + 2k D + b^2) x = e^{-kt} \sin wt$

Here the auxiliary equation is $m^2 + 2km + b^2 = 0$

$$\Rightarrow m = -k \pm i\sqrt{(b^2 - k^2)}$$

Case 1. When $w^2 \neq b^2 - k^2$

$$C.F = e^{-kt} [C_1 \cos \sqrt{(b^2 - k^2)}\ t + C_2 \sin \sqrt{(b^2 - k^2)}\ t]$$

where C_1 and C_2 are arbitrary constants

$$\& \text{ P.I.} = \frac{1}{D^2 + 2kD + b^2} (e^{-kt} \sin wt)$$

$$= e^{-kt} \frac{1}{(D-k)^2 + 2k(D-k) + b^2} (\sin wt)$$

$$= e^{-kt} \frac{1}{D^2 + (b^2 - k^2)} (\sin wt) = e^{-kt} \frac{1}{w^2 + (b^2 - k^2)} \sin wt$$

$$= \frac{e^{-kt} \sin wt}{b^2 - k^2 - w^2}$$

Here, complete solution is y = C.F + P.I

or $$y = e^{-kt} \left[C_1 \cos \sqrt{(b^2 - k^2)}\, t + C_2 \sin \sqrt{(b^2 - k^2)}\, t\right] + \frac{e^{-kt} \sin wt}{b^2 - k^2 - w^2}$$

Case 2. When $w^2 = b^2 - k^2$

$$\text{C.F.} = e^{-kt} (C_1 \cos wt + C_2 \sin wt)$$

$$\& \text{ P.I.} = \frac{1}{D^2 + 2kD + b^2} (e^{-kt} \sin wt)$$

$$= e^{-kt} \frac{1}{(D-k)^2 + 2k(D-k) + b^2} \sin wt$$

$$= e^{-kt} \frac{1}{D^2 + b^2 - k^2} \sin wt \text{ Here } f(-a^2) = 0$$

$$= e^{-kt} \left(-\frac{t}{2w} \cos wt\right)$$

$$= -\frac{t\, e^{-kt}}{2w} \cos wt$$

Hence, the required solution is y = C.F + P.I

or $$y = e^{-kt} (C_1 \cos wt + C_2 \sin wt) - \frac{t}{2w} e^{-kt} \cos wt$$

Example 17. Solve $(D^2 - 2D + 1)\, y = x\, e^x \sin x$

(Bihar P.C.S., 1997; 2007,U.P.P.C.S., 2001; L.D.A. 1995, U.P.T.U. 2005)

Solution. Here the auxiliary equation is

$m^2 - 2m + 1 = 0 \Rightarrow (m-1)^2 = 0 \therefore m = 1, 1$

$\therefore$ C.F. = $(C_1 + C_2 x)\, e^x$ where C_1 and C_2 are arbitrary constants

and $$\text{P.I.} = \frac{1}{D^2 - 2D + 1} x e^x \sin x$$

$$= e^x \frac{1}{(D+1)^2 - 2(D+1) + 1}(x \sin x)$$

$$= e^x \frac{1}{D^2 + 1 + 2D - 2D - 2 + 1} x \sin x$$

$$= e^x \frac{1}{D^2}(x \sin x)$$

$$= e^x \frac{1}{D}\int x \sin x \, dx = e^x \frac{1}{D}[-x \cos x + \sin x]$$

$$= e^x [-\int x \cos x \, dx + \int \sin x \, dx]$$

$$= e^x [-x \sin x - \cos x - \cos x]$$

$$= -e^x (x \sin x + 2 \cos x)$$

Hence, the required solution is y = C.F + P.I

$y = (C_1 + C_2 x) e^x - e^x (x \sin x + 2 \cos x)$

Example 18. Find the complete solution of the differential equation $\frac{d^2y}{dx^2} - 2\frac{dy}{dx} + y = x e^x \cos x$ **(U.P.T.U. 2009)**

Solution. Here the auxiliary equation is

$m^2 - 2m + 1 = 0 \Rightarrow (m-1)^2 = 0; \therefore m = 1, 1$

$\therefore$ C.F. $= (C_1 + C_2 x) e^x$

and $$\text{P.I.} = \frac{1}{D^2 - 2D + 1} x e^x \cos x$$

$$= e^x \frac{1}{D^2}(x \cos x)$$

$$= e^x (-x \cos x + 2 \sin x)$$

Therefore, complete solution is y =C.F + P.I

or $$y = (C_1 + C_2 x) e^x + e^x (2 \sin x - x \cos x)$$

Example 19. Solve $(D^4 + 6D^3 + 11D^2 + 6D) y = 20 e^{-2x} \sin x$

[U.P.T.U. (C.O.) 2005]

Solution. Here the auxiliary equation is

$$m^4 + 6m^3 + 11m^2 + 6m = 0$$

$$\Rightarrow \quad m(m^3 + 6m^2 + 11m + 6) = 0$$

$\Rightarrow \qquad m(m+1)(m+2)(m+3) = 0$

$\therefore m = 0, -1, -2, -3$

$\therefore$ C.F. $= C_1 + C_2 e^{-x} + C_3 e^{-2x} + C_4 e^{-3x}$

$$\& \text{ P.I} = \frac{1}{D^4 + 6D^3 + 11D^2 + 6D} 20 e^{-2x} \sin x$$

$$= 20 e^{-2x} \frac{1}{(D-2)^4 + 6(D-2)^3 + 11(D-2)^2 + 6(D-2)} \sin x$$

$$= 20 e^{-2x} \frac{1}{(D-2)(D^3 - D)} \sin x$$

$$= 20 e^{-2x} \frac{1}{(D-2)\{(D)^2 D - D\}} \sin x$$

$$= 20 e^{-2x} \frac{1}{(D-2)\{(-1)^2 D - D\}} \sin x \quad \text{replacing } D^2 \text{ by } -1^2$$

$$= 20 e^{-2x} \frac{1}{(D-2)(-D-D)} \sin x$$

$$= 20 e^{-2x} \frac{1}{4D - 2D^2} \sin x = 20 e^{-2x} \frac{1}{4D - 2(-1)^2} \sin x$$

$$= 20 e^{-2x} \frac{1}{4D + 2} \sin x$$

$$= 10 e^{-2x} \frac{2D - 1}{4D^2 - 1} \sin x = 10 e^{-2x} \frac{(2\cos x - \sin x)}{4(-1) - 1}$$

$$= 2 e^{-2x} (\sin x - 2\cos x)$$

Hence, Complete solution is y = C.F + P.I

$y = C_1 + C_2 e^{-x} + C_3 e^{-2x} + C_4 e^{-3x} + 2e^{2x} (\sin x - 2\cos x)$

Example 20. Solve $(D^2 - 4D - 5) y = e^{2x} + 3\cos(4x + 3)$ $D \equiv \frac{d}{dx}$

(U.P.T.U. 2008)

Solution. We have $(D^2 - 4D - 5) y = e^{2x} + 3\cos(4x + 3)$

Here the auxiliary equation is

$m^2 - 4m - 5 = 0 \Rightarrow (m - 5)(m + 1) = 0 \therefore m = 5, -1$

C.F. $= C_1 e^{5x} + C_2 e^{-x}$

$$\& \text{ P.I.} = \frac{1}{D^2 - 4D - 5}\{e^{2x} + 3\cos(4x+3)\}$$

$$= \frac{1}{D^2 - 4D - 5} e^{2x} + 3\frac{1}{D^2 - 4D - 5}\{\cos 4x \cos 3 - \sin 4x \sin 3\}$$

$$= \frac{1}{(2)^2 - 4(2) - 5} e^{2x} + 3\cos 3 \frac{1}{D^2 - 4D - 5}\cos 4x - 3\sin 3\frac{1}{D^2 - 4D - 5}\sin 4x$$

$$= -\frac{1}{9}e^{2x} + 3\cos 3\frac{1}{-4^2 - 4D - 5}\cos 4x - 3\sin 3\frac{1}{-4^2 - 4D - 5}\sin 4x$$

$$= -\frac{1}{9}e^{2x} + 3\cos 3\frac{1}{-16 - 4D - 5}\cos 4x - 3\sin 3\frac{1}{-16 - 4D - 5}\sin 4x$$

$$= -\frac{1}{9}e^{2x} + 3\cos 3\frac{1}{-(4D+21)}\cos 4x + 3\sin 3\frac{1}{4D+2}\sin 4x$$

$$= -\frac{1}{9}e^{2x} - 3\cos 3\frac{(4D-21)}{(6D^2 - 44)}\cos 4x + 3\sin 3\frac{(4D-21)}{16D^2 - 441}\sin 4x$$

$$= -\frac{1}{9}e^{2x} - 3\cos 3\frac{(4D-21)}{4(-4^2) - 441}\cos 4x + 3\sin 3\frac{4D-21}{16(-4^2)441}\sin 4x$$

$$= -\frac{1}{9}e^{2x} - 3\cos 3\frac{(-16\sin 4x - 21\cos 4x)}{-697} + 3\sin 3\frac{16\cos 4x - 21\sin 4x}{-697}$$

$$= -\frac{1}{9}e^{2x} - \frac{3}{697}\cos 3(16\sin 4x + 21\cos 4x) - \frac{3\sin 3}{697}(16\cos 4x - 21\sin 4x)$$

$$= -\frac{1}{9}e^{2x} - \frac{3}{697}[16\sin 4x\cos 3 + 21\cos 4x\cos 3 + 16\cos 4x\sin 3 - 21\sin 4x\sin 3]$$

$$= -\frac{1}{9}e^{2x} - \frac{3}{697}[16\sin(4x+3) + 21\cos(4x+3)]$$

Hence the required solution is y = C.F + P.I

$$\text{or } y = C_1 e^{5x} + C_2 e^{-x} - \frac{1}{9}e^{2x} - \frac{3}{697}[16\sin(4x+3) + 21\cos(4x+3)]$$

Example 21. Solve $\frac{d^2y}{dx^2} + 2\frac{dy}{dx} + 10y + 37\sin 3x = 0$, and find the value of y when $x = \pi/2$, if it is given that y = 3 and $\frac{dy}{dx} = 0$ when x = 0 **(I.A.S. 1996)**

Solution. The given equation can be written as

$$(D^2 + 2D + 10)y = -37\sin 3x$$

its auxiliary equation is $m^2 + 2m + 10 = 0$, which gives

$$m = \frac{1}{2}\left[-2 \pm \sqrt{(4-40)}\right] = -1 \pm 3i$$

$\therefore$ C.F. = e^{-x} [$C_1 \cos 3x + C_2 \sin 3x$], where C_1 & C_2 are arbitrary constants,

and $$\text{P.I.} = \frac{1}{D^2 + 2D + 10}(-37 \sin 3x)$$

$$= -37 \frac{1}{D^2 + 2D + 10} \sin 3x$$

$$= -37 \frac{1}{-3^2 + 2D + 10} \sin 3x$$

$$= -37 \frac{1}{2D + 1} \sin 3x = -37 \frac{1}{(4D^2 - 1)}(2D - 1) \sin 3x$$

$$= -37 \frac{1}{\{4(-3^2) - 1\}}(2D - 1) \sin 3x$$

$$= (2D - 1) \sin 3x = 6 \cos 3x - \sin 3x$$

Hence the required solution of the given differential equation is

$$y = e^{-x}(C_1 \cos 3x + C_2 \sin 3x) + 6 \cos 3x - \sin 3x \qquad (1)$$

Differentiating both sides of (1) with respect to x,

we get

$$\frac{dy}{dx} = e^{-x}(-3C_1 \sin 3x + 3C_2 \cos 3x) - e^{-x}(C_1 \cos 3x + C_2 \sin 3x) - 18 \sin 3x - 3 \cos 3x \qquad (2)$$

it is given that when x = 0, y = 3 and $\frac{dy}{dx} - 0$

$\therefore$ From (1) and (2) we, have

$$3 = C_1 + 6 \qquad (3)$$

and $$0 = 3C_2 - C_1 - 3 \qquad (4)$$

From (3) and (4) we get $C_1 = -3$ and $C_2 = 0$

$\therefore$ From (1), we have

$$y = -3e^{-x} \cos 3x + 6 \cos 3x - \sin 3x \qquad (5)$$

$\therefore$ when $x = \pi/2$ we have from (5)

$$y = 3e^{-\pi/2}\cos\frac{3\pi}{2} + 6\cos\frac{3\pi}{2} - \sin\frac{3\pi}{2}$$

$$y = -3e^{\frac{-\pi}{2}}.0 + 6.0 + 1$$

or $y = 1$

Example 22. Solve $(D^2 + 1)^2 = 24\,x\cos x$ given the initial conditions $x = 0$, $y = 0$, $Dy = 0$, $D^2y = 0$, $D^3y = 12$ **(I.A.S. 2001)**

Solution. Here the auxiliary equation is

$$(m^2 + 1)^2 = 0$$

or $$m = \pm i \text{ (twice)}$$

$\therefore$ C.F. = $(C_1 x + C_2)\cos x + (C_3 x + C_4)\sin x$

and $$\text{P.I.} = \frac{1}{(D^2+1)^2}(24\,x\cos x)$$

$$= \text{R.P. of } \frac{1}{(D^2+1)^2}\,24\,x\,e^{ix}$$

$$= \text{R.P. of } 24\,e^{ix}\,\frac{1}{\left[(D+i)^2+1\right]^2}\,x$$

$$= \text{R.P. of } 24\,e^{ix}\,\frac{1}{(D^2+2iD)^2}\,x$$

$$= \text{R.P. of } 24\,e^{ix}\,\frac{1}{-4D^2\left[1+\frac{D}{2i}\right]^{-2}}\,x$$

$$= \text{R.P. of } -6\,e^{ix}\,\frac{1}{D^2}\left(1 + iD - \frac{3}{4}D^2 + \ldots\ldots\right)x$$

$$= \text{R.P. of } -6\,e^{ix}\,\frac{1}{D^2}(x+i)$$

$$= \text{R.P. of } -6\,e^{ix}\left(\frac{1}{6}x^3 + \frac{1}{2}i\,x^2\right)$$

$$= \text{R.P. of } -(\cos x + i\sin x)(x^3 + 3i\,x^2)$$

$$= -x^2\cos x + 3x^2\sin x$$

$\therefore$ The solution of the given differential equation is

$$y = (C_1 x + C_2) \cos x + (C_3 x + C_4) \sin x - x^3 \cos x + 3x^2 \sin x$$

or $\quad y = (C_1 x + C_2 - x^3) \cos x + (C_3 x + C_4 + 3x^2) \sin x \qquad (1)$

$\therefore$ $Dy = (C_1 - 3x^2) \cos x - (C_1 x + C_2 - x^3) \sin x + (C_3 x + C_4 + 3x^2) \cos x + (C_3 + 6x) \sin x$

or $\quad Dy = (C_1 + C_4 + C_3 x) \cos x + (C_3 - C_2 + 6x - C_1 x + x^3) \sin x \qquad (2)$

$\therefore$ $D^2y = -(C_1 + C_4 + C_3 x) \sin x + C_3 \cos x + (C_3 - C_2 + 6x - C_1 x + x^3) \cos x + (6 - 4 + 3x^2) \sin x$

or $D^2y = (6 - 2 C_1 - C_4 - C_3 x + 3x^2) \sin x + (2 C_3 - C_2 + 6x - C_1 x + x^3) \cos x \qquad (3)$

$\therefore$ $D^2y = (6 - 2 C_1 - C_4 - C_3 x + 3x^2) \cos x + (-C_3 + 6x) \sin x - (2 C_3 - C_2 + 6x - C_1 x + x^3) \sin x + (6 - C_1 + 3x^2) \cos x$

or $D^3y = (12 - 3 C_1 - C_4 - C_3 x + 6x^2) \cos x + (C_2 - 3 C_3 - 6x + C_1 x + 6x - x^2) \sin x \qquad (4)$

Since we are given that $x = 0, y = 0, Dy = 0, D^2y = 0, D^3y = 12$ so from (1), (2), (3) and (4) we get

$0 = C_2 \qquad (5)$

$0 = C_1 + C_4 \qquad (6)$

$0 = 2C_3 - C_2 \qquad (7)$

$12 = 12 - 3 C_1 - C_4 \qquad (8)$

From (5) and (7) we get $C_2 = 0 = C_3$

From (6) and (8) we get $C_1 = 0 = C_4$

$\therefore$ From (1) the required solution is

$$y = 3x^2 \sin x - x^3 \cos x$$

(D) P.I when Q is of the form x. V, where V is any function of x

$$\frac{1}{f(D)} (x. V) = x. \frac{1}{f(D)} V - \frac{f'(D)}{\{f(D)\}^2} V$$

Example 23. Solve $(D^2 - 2D + 1) y = x \sin x$

Solution. Here the auxiliary equation is $m^2 - 2m + 1 = 0$

or $\quad (m - 1)^2 = 0$ or $m = 1, 1$

$\therefore$ C.F. $= (C_1 x + C_2) e^x$, where C_1 and C_2 are arbitrary constants

and $\quad$ P.I $= \dfrac{1}{D^2 - 2D + 1} x \sin x$

$$= x \frac{1}{D^2 - 2D + 1} \sin x - \frac{(2D - 2)}{(D^2 - 2D + 1)^2} \sin x$$

$$= x \frac{1}{-1^2 - 2D + 1} \sin x - \frac{(2D - 2)}{(-1^2 - 2D + 1)^2} \sin x$$

$$= -\frac{x}{2} \frac{1}{D} \sin x - \frac{1}{4 D^2} (2D - 2) \sin x$$

$$= -\frac{x}{2} \int \sin x \, dx - \frac{1}{2} \frac{1}{D^2} (\cos x - \sin x)$$

$$= \frac{x}{2} \cos x - \frac{1}{2} \frac{1}{D} \int (\cos x - \sin x) \, dx$$

$$= \frac{x}{2} \cos x - \frac{1}{2} \frac{1}{D} (\sin x + \cos x)$$

$$= \frac{x}{2} \cos x - \frac{1}{2} \int (\sin x + \cos x) \, dx$$

$$= \frac{x}{2} \cos x - \frac{1}{2} (-\cos x + \sin x)$$

$$= \frac{1}{2} (x \cos x + \cos x - \sin x)$$

∴ The required solution is y = C.F + P.I

or $$y = (C_1 x + C_2) e^x + \frac{1}{2} (x \cos x + \cos x - \sin x)$$

(E) To show that $\frac{1}{D - a} Q = e^{ax} \int Q e^{-ax} \, dx$, where Q is a function of x.

Proof Let $y = \frac{1}{D - a} Q$

Then (D – a) y = Q, operating both sides with D – a

or $\frac{dy}{dx} - ay = Q$, Which is a linear equation in y whose integrating factor is e^{-ax} its solution is $y e^{-ax} = \int Q e^{-ax} \, dx$, neglecting the constant of integration as P.I. is required

or $$y = e^{ax} \int Q e^{-ax} \, dx$$

or $$\frac{1}{D - a} Q = e^{ax} \int Q e^{-ax} \, dx$$

Example 24. Solve $(D^2 + a^2)\, y = \sec ax$ **(U.P.PC.S. 1971, 1973, 1977)**

Solution. The auxiliary equation is $m^2 + a^2 = 0$ or $m = \pm ai$

$\therefore$ C.F. $= C_1 \cos ax + C_2 \sin ax$, where C_1 and C_2 are arbitrary constants

and $$\text{P.I} = \frac{1}{D^2 + a^2} \sec ax = \frac{1}{2\,i\,a}\left[\frac{1}{D - ia} - \frac{1}{D + ia}\right] \sec ax \qquad (1)$$

Now $$\frac{1}{D - i\,a} \sec ax = e^{iax} \int \sec ax.\, e^{-iax}\, dx$$

$$= e^{iax} \int \sec ax.\, (\cos ax - i \sin ax)\, dx$$

$$= e^{iax} \int (1 - i \tan ax)\, dx$$

$$= e^{iax} \left[x + \left(\frac{i}{a}\right) \log \cos ax\right]$$

similarly,

$$\frac{1}{D + ia} \sec ax = e^{-i\,ax} \int \sec ax.\, e^{iax}\, dx$$

$$= e^{-i\,ax} \int \sec ax\, (\cos ax + i \sin ax)\, dx$$

$$= e^{-i\,ax} \int (1 + i \tan ax)\, dx$$

$$= e^{-i\,ax} \left[x - \left(\frac{i}{a}\right) \log \cos ax\right]$$

$\therefore$ From (1) we, have

$$\text{P.I} = \frac{1}{2i\,a}\left[e^{i\,ax}\left\{x + \frac{i}{a} \log \cos ax\right\} - e^{-i\,ax}\left\{x - \frac{i}{a} \log \cos ax\right\}\right]$$

$$= \left[x\left(\frac{e^{i\,ax} - e^{-i\,ax}}{2ia}\right) + \frac{i}{a} (\log \cos ax).\left(\frac{e^{i\,ax} + e^{-i\,ax}}{2ia}\right)\right]$$

$$= \left(\frac{x}{a}\right) \sin ax + \frac{1}{a^2} \cos ax.\, (\log \cos ax)$$

$\therefore$ The required solution is

$$y = \text{C.F.} + \text{P.I.}$$

or $y = C_1 \cos ax + C_2 \sin ax + \left(\frac{x}{a}\right) \sin ax + \left(\frac{1}{a^2}\right) \cos ax . \log \cos ax$

Example 25. $(D^2 + a^2)\, y = \text{cosec}\, ax$

Solution. Here the auxiliary equation is $m^2 + a^2 = 0$

or $m = \pm ai$

$\therefore$ C.F. = $C_1 \cos ax + C_2 \sin ax$ and

$$\text{P.I} = \frac{1}{D^2 + a^2} \text{cosec}\, ax = \frac{1}{2ia}\left[\frac{1}{D - ia} - \frac{1}{D + ia}\right] \text{cosec}\, ax \qquad (1)$$

Now $\frac{1}{D - ia} \text{cosec}\, ax = e^{i\,ax} \int (\text{cosec}\, ax)\, e^{-i\,ax}\, dx$

$$= e^{i\,ax} \int \text{cosec}\, ax\, (\cos ax - i \sin ax)\, dx$$

$$= e^{i\,ax} \int (\cot\, ax - i)\, dx$$

$$= e^{i\,ax} \left[\frac{1}{a} (\log \sin ax) - ix\right]$$

and $\frac{1}{D + ia} \text{cosec}\, ax = e^{-i\,ax} \int (\text{cosec}\, ax)\, e^{i\,ax}\, dx$

$$= e^{-i\,ax} \int (\text{cosec}\, ax)\, (\cos ax + i \sin ax)\, dx$$

$$= e^{-i\,ax} \int (\cot\, ax + i)\, dx$$

$$= e^{-i\,ax} \left[\frac{1}{a} (\log \sin ax) + ix\right]$$

$\therefore$ From (1) we have

$$\text{P.I.} = \frac{1}{2ia}\left[e^{i\,ax} \left\{\frac{1}{a} (\log \sin ax) - i\,x\right\} - e^{-i\,ax} \left\{\frac{1}{a} (\log \sin ax) + ix\right\}\right]$$

$$= \left[\frac{1}{a^2} (\log \sin ax) \left(\frac{e^{i\,ax} - e^{-i\,ax}}{2i}\right) - \frac{x}{a}\left(\frac{e^{i\,ax} + e^{-i\,ax}}{2}\right)\right]$$

$$= \left(\frac{1}{a^2}\right) \sin ax\, (\log \sin ax) - \left(\frac{x}{a}\right) (\cos ax)$$

$\therefore$ The required solution is y = C.F + P.I

or $y = C_1 \cos ax + C_2 \sin ax + \left(\frac{1}{a^2}\right) \sin ax \log(\sin ax) - \left(\frac{x}{a}\right) \cos ax$

Example 26. $(D^2 + a^2)\, y = \tan ax$

Solution. Here C.F. = $C_1 \cos ax + C_2 \sin ax$

and P.I. = $\frac{1}{D^2 + a^2} \tan ax = \frac{1}{2ia}\left[\frac{1}{D - ia} - \frac{1}{D + ia}\right] \tan ax$ (1)

Now $\frac{1}{D - ia} \tan ax = e^{i\,ax} \int e^{-i\,ax} \tan ax\, dx$

$= e^{i\,ax} \int (\cos ax - i \sin ax) \tan ax\, dx$

$= e^{i\,ax} \int \left[\sin ax - i\frac{\sin^2 ax}{\cos ax}\right] dx$

$= e^{i\,ax} \int \sin ax\, dx - i\, e^{i\,ax} \int \left(\frac{1 - \cos^2 ax}{\cos ax}\right) dx$

$= e^{i\,ax} \{(-\cos ax)/a\} - i\, e^{i\,ax} \int (\sec ax - \cos ax)\, dx$

$= -\frac{1}{a}\left\{\left(e^{i\,ax} \cos ax\right)\right\} - i e^{i\,ax}\left[\frac{1}{a} \log \tan\left(\frac{\pi}{4} + \frac{ax}{2}\right) - \frac{1}{a} \sin ax\right]$

$= -\left(\frac{1}{a}\right) e^{i\,ax}\left[\cos ax + i \log \tan\left(\frac{\pi}{4} + \frac{ax}{2}\right) - i \sin ax\right]$

$= -\frac{1}{a} e^{i\,ax}\left[(\cos ax - i \sin ax) + i \log \tan\left(\frac{\pi}{4} + \frac{ax}{2}\right)\right]$

$- \frac{1}{a} e^{i\,ax}\left[e^{-i\,ax} + i \log \tan\left(\frac{\pi}{4} + \frac{ax}{2}\right)\right]$

$= -\frac{1}{a}\left[1 + i\, e^{i\,ax} \log \tan\left(\frac{\pi}{4} + \frac{ax}{2}\right)\right]$ (2)

Similarly replacing i by –i we get

$\frac{1}{D + ia} \tan ax = -\left(\frac{1}{a}\right)\left[1 - i\, e^{-i\,ax} \log \tan\left(\frac{\pi}{4} + \frac{ax}{2}\right)\right]$ (3)

∴ From (1), (2) and (3), we have

$$\text{P.I} = \frac{1}{2ia}\left[-\frac{1}{a}\left\{1 + i\,e^{i\,ax}\log\tan\left(\frac{\pi}{4}+\frac{ax}{2}\right)\right\} + \frac{1}{a}\left\{1 - i\,e^{-i\,ax}\log\tan\left(\frac{\pi}{4}+\frac{ax}{2}\right)\right\}\right]$$

$$= \frac{1}{2i\,a}\left[-\frac{i}{a}\left(e^{i\,ax}+e^{-i\,ax}\right)\log\tan\left(\frac{\pi}{4}+\frac{ax}{2}\right)\right]$$

$$= -\frac{1}{a^2}(\cos ax)\log\tan\left(\frac{\pi}{4}+\frac{ax}{2}\right)$$

∴ The required solution is y = C.F + P.I

$$\text{or } y = C_1\cos ax + C_2\sin ax - \frac{1}{a^2}\cos ax\log\tan\left(\frac{\pi}{4}+\frac{ax}{2}\right)$$

Miscellaneous solved Examples

Example 27. Solve $(D^2 - 2D + 1)\,y = x^2 e^{3x}$

Solution. Here the auxiliary equation is

$m^2 - 2m + 1 = 0$

or $(m-1)^2 = 0$

or $m = 1, 1$

∴ C.F. = $(C_1 x + C_2)\,e^x$

$$\text{and P.I.} = \frac{1}{D^2 - 2D + 1}x^2 e^{3x} = e^{3x}\frac{1}{(D+3)^2 - 2(D+3) + 1}x^2$$

$$= e^{3x}\frac{1}{D^2 + 4D + 4}x^2 = e^{3x}\frac{1}{4\left(1+\frac{D}{2}\right)^2}x^2$$

$$= \frac{e^{3x}}{4}\left(1+\frac{1}{2}D\right)^{-2}x^2 = \frac{e^{3x}}{4}\left[\left(1 - D + \frac{3}{4}D^2 + \ldots\ldots\right)x^2\right]$$

$$= \frac{1}{4}e^{3x}\left(x^2 - Dx^2 + \frac{3}{4}D^2x^2\right)$$

$$= \frac{1}{4}e^{3x}\left(x^2 - 2x + \frac{3}{2}\right)$$

$$= \frac{1}{8}e^{3x}\left(2x^2 - 4x + 3\right)$$

∴ The required solution is y = C.F + P.I

or $\quad y = (C_1 x + C_2) e^x + \frac{1}{8} e^{3x} (2x^2 - 4x + 3)$

Example 28. Solve $(D^2 - 1) y = \cos hx$

Solution. The given equation can be written as

$(D^2 - 1) y = \frac{1}{2} (e^x + e^{-x})$

its auxiliary equation is $m^2 - 1 = 0$

or $m = \pm 1$

$\therefore$ C.F. $= C_1 e^x + C_2 e^{-x}$, where C_1, C_2 are arbitrary constants

and $\quad$ P.I $= \frac{1}{D^2 - 1} \frac{1}{2} (e^x + e^{-x})$

$$= \frac{1}{2} \frac{1}{(D^2 - 1)} e^x + \frac{1}{2} \frac{1}{D^2 - 1} e^{-x}$$

$$= \frac{1}{2} e^x \frac{1}{(D+1)^2 - 1} 1 + \frac{1}{2} \frac{1}{(-1)^2 - 1} e^{-x}$$

$$= \frac{1}{2} e^x \frac{1}{D^2 + 2D} (1) - \frac{1}{4} e^{-x}$$

$$= \frac{1}{2} e^x \frac{1}{2D\left(1 + \frac{D}{2}\right)} (1) - \frac{1}{4} e^{-x}$$

$$= \frac{1}{4} e^x \frac{1}{D} \left(1 - \frac{D}{2} + \ldots\ldots\right) 1 - \frac{1}{4} e^{-x}$$

$$= \frac{1}{4} e^x \frac{1}{D} (1) - \frac{1}{4} e^{-x}$$

$$= \frac{1}{4} e^x \int 1\, dx - \frac{1}{4} e^{-x} = \frac{1}{4} e^x x - \frac{1}{4} e^{-x}$$

$\therefore$ The required solution is y = C.F + P.I

or $y = C_1 e^x + C_2 e^{-x} + \frac{1}{4} (x e^x - e^{-x})$

Example 29. Solve $(D^2 - 4) y = \cos^2 x$

Solution. Its auxiliary equation is $m^2 - 4 = 0$

or $m = \pm 2$

$\therefore$ C.F. = $C_1 e^{2x} + C_2 e^{-2x}$, where C_1 and C_2 are arbitrary constants

and $P.I = \frac{1}{D^2 - 4} \cos^2 x = \frac{1}{D^2 - 4}\left[\frac{1}{2}(1 + \cos 2x)\right]$

$$= \frac{1}{D^2 - 4}\,\frac{1}{2} + \frac{1}{D^2 - 4}\left(\frac{1}{2}\cos 2x\right)$$

$$= -\frac{1}{4}\left(1 - \frac{1}{4}D^2\right)^{-1}\frac{1}{2} + \frac{1}{2}\,\frac{1}{-2^2 - 4}\cos 2x$$

$$= -\frac{1}{4}\left[1 + \frac{1}{4}D^2 + ...\right]\frac{1}{2} - \frac{1}{16}\cos 2x$$

$$= -\frac{1}{4}\left(\frac{1}{2}\right) - \frac{1}{1}\cos 2x$$

$$= -\frac{1}{16}(2 + \cos 2x)$$

$\therefore$ The required solution is y = C.F + P.I

or $y = C_1 e^{2x} + C_2 e^{-2x} - \frac{1}{16}(2 + \cos 2x)$

Example 30. Solve $(D^2 + 1)\, y = x^2 \sin 2x$

Solution. Here the auxiliary equation is $m^2 + 1 = 0$

or $m = \pm i$

$\therefore$ C.F = $C_1 \cos x + C_2 \sin x$, where C_1 and C_2 are arbitrary constants

and $P.I = \frac{1}{D^2 + 1} x^2 \sin 2x$ = I.P of $\frac{1}{D^2 + 1} x^2 e^{2ix}$

$$= \text{I.P of } e^{2ix}\,\frac{1}{\{(D + 2i)^2 + 1\}}\,x^2$$

$$= \text{I.P of } e^{2ix}\,\frac{1}{(D^2 + 4iD - 3)}\,x^2$$

$$= \text{I.P of } e^{2ix}\,\frac{1}{-3\left(1 - \frac{4iD}{3} - \frac{D^2}{3}\right)}\,x^2$$

$$= \text{I.P. of } \frac{e^{2ix}}{-3}\left[1 - \left(\frac{4iD + D^2}{3}\right)\right]^{-1} x^2$$

$$= \text{I.P of } \frac{e^{2ix}}{-3}\left[1+\left(\frac{4iD+D^2}{3}\right)+\left(\frac{16\,i^2\,D^2}{9}+\ldots\right)+\ldots\right]x^2$$

$$= \text{I.P of } -\frac{1}{3}\,e^{2ix}\left(1+\frac{4}{3}\,iD-\frac{13}{9}\,D^2+\ldots\ldots\right)x^2$$

$$= \text{I.P of } -\frac{1}{3}\,e^{2ix}\left[x^2+\frac{4}{3}\,i\,(2x)-\frac{13}{9}\,(2)\right]$$

$$= \text{I.P of } -\frac{1}{3}\,(\cos 2x + i\sin 2x)\left[\left(x^2-\frac{26}{9}\right)+\frac{8}{3}\,xi\right]$$

$$= -\frac{1}{3}\left(\frac{8}{3}\,x\right)\cos 2x - \frac{1}{3}\left(x^2-\frac{26}{9}\right)\sin 2x$$

$$= -\frac{1}{27}\,[24\,x\cos 2x + (9x^2-26)\sin 2x]$$

∴ The required solution is

$$y = C_1\cos x + C_2\sin x - \frac{1}{27}\,[24\,x\cos 2x + (9x^2-26)\sin 2x]$$

EXERCISE

Solve the following differential Equations

1. $(D^3 + 6D^2 + 11\,D + 6)\,y = 0$

Ans. $y = C_1\,e^{-x} + C_2\,e^{-2x} + C_3\,e^{-3x}$

2. $\dfrac{d^2x}{dt^2} - 3\,\dfrac{dx}{dt} + 2x = 0$ given that when $t = 0$, $x = 0$ and $\dfrac{dx}{dt} = 0$

Ans. $x = 0$

3. $(D^2 + D + 1)\,y = e^{-x}$, where $D \equiv \dfrac{d}{dx}$

Ans. $y = e^{-x/2}\left[C_1\cos\left(\frac{1}{2}\,x\sqrt{3}\right) + C_2\sin\left(\frac{1}{2}\,x\sqrt{3}\right)\right] + e^{-x}$

4. $(D-1)^2\,(D^2+1)^2\,y = e^x$

Ans. $y = (C_1x + C_2)\,e^x + (C_3x + C_4)\cos x + (C_5x + C_6)\sin x + \frac{1}{8}\,x^2\,e^x$

5. $\dfrac{d^2y}{dx^2} - 3\,\dfrac{dy}{dx} + 2y = e^x$, $y = 3$ and $\dfrac{dy}{dx} = 3$, when $x = 0$

Ans. $y = 2e^{x} + e^{2x} - x e^{x}$

6. $\frac{d^2y}{dx^2} - \frac{dy}{dx} - 2y = \sin 2x$

Ans. $y = C_1 e^{2x} + C_2 e^{-x} + \frac{1}{20}(\cos 2x - 3 \sin 2x)$

7. $(D^2 + 16)\, y = \sin 2x$, given that $y = 0$ and $\frac{dy}{dx} = \frac{5}{6}$ where $x = 0$

Ans. $12\, y = 2 \sin 4x + \sin 2x$

8. $\frac{d^2y}{dx^2} - 4y = e^{x} + \sin 2x$

Ans. $y = C_1 e^{2x} + C_2 e^{-2x} - \frac{1}{3} e^{x} - \frac{1}{8} \sin 2x$

9. $(4D^2 + 9)\, y = \sin x$, given that $y = \frac{1}{2}, \frac{dy}{dx} = \frac{4}{5}$ when $x = \pi$

Ans. $y = \frac{2}{3} \cos \frac{3x}{2} - \frac{1}{2} \sin \frac{3x}{2} + \frac{1}{5} \sin x$

10. Find the integral of the equation $\frac{d^2x}{dt^2} + 2n \cos x \frac{dx}{dt} + n^2 x = a \cos nt$ which is such that when $t = 0$, $x = 0$ and $\frac{dx}{dt} = 0$

Ans. $x = e^{-nt\cos\alpha}\left\{-\frac{a}{n^2 \sin 2\alpha} \sin (n \sin \alpha)\, t\right\} + \frac{a \sin nt}{2n^2 \cos \alpha}$

11. Solve $\frac{d^2y}{dx^2} + a^2y = \sin ax$

Ans. $y = C_1 \cos ax + C_2 \sin ax - \frac{x}{2a} \cos ax$

12. $(D^2 + a^2)\, y = \cos ax$

Ans. $y = C_1 \cos ax + C_2 \sin ax + \frac{1}{2}\left(\frac{x}{a}\right) \sin ax$

13. $(D^4 - 1)\, y = \sin x$

Ans. $y = C_1 e^{x} + C_2 e^{-x} + C_3 \cos x + C_4 \sin x + \frac{x}{4} \cos x$

14. $(D^4 + D^2 + 1)\, y = e^{-x/2} \cos\left(\frac{1}{2} x \sqrt{3}\right)$

Ans. $y = e^{-x/2}\left[C_1 \cos \frac{1}{2} x\sqrt{3} + C_2 \sin \frac{1}{2} x\sqrt{3}\right] + e^{x/2}$

$\left[C_3 \cos \frac{x}{2}\sqrt{3} + C_4 \sin \frac{x}{2}\sqrt{3}\right] - 16e^{-x/2} \cos\left(\frac{x}{2}\sqrt{3}\right)$

15. $(D^2 + 1)\, y = \sin x \sin 2x$

Ans. $y = C_1 \cos x + C_2 \sin x + \frac{x}{4} \sin x + \frac{1}{16} \cos 3x$

16. $\frac{d^2y}{dx^2} + 4y = \sin^2 x$

Ans. $y = C_1 \cos 2x + C_2 \sin 2x + \frac{1}{8} - \frac{x}{8} \sin 2x$

17. $(D^2 - 4)\, y = x^2$

Ans. $y = C_1 e^{2x} + C_2 e^{-2x} - \frac{1}{4}\left(x^2 + \frac{1}{2}\right)$

18. Solve $(D^3 - 8)\, y = x^3$

Ans. $y = C_1 e^{2x} + e^{-x} (C_2 \cos x\sqrt{3} + C_3 \sin x\sqrt{3}) - \frac{1}{8}\left(x^3 + \frac{3}{4}\right)$

19. $(D^3 + 2D^2 + D)\, y = e^{2x} + x^2 + x$

Ans. $y = C_1 + (C_2 x + C_3)\, e^{-x} + \frac{1}{18} e^{2x} + \frac{1}{3} x^3 - \frac{3}{2} x^2 + 4x$

20. $\frac{d^2y}{dx^2} = a + bx + cx^2$, given that $\frac{dy}{dx} = 0$, when $x = 0$ and $y = d$, when $x = 0$

Ans. $y = d + \frac{ax^2}{2} + \frac{bx^3}{6} + \frac{cx^4}{12}$

21. $(D^2 + 4D - 12)\, y = (x - 1)\, e^{2x}$

Ans. $y = C_1 e^{2x} + C_2 e^{-6x} + \frac{1}{64} (4x^2 - 9x)\, e^{2x}$

22. $\left(\frac{d^2y}{dx^2} - 2\frac{dy}{dx} + 5\right) y = e^{2x} \sin x$

Ans. $y = e^x (C_1 \cos 2x + C_2 \sin 2x) - \frac{1}{10} e^{2x} (\cos x - 2 \sin x)$

23. $(D^2 - 4D + 4)\, y = e^{2x} \sin 3x$

Ans. $y = (C_1 x + C_2)\, e^{2x} - \frac{1}{9} e^{2x} \sin 3x$

24. $\left(\frac{d}{dx} + 1\right)^3 y = x^2 e^{-x}$

Ans. $y = (C_1 x^2 + C_2 x + C_3)\, e^{-x} + \frac{1}{60} x^5 e^{-x}$

25. $\frac{d^2y}{dx^2} + 2\frac{dy}{dx} + y = x \cos x$

Ans. $y = (C_1 x + C_2)\, e^{-x} + \frac{1}{2} x \sin x - \frac{1}{2} \sin x + \frac{1}{2} \cos x$

26. $\frac{d^4y}{dx^4} - y = x \sin x$

Ans. $y = C_1 e^x + C_2 e^{-x} + C_3 \cos x + C_4 \sin x + \frac{1}{8} (x^2 \cos x - 3x \sin x)$

27. $\frac{d^2y}{dx^2} + 2\frac{dy}{dx} + y = x \sin x$ **(I.A.S. 1998)**

Ans. $y = (C_1 + C_2 x)\, e^{-x} + \frac{1}{2} (\sin x + \cos x - x \cos x)$

28. Solve $(D^2 + 1)^2 y = 24 x \cos x$, given that $y = Dy = D^2y = 0$ and $D^3y = 12$ when $x = 0$

Ans. $y = 3x^2 \sin x - x^3 \cos x$

29. $(D^2 - 4D + 4)\, y = 8x^2 e^{2x} \sin 2x$ **(U.P.T.U. 2004, 2005)**

Ans. $y = (C_1 + C_2 x)\, e^{2x} + e^{2x} (-2x^2 \sin 2x - 4x \cos 2x + 3 \sin 2x)$

Objective Type of Questions

Choose a correct answer from the four answers given in each of the following questions.

1. The solution of the differential equation $\frac{d^2y}{dx^2} + (3i - 1)\frac{dy}{dx} - 3iy = 0$ is

(I.A.S. 1998)

(a) $y = C_1 e^x + C_2 e^{3ix}$ (b) $y = C_1 e^{-x} + C_2 e^{3ix}$
(c) $y = C_1 e^x + C_2 e^{-3ix}$ (d) $y = C_1 e^{-x} + C_2 e^{-3ix}$

Ans. (c)

2. A particular integral of the differential equation $(D^2 + 4)\, y = x$ is

(I.A.S. 1998)

(a) xe^{-2x} (b) $x \cos 2x$
(c) $x \sin 2x$ (d) $x/4$

Ans. (d)

3. The particular integral of $(D^2 + 1)\, y = e^{-x}$ is

(I.A.S. 1999)

(a) $\left(\frac{1}{4} - \frac{x}{2}\right) e^{-x}$ (b) $\left(\frac{1}{4} + \frac{x}{2}\right) e^{-x}$

(c) $\frac{e^{-x}}{2}$ (d) $\frac{e^{-x}}{-2}$

Ans. (c)

4. For the differential equation $(D + 2)(D - 1)^3 y = e^x$ the particular integral is

(I.A.S. 1990, U.P.P.C.S. 2000)

(a) $\frac{1}{18} x^4 e^x$ (b) $\frac{1}{18} x^3 e^x$

(c) $\frac{1}{18} x\, e^{3x}$ (d) $\frac{1}{36} x\, e^{3x}$

Ans. (b)

5. The particular integral of the differential equation $\frac{d^2y}{dx} + 9y = \sin 3x$ is

(a) $\frac{x \sin 3x}{18}$ (b) $\frac{x \sin 3x}{6}$

(c) $\frac{-x \cos 3x}{6}$ (d) $\frac{x \cos 3x}{18}$

Ans. (c)

6. The solution of the differential equation $\frac{d^2y}{dx^2} - 3\frac{dy}{dx} - 4y = 0$ is

(U.P.P.C.S. 2001)

(a) $y = C_1 e^{-x} + C_2 e^{-4x}$ (b) $y = C_1 e^{x} + C_2 e^{4x}$

(c) $y = C_1 e^{-x} + C_2 e^{4x}$ (d) $y = C_1 e^{x} + C_2 e^{-4x}$

Ans. (c)

7. The general solution of the differential equation $(D^2 - 1)\, y = x^2$ is

(a) $y = C_1 e^{x} + C_2 e^{-x} - x^2$ (b) $y = C_1 e^{x} + C_2 e^{-x} + (x^2 + 2)$

(c) $y = C_1 e^{x} + C_2 e^{-x} - 2$ (d) $y = C_1 e^{x} + C_2 e^{-x} - (x^2 + 2)$

Ans. (d)

8. The P.I. of $(D^2 - 2D)\, y = e^x \sin x$ is

(a) $-\frac{1}{2} e^x \sin x$ (b) $e^x \cos x$

(c) $-\frac{1}{2} e^x \cos x$ (d) none of these

Ans. (a)

9. The general solution of the differential equation $D^2 (D + 1)^2 y = e^x$ is

(I.A.S. 1990)

(a) $y = C_1 + C_2 x + (C_3 + C_4 x)\, e^x$

(b) $y = C_1 + C_2 x + (C_3 + C_4 x)\, e^{-x} + \frac{1}{4} e^x$

(c) $y = (C_1 + C_2 e^{-x}) + (C_3 + C_4 x)\, e^{-x} + \frac{1}{4} e^x$

(d) none of these

Ans. (b)

10. $y = e^{-x} (C_1 \cos\sqrt{3}\, x + C_2 \sin\sqrt{3}\, x) + C_3 e^{2x}$ is the solution of

(I.A.S. 1994)

(a) $\frac{d^3y}{dx^3} + 4y = 0$ (b) $\frac{d^3y}{dx^3} + 8y = 0$

(c) $\frac{d^3y}{dx^3} - 8y = 0$ (d) $\frac{d^3y}{dx^3} - 2\frac{d^2y}{dx^2} + \frac{dy}{dx} - 2 = 0$

Ans. (c)

11. The P.I of the differential equation $(D^3 - D)\ y = e^x + e^{-x}$ is

(I.A.S. 1993)

(a) $\frac{1}{2}(e^x + e^{-x})$ (b) $\frac{1}{2}x(e^x + e^{-x})$

(c) $\frac{1}{2}x^2(e^x + e^{-x})$ (d) $\frac{1}{2}x^2(e^x - e^{-x})$

Ans. (b)

12. Given y = 1 + cos x and y = 1 + sin x are solutions of the differential equation $\frac{d^2y}{dx^2} + y = 1$, its solution will be also

(R.A.S. 1994)

(a) y = 2 (1 + cos x) (b) y = 2 + cos x + sin x

(c) y = cos x – sin x (d) y = 1 + cos x + sin x

Ans. (d)

13. The solution of the differential equation $(D^3 - 6D^2 + 11D - 6)\ y = 0$ is

(R.A.S. 1994)

(a) $y = C_1 e^x + C_2 e^{2x} + C_3 e^{4x}$ (b) $y = C_1 e^{2x} + C_2 e^{3x} + C_3 e^{4x}$

(c) $y = C_1 e^{-x} + C_2 e^{2x} + C_3 e^{4x}$ (d) $y = C_1 e^x + C_2 e^{2x} + C_3 e^{3x}$

Ans. (d)

14. The solution of the differential equation $\frac{d^2y}{dx^2} - \frac{dy}{dx} - 2y = 3\,e^{2x}$, when y(0) = 0 = and y(0) = –2 is

(R.A.S. 1994)

(a) $y = e^{-x} - e^{2x} + x\,e^{2x}$ (b) $y = e^x - e^{-2x} - x\,e^{2x}$

(c) $y = e^{-x} + e^{2x}\ \frac{x}{2}e^{2x}$ (d) $y = e^x - e^{-2x} - \frac{x}{2}e^{2x}$

Ans. (a)

15. The solution of the differential equation $\frac{d^2y}{dx^2} + y = \cos 2x$ is

(U.P.P.C.S. 1995)

(a) $A \cos x + B \sin x + \frac{1}{3} \cos 2x$ (b) $A \cos x + B \sin x + \frac{1}{3} \sin 2x$

(c) $A \cos x + B \sin x - \frac{1}{3} \cos 2x$ (d) $A \cos x + B \sin x - \frac{1}{3} \sin 2x$

Ans. (c)

16. The general solution of the differential equation $\frac{d^2y}{dx^2} + n^2 y = 0$ is

(R.A.S. 1995)

(a) $C_1 \sqrt{\cos nx} + C_2 \sqrt{\sin nx}$ (b) $C_1 \cos nx + C_2 \sin nx$

(c) $C_1 \cos^2 nx + C_2 \sin^2 nx$ (d) $C_1 \cos^3 nx + C_2 \sin^3 nx$

Ans. (b)

17. The particular integral of $\frac{d^2y}{dx^2} + \frac{dy}{dx} = x^2 + 2x + 4$ is

(I.A.S. 1996)

(a) $\frac{x^2}{3} + 4x$ (b) $\frac{x^3}{3} + 4x$

(c) $\frac{x^3}{3} + 4$ (d) $\frac{x^3}{3} + 4x^2$

Ans. (b)

18. The general solution of the differential equation $\frac{d^4y}{dx^4} - 6\frac{d^3y}{dx^3} + 12\frac{d^2y}{dx^2} - 8\frac{dy}{dx} = 0$ is

(I.A.S. 1996)

(a) $y = C_1 + (C_2 + C_3 x + C_4 x^2) e^{2x}$ (b) $y = (C_1 + C_2 x + C_3 x^2) e^{2x}$

(c) $y = C_1 + C_2 x + C_3 x^2 + C_4 x^3$ (d) $y = C_1 x + C_2 x^2 + C_3 x^3 + C_4 x^4$

Ans. (a)

19. The primitive of the differential equation $(D^2 - 2D + 5)^2 y = 0$ is

(I.A.S. 1995)

(a) $e^{x}\{(C_1 + C_2 x)\cos 2x + (C_3 + C_4 x)\sin 2x\}$

(b) $e^{2x}\{(C_1 + C_2 x)\cos x + (C_3 + C_4 x)\sin x\}$

(c) $(C_1 e^{x} + C_2 e^{2x})\cos x + (C_3 e^{x} + C_4 e^{2x})\sin x$

(d) $e^{x}\{C_1 \cos x + C_2 \cos 2x + C_3 \sin x + C_4 \sin 2x\}$

Ans. (a)

20. Which one of the following does not satisfy the differential equation $\frac{d^3y}{dx^3} - y = 0$?

(U.P.P.C.S. 1994)

(a) e^{x} (b) e^{-x}

(c) $e^{-x/2}\sin\left(\frac{\sqrt{3}}{2}x\right)$ (d) $e^{-x/2}\cos\left(\frac{\sqrt{3}}{2}x\right)$

Ans. (b)

21. The particular integral of $(D^2 + a^2)\,y = \sin ax$ is

(I.A.S. 1995)

(a) $-\frac{x}{2a}\cos ax$ (b) $\frac{x}{2a}\cos ax$

(c) $-\frac{ax}{2}\cos ax$ (d) $\frac{ax}{2}\cos ax$

Ans. (a)

Chapter 4

Equations Reducible To Linear Equations with Constant Coefficients

INTRODUCTION

Now the shall study two such forms of linear differential equations with variable coefficients which can be reduced to linear differential equations with constant coefficients by suitable substituitions.

1. Cauchy's Homogeneous Linear Equations

A differential equation of the form

$$x^n \frac{d^n y}{dx^n} + P_1 x^{n-1} \frac{d^{n-1} y}{dx^{n-1}} + \ldots\ldots\ldots\ldots + P_n y = X$$

where $P_1, P_2, \ldots\ldots\ldots\ldots P_n$ are constants and X is either a function of x or a constant is called Cauchy-Euler homogeneous linear differential equation.

The solution of the above homogenous linear equation may be obtained after transforming it into linear equation with constant coefficients by using the substitution.

By the substitution $x = e^z$ or $z = \log_e x$; $\therefore \frac{dz}{dx} = \frac{1}{x}$

$$\text{Now } \frac{dy}{dx} = \frac{dy}{dz}\frac{dz}{dx} = \frac{1}{x}\cdot\frac{dy}{dz}$$

$$\therefore x\frac{dy}{dx} = \frac{dy}{dz} \tag{1}$$

$$\text{Again } \frac{d^2y}{dx^2} = \frac{d}{dx}\left(\frac{dy}{dx}\right) = \frac{d}{dx}\left(\frac{1}{x}\frac{dy}{dz}\right)$$

$$= \frac{x\frac{d^2y}{dz^2}\frac{dz}{dx} - \frac{dy}{dz}}{x^2} = \frac{x\frac{d^2y}{dz^2}\frac{1}{x} - \frac{dy}{dz}}{x^2}$$

$$\therefore x^2 \frac{d^2y}{dx^2} = \frac{d^2y}{dz^2} - \frac{dy}{dz} \qquad (2)$$

Also $$\frac{d^3y}{dx^3} = \frac{d}{dx}\left(\frac{d^2y}{dx^2}\right) = \frac{d}{dx}\left[\frac{1}{x^2}\left(\frac{d^2y}{dz^2} - \frac{dy}{dz}\right)\right]$$

$$= \frac{x^2\left[\frac{d^3y}{dz^3}\frac{dz}{dx} - \frac{d^2y}{dz^2}\frac{dz}{dx}\right] - 2x\left(\frac{d^2y}{dz^2} - \frac{dy}{dz}\right)}{x^4}$$

Substituting $\frac{dz}{dx} = \frac{1}{x}$ and simplifying, we get

$$x^3 \frac{d^3y}{dx^3} = \frac{d^3y}{dz^3} - 3\frac{d^2y}{dz^2} + 2\frac{dy}{dz} \qquad (3)$$

Using $x\frac{d}{dx} \equiv \frac{d}{dz} \equiv D$, in (1), (2) and (3)

we get $$x\frac{dy}{dx} = Dy$$

$$x^2\frac{d^2y}{dx^2} = D(D-1)y$$

$$x^3\frac{d^3y}{dx^3} = D(D-1)(D-2)y$$

In general, we have

$$x^n\frac{d^ny}{dx^n} = D(D-1)(D-2)\ldots\ldots\ldots\ldots(D-n+1)y$$

Using, these results in homogeneous linear equation, it will be transformed into a linear differential equation with constant coefficients.

Example 1. Solve $x^2\frac{d^2y}{dx^2} - x\frac{dy}{dx} - 3y = x^2 \log x$ (I.A.S. 2001)

Solution. On changing the independent variable by substituting

$$x = e^z \text{ or } z = \log_e x \text{ and } \frac{d}{dz} \equiv D$$

The differential equation becomes

$$[D(D-1)-D-3]\,y = ze^{2z}$$

or $$(D^2-2D-3)\,y = ze^{2Z}$$

Now the auxiliary equation is $m^2 - 2m - 3 = 0$

$\Rightarrow$ $$m = 3, -1$$

Hence, the C.F. $= C_1 e^{3Z} + C_2 e^{-Z} = C_1 x^3 + \dfrac{C_2}{x}$

and $$\text{P.I.} = \frac{1}{D^2-2D-3} ze^{2z}$$

$$= e^{2z} \frac{1}{(D+2)^2 - 2(D+2) - 3} z = e^{2z} \frac{1}{D^2+4+4D-2D-4-3} z$$

$$= e^{2z} \frac{1}{D^2+2D-3} z$$

$$= e^{2z} \frac{1}{-3\left(1 - \frac{2D}{3} - \frac{D^2}{3}\right)} z$$

$$= \frac{e^{2z}}{-3}\left[1 - \left(\frac{2D}{3} + \frac{D^2}{3}\right)\right]^{-1} z$$

$$= \frac{e^{2z}}{-3}\left[1 + \frac{2D}{3} + \frac{D^2}{3} + \dots\dots\dots\right] z$$

$$= \frac{e^{2z}}{-3}\left(z + \frac{2}{3}\right) = e^{2z}\left(-\frac{z}{3} - \frac{2}{9}\right)$$

$$= x^2\left(-\frac{1}{3}\log_e x - \frac{2}{9}\right)$$

Hence solution of the given differential equation is

$$y = C_1 x^3 + \frac{C_2}{x} + x^2\left(-\frac{1}{3}\log_e x - \frac{2}{9}\right)$$

Example 2. Solve $x^2 \dfrac{d^2y}{dx^2} - 2x\dfrac{dy}{dx} - 4y = x^4$

Solution. On changing the independent variable by substituting $x = e^z$ or $z = \log_e x$ and $\frac{d}{dz} \equiv D$

The differential equation becomes

$$[D(D-1) - 2D - 4]\, y = e^{4z}$$

Now, the auxiliary equation is

$$m^2 - 3m - 4 = 0 \text{ or } m = 4, -1$$

$\therefore$ C.F. $= C_1 e^{4z} + C_2 e^{-z}$

$= C_1 x^4 + C_2 \frac{1}{x}$ $\qquad \because e^z = x$

and $\quad$ P.I. $= \frac{1}{D^2 - 3D - 4} e^{4z}$

$$= e^{4z} \frac{1}{(D+4)^2 - 3(D+4) - 4} 1$$

$$= e^{4z} \frac{1}{D^2 + 16 + 8D - 3D - 12 - 4} 1$$

$$= e^{4z} \frac{1}{D^2 + 5D} 1$$

$$= e^{4z} \frac{1}{5D\left(1 + \frac{D}{5}\right)} 1 = e^{4z} \frac{1}{5D}\left(1 + \frac{D}{5}\right)^{-1} 1$$

$$= e^{4z} \frac{1}{5D}\left[1 - \frac{D}{5} + \dots\dots\dots\dots\right] 1$$

$$= e^{4z} \frac{1}{5D} 1 = e^{4z} \frac{1}{5} z = \frac{1}{5} z\, e^{4z}$$

$$= \frac{1}{5} x^4 \log_e x$$

Hence the required solution is

$$y = C_1 x^4 + C_2 \frac{1}{x} + \frac{1}{5} x^4 \log_e x$$

Example 3. Solve $x^3 \frac{d^3y}{dx^3} + 2x^2 \frac{d^2y}{dx^2} + 2y = 10\left(x + \frac{1}{x}\right)$

(I.A.S. 2006, 1998, U.P.P.C.S. 1973)

Solution. On changing the independent variable by substituting $x = e^z$ or $z = \log_e x$ and $\frac{d}{dz} \equiv D$ the given differential equation becomes

$$[D(D-1)(D-2) + 2D(D-1) + 2]\, y = 10\left(e^z + e^{-z}\right)$$

or $$(D^3 - D^2 + 2)\, y = 10\,(e^z + e^{-z})$$

The auxiliary equation is

$m^3 - m^2 + 2 = 0$ or $(m+1)(m^2 - 2m + 2) = 0$

or $$D = -1, 1 \pm i$$

$\therefore$ $$C.F = C_1 e^{-z} + C_2 e^z \cos(z + C_3)$$

$$= C_1 x^{-1} + C_2 x \cos(\log_e x + C_3)$$

and $$P.I. = \frac{1}{(D+1)(D^2 - 2D + 2)} 10\, e^z + \frac{1}{(D+1)(D^2 - 2D + 2)} 10\, e^{-z}$$

$$= \frac{1}{(1+1)(1^2 - 2.1 + 2)} 10\, e^z + e^{-z} \frac{1}{\{(D-1+1)\}\{(D-1)^2 - 2(D-1) + 2\}} 10$$

$$= \frac{1}{2} 10\, e^z + e^{-z} \frac{1}{D(D^2 + 1 - 2D - 2D + 2 + 2)} 10$$

$$= 5\, e^z + e^{-z} \frac{1}{D(D^2 - 4D + 5)} 10$$

$$= 5e^z + e^{-z} \frac{1}{5D\left(1 - \frac{4D}{5} + \frac{D^2}{5}\right)} 10$$

$$= 5e^z + e^{-z} \frac{1}{5D} \left\{1 - \left(\frac{4D}{5} - \frac{D^2}{5}\right)\right\}^{-1} 10$$

$$= 5e^z + e^{-z} \frac{1}{5D} 10$$

$$= 5e^z + 2e^{-z}\,\frac{1}{D}\,1 = 5e^z + 2e^{-z}\,z$$

$$= 5e^z + 2z\,e^{-z}$$

$$= 5x + 2\,(\log_e x)\,\frac{1}{x}$$

Hence the required solution is

$$y = C_1\,x^{-1} + C_2\,x\,(\log_e x + C_3) + 5x + (2\log_e x)\,\frac{1}{x}$$

Example 4. Solve $x^3\,\frac{d^3y}{dx^3} + 3x^2\,\frac{d^2y}{dx^2} + x\,\frac{dy}{dx} + y = x + \log x$

(Bihar P.C.S. 2002, U.P.T.U. 2001)

Solution. On changing the independent variable by substituting

$x = e^z$ or $z = \log_e x$ and $\frac{d}{dz} \equiv D$, we have

$$[D\,(D-1)\,(D-2) + 3\,D\,(D-1) + D + 1]\,y = e^z + z$$

or $\quad (D^3 + 1)\,y = e^z + z$

The auxiliary equation is $m^3 + 1 = 0$

or $\quad (m + 1)\,(m^2 - m + 1) = 0$

$\Rightarrow \quad m = -1, \dfrac{1 \pm \sqrt{3}\,i}{2}$

so C.F. $= C_1\,e^{-z} + e^{z/2}\left(C_2\cos\frac{\sqrt{3}}{2}\,z + C_3\sin\frac{\sqrt{3}}{2}\,z\right)$

and $\quad$ P.I. $= \dfrac{1}{D^3 + 1}\,(e^z + z)$

$$= \frac{1}{D^3 + 1} e^z + \frac{1}{1 + D^3} z$$

$$= \frac{1}{1^3 + 1} e^z + (1 + D^3)^{-1} (z)$$

$$= \frac{1}{2} e^z + (1 - D^3 + \ldots\ldots..) z$$

$$= \frac{e^z}{2} + z$$

Therefore required solution is

$$y = C_1 e^z + e^{z/2} \left(C_2 \cos \frac{\sqrt{3}}{2} z + C_3 \sin \frac{\sqrt{3}}{2} z \right) + \frac{e^Z}{2} + z$$

or $$y = C_1 x^{-1} + \sqrt{x} \left[C_2 \cos \frac{\sqrt{3}}{2} (\log x) + C_3 \sin \frac{\sqrt{3}}{2} (\log x) \right] + \frac{x}{2} + \log x$$

Example 5. Solve $x^2 \frac{d^2y}{dx^2} + x \frac{dy}{dx} + y = (\log x) \sin (\log x)$

(U.P.T.U. 2002)

Solution. On changing the independent variable by substituting $x = e^z$ or $z = \log_e x$ and $\frac{d}{dz} = D$, we have

$$[D (D - 1) + D + 1]\, y = z \sin z$$

or $$(D^2 + 1)\, y = z \sin z$$

The auxiliary equation is

$$m^2 + 1 = 0$$

or $$m = \pm i$$

Thus $$\text{C.F.} = C_1 \cos z + C_2 \sin z$$

$$= C_1 \cos (\log x) + C_2 \sin (\log x)$$

& $$\text{P.I.} = \frac{1}{D^2 + 1} z \sin z$$

$$= \text{imaginary part of } \frac{1}{D^2 + 1} z\, e^{iz}$$

$$= \text{I.P. of } e^{iz} \frac{1}{(D+i)^2+1} z$$

$$= \text{I.P. of } e^{iz} \frac{1}{D^2+2iD-1+1} z$$

$$= \text{I.P. of } e^{iz} \frac{1}{2iD}\left(1+\frac{D}{2i}\right)^{-1} z$$

$$= \text{I.P. of } e^{iz} \frac{1}{2iD}\left(1-\frac{D}{2i}+\ldots\ldots\right) z$$

$$= \text{I.P. of } e^{iz} \frac{1}{2iD}\left(z-\frac{1}{2i}\right)$$

$$= \text{I.P. of } e^{iz} \frac{1}{2i}\int\left(z+\frac{i}{2}\right) dz$$

$$= \text{I.P. of } \frac{e^{iz}}{2i}\left(\frac{z^2}{2}+\frac{iz}{2}\right)$$

$$= \text{I.P. of } \frac{-i}{2}(\cos z + i \sin z)\left(\frac{z^2}{2}+\frac{i}{2}z\right)$$

$$= \text{I.P. of} -\frac{1}{2}(i\cos z - \sin z)\left(\frac{z^2}{2}+\frac{i}{2}z\right)$$

$$= -\frac{z^2}{4}\cos z + \frac{1}{4} z \sin z = \frac{z}{4}(\sin z - z\cos z)$$

$$= \frac{\log x}{4}[\sin(\log x) - \log x \cos(\log x)]$$

Hence required solution is

$$y = C_1 \cos(\log x) + C_2 \sin(\log x) + \frac{\log x}{4}[\sin(\log x) - \log x \cos(\log x)]$$

Example 6. Solve $x^2\dfrac{d^2y}{dx^2} - x\dfrac{dy}{dx} + 4y = \cos(\log x) + x\sin(\log x)$

Solution. On changing the independent variable by substituting $x = e^z$ or $z = \log x$ and $\dfrac{d}{dz} \equiv D$ we have

$$[D(D-1) - D + 4]\, y = \cos z + e^z \sin z$$

or $(D^2 - 2D + 4)\, y = \cos z + e^z \sin z$

The auxiliary equation is

$$m^2 - 2m + 4 = 0$$

$$\Rightarrow \quad m = 1 \pm i\sqrt{3}$$

Therefore C.F. $= e^z (C_1 \cos \sqrt{3}\, z + C_2 \sin \sqrt{3}\, z)$

$$= x\left[C_1 \cos\left\{\sqrt{3} \log x + C_2 \sin\left(\sqrt{3} \log x\right)\right\}\right]$$

and

$$\text{P.I} = \frac{1}{D^2 - 2D + 4} \cos z + \frac{1}{D^2 - 2D + 4} e^z \sin z$$

$$= \frac{1}{-1^2 - 2D + 4} \cos z + e^z \frac{1}{(D+1)^2 - 2(D+1) + 4} \sin z$$

$$= \frac{1}{3 - 2D} \cos z + e^z \frac{1}{D^2 + 3} \sin z$$

$$= \frac{3 + 2D}{9 - 4D^2} \cos z + e^z \frac{1}{-1^2 + 3} \sin z$$

$$= \frac{(3 + 2D)}{9 - 4(-1)^2} \cos z + \frac{e^z}{2} \sin z$$

$$= \frac{1}{13}(3 + 2D) \cos z + \frac{1}{2} e^z \sin z$$

$$= \frac{1}{13}(3 \cos z - 2 \sin z) + \frac{1}{2} e^z \sin z$$

$$= \frac{1}{13}[3 \cos(\log x) - 2 \sin(\log x)] + \frac{1}{2} x \sin(\log x)$$

Therefore required solution is

$$y = x\left[C_1 \cos\left(\sqrt{3} \log x\right) + C_2 \sin\left(\sqrt{3} \log x\right)\right] + \frac{1}{13}[3 \cos(\log x) - 2 \sin(\log x)] + \frac{1}{2} x \sin(\log x)$$

Example 7. Solve $x^2 \dfrac{d^2y}{dx^2} + 4x \dfrac{dy}{dx} + 2y = e^x$

[U.P.T.U. (CO) 2005]

Solution. Substituting $x = e^z$ or $z = \log_e x$ and putting $\dfrac{d}{dz} \equiv D$, we have

$$[D(D - 1) + 4D - 2]\, y = e^{e^z}$$

or $\quad (D + 2)(D + 1)\, y = e^{e^z}$

The auxiliary equation is $(m + 2)(m + 1) = 0$

$\Rightarrow \quad m = -2, -1$

$\therefore \quad$ C.F. $= C_1 e^{-2Z} + C_2 e^{-Z} = C_1 x^{-2} + C_2 x^{-1}$

and $\quad$ P.I $= \dfrac{1}{(D+2)(D+1)} e^{e^z} = \left[\dfrac{1}{D+1} - \dfrac{1}{D+2}\right] e^{e^z}$

Let $\quad \dfrac{1}{D+1} e^{e^z} = u \quad \therefore (D + 1)\, u = e^{e^z}$

or $\quad \dfrac{du}{dz} + u = e^{e^z}$, which is linear

Integrating factor = e^z, Hence its solution is

$$u e^z = \int e^z e^{e^z} dz$$

$$= \int e^x dx \qquad \because e^Z = x \therefore e^Z dz = dx$$

$$= e^x$$

or $\quad u = e^x \dfrac{1}{e^z} = \dfrac{1}{x} e^x \qquad \because e^z = x$

Further, let $\dfrac{1}{D+2} e^{e^z} = v$

$\therefore (D+2)\, v = e^{e^z}$

or $\dfrac{dv}{dz} + 2v = e^{e^z}$, which is linear

Integrating factor = e^{2z}, Hence its solution is

$$ve^{2z} = \int e^{2z} e^{e^z} dz$$

$$= \int e^z e^{e^z} e^z dz$$

$$= \int x e^x dx \qquad \because e^z = x, \therefore e^z dz = dx$$

$$= e^x (x - 1)$$

$$\therefore \qquad v = \frac{e^x (x-1)}{e^{2z}} = \frac{e^x (x-1)}{x^2} = \frac{e^x}{x} - \frac{e^x}{x^2}$$

Hence $\qquad \text{P.I.} = u - v = \frac{1}{x} e^x - \left(\frac{e^x}{x} - \frac{e^x}{x^2}\right) = \frac{e^x}{x^2}$

Hence the required solution is

$$y = C_1 x^{-2} + C_2 x^{-1} + \frac{e^x}{x^2}$$

2. Legendre's linear differential equation (Equation reducible to homogeneous form)

An equation of the form

$$(ax+b)^n \frac{d^n y}{dx^n} + k_1 (ax+b)^{n-1} \frac{d^{n-1} y}{dx^{n-1}} + ... + k_n y = X \qquad (1)$$

Where a, b, k_1, k_2,..........k_n are all constants and X is a function of x, is called Legendre's linear equation.

Such equations can be reduced to linear equations with constant coefficients by substituting $ax + b = e^z$ i.e. $z = \log(ax+b)$

Then if $D = \frac{d}{dz}$, $\frac{dy}{dx} = \frac{dy}{dz}\frac{dz}{dx} = \frac{a}{ax+b}\frac{dy}{dz}$

i.e. $(ax+b)\frac{dy}{dx} = aDy$

$$\frac{d^2y}{dx^2} = \frac{d}{dx}\left(\frac{a}{ax+b}\frac{dy}{dz}\right) = \frac{-a^2}{(ax+b)^2}\frac{dy}{dz} + \frac{a}{ax+b}\frac{d}{dz}\left(\frac{dy}{dz}\right)\frac{dz}{dx}$$

$$= \frac{a^2}{(ax+b)^2}\left(\frac{d^2y}{dz^2} - \frac{dy}{dz}\right)$$

i.e. $(ax+b)^2 \frac{d^2y}{dx^2} = a^2 D(D-1)y$

Similarly $(ax+b)^3 \frac{d^3y}{dx^3} = a^3 D(D-1)(D-2)y$ and so on.

After making these replacements in (1), there results a linear equation with constant coefficients.

Example 8. Solve $(1+x)^2 \frac{d^2y}{dx^2} + (1+x)\frac{dy}{dx} + y = 4\cos\log(1+x)$

Solution. put $1 + x = e^z$ and $\frac{d}{dz} = D$

Hence the given differential equation becomes

$[D(D-1) + D + 1]\, y = 4\cos z$

$\therefore$ Auxiliary equation is

$$D^2 + 1 = 0 \text{ or } D = \pm i$$

$\therefore$ C.F. $= C_1\cos(z + C_2) = C_1\cos[\log(1+x) + C_2]$

and P.I $= \frac{1}{D^2+1}\, 4\cos z = 4.\,\frac{z}{2}\sin z$

$$= 2z\sin z$$

$$= 2\log(1+x)\sin\log(1+x)$$

Hence the required solution is

$y = C_1\cos[\log(1+x) + C_2] + 2\log(1+x)\sin\log(1+x)$

Example 9 : Solve

$(1+2x)^2 \frac{d^2y}{dx^2} - 6(1+2x)\frac{dy}{dx} + 16y = 8(1+2x)^2$

and $y(0) = 0$, $y'(0) = 2$ (I.A.S. 1997)

Solution : Let $1 + 2x = z$ then

$$\frac{dy}{dx} = \frac{dy}{dz}\cdot\frac{dz}{dx} = 2\frac{dy}{dz}$$

$$\because \frac{dz}{dx} = 2$$

and $\frac{d^2y}{dx^2} = \frac{d}{dz}\left(2\frac{dy}{dz}\right)\frac{dz}{dx} = 4\frac{d^2y}{dz^2}$

Substituting these in given differential equation we have

$$4z^2\frac{d^2y}{dz^2}-6\cdot 2z\frac{dy}{dz}+16y=8z^2$$

$$\text{or } z^2\frac{d^2y}{dz^2}-3z\frac{dy}{dz}+4y=2z^2$$

putting $z = e^t$, we have

$\{\theta(\theta-1)-2\theta+4\}\, y = 2e^{2t}$

or $(\theta^2-4\theta+4)\, y = 2e^{2t}$

its auxiliary equation is

$m^2 - 4m + 4 = 0$

i.e. $(m-2)^2 = 0$

or $m = 2$ (twice)

$\therefore$ C.F $= (C_1 + C_2t)e^{2t}$

$= (C_1+C_2\log z)z^2$

$= \{C_1 + C_2 \log (1+2x)\}\, (1+2x)^2$

$$\text{and P.I}=\frac{1}{\theta^2-4\theta+4}2e^{2t}=2e^{2t}\frac{1}{(\theta+2)^2-4(\theta+2)+4}\cdot 1$$

$$=2e^{2t}\frac{1}{\theta^2+4\theta-4-4\theta-8+4}\cdot 1=2e^{2t}\frac{1}{\theta^2}\cdot 1$$

$$=2e^{2t}\frac{t^2}{2}=z^2(\log z)^2$$

$$=(1+2x)^2\{\log(1+2x)\}^2$$

Hence the complete solution is

$y = \{C_1+C_2 \log (1+2x)\}\, (1+2x)^2+(1+2x)^2\{\log (1+2x)\}^2$

METHOD OF VARIATION OF PARAMETERS

Method of variation of parameters enables to find solution of any linear non homogeneous differential equation of second order even (with variable coefficients also) provided its complimentary function is given (known). The particular integral of the non-homogeneous equation is obtained by varying the parameters i.e. by replacing the arbitrary constants in the C.F. by variable functions.

Consider a linear non-homogeneous second order differential equation with variable coefficients

$$\frac{d^2y}{dx^2} + P(x)\frac{dy}{dx} + Q(x)\,y = X(x) \qquad (1)$$

Suppose the complimentary function of (1) is = $C_1y_1(x) + C_2y_2(x)$ (2)

so that y_1 and y_2 satisfy

$$\frac{d^2y}{dx^2} + P(x)\frac{dy}{dx} + Q(x)\,y = 0$$

In method of variation of parameters the arbitrary constants C_1 and C_2 in (2) are replaced by two unknown functions u (x) and v (x).

Let us assume particular integral is = $u(x)\,y_1(x) + v(x)\,y_2(x)$ (3)

where $$u = \int \frac{-X\,y_2}{y_1\,y_2' - y_1'\,y_2}\,dx$$

and $$v = \int \frac{X\,y_1}{y_1\,y_2' - y_1'\,y_2}\,dx$$

on putting the values of u and v in (3) we get P.I

Thus, required general solution = C.F + P.I

Example 10. Apply the method of variation of parameters to solve

(U.P.T.U. 2009)

$$\frac{d^2y}{dx^2} + y = \tan x$$

Solution. The auxiliary equation is $m^2 + 1 = 0 \Rightarrow m = \pm i$

$\therefore$ C.F. = $C_1 \cos x + C_2 \sin x$ (1)

Here $y_1 = \cos x$, $y_2 = \sin x$

Therefore $y_1\,y_2' - y_1'\,y_2 = \cos^2 x + \sin^2 x = 1$

Let us suppose P.I = $u.y_1 + v\,y_2$ (2)

where

$$u = \int \frac{-Xy_2}{y_1\,y_2' - y_1'\,y_2}\,dx = -\int \frac{\sin x \tan x}{1}\,dx$$

$$= -\int \frac{\sin^2 x}{\cos x}\, dx = -\int \frac{1 - \cos^2 x}{\cos x}\, dx$$

$$= \int (\cos x - \sec x)\, dx$$

$$= \sin x - \log (\sec x + \tan x)$$

& $$v = \int \frac{X\, y_1}{y_1\, y_2' - y_1'\, y_2}\, dx = \int \frac{\tan x \cos x}{1}\, dx$$

$$= \int \sin x\, dx = -\cos x$$

Putting the values of u and v in (2), we get

$$\text{P.I} = u\, y_1 + v\, y_2$$

$$= [\sin x - \log (\sec x + \tan x)] \cos x - \cos x \sin x$$

$$= -\cos x \log (\sec x + \tan x)$$

Therefore, complete solution is

$$y = C_1 \cos x + C_2 \sin x - \cos x \log (\sec x + \tan x)$$

Example 11. Use variation of parameters to solve

$$\frac{d^2y}{dx^2} + y = \sec x$$

(U.P.T.U. 2002)

Solution. The auxiliary equation is

$$m^2 + 1 = 0$$

$\Rightarrow$ $$m = \pm i$$

$\therefore$ $$\text{C.F} = C_1 \cos x + C_2 \sin x \qquad (1)$$

Here $y_1 = \cos x$, $y_2 = \sin x$

Let us suppose P.I. $= u\, y_1 + v\, y_2$ (2)

where $$u = \int \frac{-\sec x \sin x}{1}\, dx \quad \because u = \int \frac{-X\, y_2}{y_1\, y_2' - y_1'\, y_2}\, dx$$

As $y_1\, y_2' - y_1'\, y_2 = \cos x \cos x - (-\sin x) \sin x$

$= \cos^2 x + \sin^2 x$

$= 1$

$= -\int \tan x \, dx$

$= \log \cos x$

and $\qquad v = \int \frac{X y_1}{y_1 y_2' - y_1' y_2} dx$

$= \int \frac{\cos x \sec x}{1} dx = \int dx = x$

putting the values of u and v in (2), we get

$$P.I = \log \cos x . \cos x + x . \sin x$$

Therefore, complete solution is

$$y = C_1 \cos x + C_2 \sin x + \cos x . \log \cos x + x \sin x$$

Example 12. Using the method of variation of parameters solve

$$\frac{d^2 y}{dx^2} + 4y = 4 \tan 2x$$

(I.A.S. 2001, U.P.T.U. 2006)

Solution. Here the auxiliary equations

$$m^2 + 4 = 0 \quad \Rightarrow \quad m = \pm 2i$$

$\therefore \qquad C.F = C_1 \cos 2x + C_2 \sin 2x \qquad (1)$

Here $\qquad y_1 = \cos 2x, y_2 = \sin 2x$

Let us suppose P.I. $= u y_1 + v y_2 \qquad (2)$

where $\qquad u = \int \frac{-X y_2 dx}{y_1 y_2' - y_1' y_2} = \int \frac{-4 \tan 2x \sin 2x}{2} dx$

$\because$ As $y_1 y_2' - y_1' y_2 = 2 \cos 2x \cos 2x + 2 \sin 2x \sin 2x = 2$

$$= -\int 2 \frac{\sin^2 2x}{\cos 2x} dx = -\int \frac{1 - \cos^2 2x}{\cos 2x} dx$$

$= 2 \int (\cos 2x - \sec 2x) \, dx$

$= \sin 2x - \log (\sec 2x + \tan 2x)$

and $$v = \int \frac{X y_1}{y_1 y_2' - y_1' y_2} dx = \int \frac{4 \tan 2x \cos 2x}{2} dx$$

$$= 2 \int \sin 2x \, dx = - \cos 2x$$

putting the values of u and v in (2) we get

$$P.I = \{\sin 2x - \log (\sec 2x + \tan 2x)\} \cos 2x - \cos 2x \sin 2x$$

$$= - \cos 2x \log (\sec 2x + \tan 2x)$$

Hence, the complete solution is

$y = C_1 \cos 2x + C_2 \sin 2x - \cos 2x \log (\sec 2x + \tan 2x)$

Example 13. Obtain general solution of the differential equation $x^2 \frac{d^2y}{dx^2} + x\frac{dy}{dx} - y = x^3 e^x$

(U.P.T.U. 2002)

Solution. On changing the independent variable by substituting $x = e^Z$ or $z = \log_e x$ and $\frac{d}{dz} \equiv D$ the differential equation becomes

$[D (D - 1) + D - 1] y = e^{3Z} e^{e^Z}$

or $$(D^2 - 1) y = e^{3Z} e^{e^Z}$$

Here auxiliary equation is $m^2 - 1 = 0 \Rightarrow m = \pm 1$

$\therefore$ $$C.F = C_1 e^Z + C_2 e^{-Z}$$

$\Rightarrow$ $$C.F = C_1 x + \frac{C_2}{x}$$

Let $P.I = u y_1 + v y_2$

Here $y_1 = x$ and $y_2 = \frac{1}{x}$

Also $$u = \int \frac{-Xy_2 dx}{y_1 y_2' - y_1' y_2} = \int \frac{-x^3 e^x \frac{1}{x} dx}{x\left(-\frac{1}{x^2}\right) - \frac{1}{x}(1)} = \int \frac{-x^2 e^x \, dx}{\frac{-2}{x}}$$

$$= \frac{1}{2} \int x^3 e^x \, dx$$

$$u = \frac{1}{2}\left[x^3 e^x - 3x^2 e^x + 6xe^x - 6e^x\right]$$

$$= \frac{1}{2}\left[x^3 - 3x^2 + 6x - 6\right] e^x$$

$$\& \;\; v = \int \frac{X\, y_1 dx}{y_1 y_2' - y_1' y_2} = \int \frac{x^3\, e^x\, x\, dx}{x\left(-\frac{1}{x^2}\right) - \frac{1}{x}(1)}$$

$$= \int \frac{x^4\, e^x}{-\frac{2}{x}} dx = -\frac{1}{2}\int x^5\, e^x\, dx$$

or $$v = -\frac{1}{2}\left[x^5 e^x - 5x^4 e^x + 20x^3 e^x - 60x^2 e^x + 120xe^x - 120e^x\right]$$

putting the values of u and v in equation (2) we get

$$P.I = \frac{1}{2}(x^3 - 3x^2 + 6x - 6)\, e^x\, x - \frac{1}{2}(x^5 - 5x^4 + 20x^3 - 60x^2 + 120x - 120)\, e^x \frac{1}{x}$$

$$= \frac{e^x}{2}\left[x^4 - 3x^3 + 6x^2 - 6x - x^4 + 5x^3 - 20x^2 + 60x - 120 + \frac{120}{x}\right]$$

$$= \frac{e^x}{2}\left[2x^3 - 14x^2 + 54x - 120 + \frac{120}{x}\right]$$

Hence the required solution is y = C.F + P.I

or $$y = C_1 x + \frac{C_2}{x} + \left(x^3 - 7x^2 + 27x - 60 + \frac{60}{x}\right) e^x$$

Example 14. Solve by method of variation of parameters

$$\frac{d^2y}{dx^2} - y = \frac{2}{1+e^x}$$ **(U.P.T.U 2001)**

Solution. Here auxiliary equation is $m^2 - 1 = 0$

$\Rightarrow$ $m = \pm 1$

$\therefore$ $C.F = C_1 e^x + C_2 e^{-x}$

Here $y_1 = e^x$, $y_2 = e^{-x}$

Let $P.I = uy_1 + vy_2$

where $$u = \int \frac{-Xy_2}{y_1\, y_2' - y_1'\, y_2} dx = \int \frac{\frac{2}{1+e^x} e^{-x}}{-2} dx$$

$\because y_1\, y_2' - y_1'\, y_2 = -e^x\, e^{-x} - e^x\, e^{-x} = -2$

$$= \int \frac{e^{-x}}{1+e^x} dx = \int \frac{1}{e^x\,(1+e^x)} dx = \int \left(\frac{1}{e^x} - \frac{1}{1+e^x}\right) dx$$

$$= \int e^{-x}\, dx - \int \frac{e^{-x}}{e^{-x}+1} dx$$

$$= -e^{-x} + \log(e^{-x}+1)$$

$$v = \int \frac{X\, y_1}{y_1\, y_2' - y_1'\, y_2} dx = \int \frac{e^x}{-2} \frac{2}{1+e^x} dx$$

$$= -\int \frac{e^x}{1+e^x} dx = -\log(1+e^x)$$

$P.I = u\, y_1 + v\, y_2$

$= [-e^{-x} + \log(e^{-x}+1)]\, e^x - e^{-x} \log(1+e^x)$

$= -1 + e^x \log(e^{-x}+1) - e^{-x} \log(e^x+1)$

$\therefore\ y = C_1 e^x + C_2 e^{-x} - 1 + e^x \log(e^{-x}+1) - e^{-x} \log(e^x+1)$

Example 15. Apply the method of variation of parameters to solve

$$\frac{d^2y}{dx^2} - 3\frac{dy}{dx} + 2y = \frac{e^x}{1+e^x}$$

(U.P.T.U. 2005)

Solution. Here auxiliary equation is

$$m^2 - 3m + 2 = 0$$

$\Rightarrow\quad (m-1)(m-2) = 0$

$\Rightarrow\quad m = 1, 2$

$\therefore$ C.F. $= C_1 e^x + C_2 e^{2x}$

Here $y_1 = e^x$, $y_2 = e^{2x}$

$P.I = u\, y_1 + v\, y_2$

where $\quad u = \int -\frac{Xy_2}{y_1 y_2' - y_1' y_2} dx$

$$= \int -\frac{\frac{e^x}{1+e^x} e^{2x}}{e^x (2e^{2x}) - e^{2x} (e^x)} dx = \int \frac{\frac{-e^{-3x}}{1+e^x}}{2e^{3x} - e^{3x}} dx$$

$$= \int \frac{-e^{-3x}}{e^{3x}\left(1+e^x\right)} dx = -\int \frac{1}{1+e^x} dx$$

$$= -\int \frac{e^{-x} dx}{e^{-x}+1} = \log\left(e^{-x}+1\right)$$

and $\quad v = \int \frac{X y_1}{y_1 y_2' - y_1' y_2} dx = \int \frac{\frac{e^x}{1+e^x} e^x}{e^x (2e^{2x}) - e^{2x} (e^x)} dx$

$$= \int \frac{\frac{e^{2x}}{1+e^x}}{2e^{3x} - e^{3x}} dx = \int \frac{e^{2x}}{e^{3x} (1+e^x)} dx$$

$$= \int \frac{1}{e^x (1+e^x)} dx = \int \left(\frac{1}{e^x} - \frac{1}{1+e^x}\right) dx$$

$$= \int \left(e^{-x} - \frac{e^{-x}}{e^{-x}+1}\right) dx = -e^{-x} + \log (e^{-x}+1)$$

Therefore $\quad$ P.I $= e^x \log (e^{-x}+1) + e^{2x} \{-e^{-x} + \log (e^{-x}+1)\}$

$$= e^x \log (e^{-x}+1) - e^x + e^{2x} \log (e^{-x}+1)$$

Therefore, complete solution is y = C.F + P.I

or $\quad y = C_1 e^x + C_2 e^{2x} + e^x \log (e^{-x}+1) - e^x + e^{2x} \log (e^{-x}+1)$

Example 16. Solve by the method of variation of parameters

$$\frac{d^2y}{dx^2} - 2\frac{dy}{dx} = e^x \sin x$$

(U.P.T.U. 2003)

Solution. Here auxiliary equation is $m^2 - 2m = 0$

$\Rightarrow \quad m (m-2) = 0$

$\Rightarrow \qquad m = 0, 2$

$\therefore \qquad \text{C.F.} = C_1 + C_2\, e^{2x}$ (1)

Here $y_1 = 1$, $y_2 = e^{2x}$

$\text{P.I} = uy_1 + vy_2$ (2)

where $u = \int \frac{-X\, y_2}{y_1\, y_2' - y_1'\, y_2}\, dx = \int \frac{-e^x \sin x \cdot e^{2x}}{1\,(2e^{2x}) - 0\,(e^{2x})}\, dx$

$$= -\frac{1}{2} \int e^x \sin x\, dx$$

$$= -\frac{1}{2} \frac{e^x}{(1)^2 + (1)^2} (\sin x - \cos x)$$

$$= -\frac{1}{4} e^x (\sin x - \cos x)$$

and $v = \int \frac{X\, y_1}{y_1\, y_2' - y_1'\, y_2}\, dx$

$$= \int \frac{e^x \sin x\, 1}{1\,(2e^{2x}) - 0\,(e^{2x})} = \frac{1}{2} \int e^{-x} \sin x\, dx$$

$$= \frac{1}{2} \frac{e^{-x}}{(-1)^2 + (1)^2} (-\sin x - \cos x)$$

$$= -\frac{1}{2} \frac{e^{-x}}{2} (\sin x + \cos x) = -\frac{e^{-x}}{4} (\sin x + \cos x)$$

putting the values of u and v in (2), we get

$$\text{P.I} = \frac{e^x}{-4}(\sin x - \cos x)1 - \frac{e^{-x}}{4}(\sin x + \cos x)e^{2x}$$

$$= \frac{e^x}{-4}(\sin x - \cos x + \sin x + \cos x) = \frac{e^{-x}}{-2} \sin x$$

Hence, the complete solution is y = C.F + P.I

$$y = C_1 + C_2\, e^{2x} - \frac{e^x}{2} \sin x$$

Example 18. Solve by method of variation of parameters

$$\frac{d^2y}{dx^2} + 2\frac{dy}{dx} + y = e^{-x}\log x$$

(U.P.T.U. 2008)

Solution. Here auxiliary equation is $m^2 + 2m + 1 = 0$

$\Rightarrow \quad (m + 1)^2 = 0$

$\Rightarrow \quad m = -1, -1$

$\therefore \quad C.F = (C_1 + C_2 x)\, e^{-x}$ (1)

Here $y_1 = e^{-x}$, $y_2 = x\,e^{-x}$

$P.I = uy_1 + vy_2$ (2)

where $u = \int \frac{-Xy_2}{y_1 y_2' - y_1' y_2}\, dx$

$$= \int \frac{-e^{-x} \log x.\, x\, e^{-x}}{-xe^{-2x} + e^{-2x} + x\,e^{-2x}}\, dx$$

$$= \int \frac{-xe^{-2x} \log x}{e^{-2x}}\, dx = -\int x \log x\, dx$$

$$= -\frac{x^2}{2}\log x + \int \frac{x^2}{2}\frac{1}{x}\, dx$$

$$= -\frac{x^2}{2}\log x + \frac{x^2}{4}$$

and $v = \int \frac{X y_1}{y_1 y_2' - y_1' y_2}\, dx$

$$= \int \frac{e^{-x} \log x \cdot e^{-x}}{e^{-2x}}\, dx$$

$$= \int \log x\, dx$$

$$= x \log x - \int \frac{1}{x}\, x\, dx$$

$$= x \log x - x$$

Putting these values of u and v in (2) we get

$$\text{P.I} = -\frac{x^2e^{-x}}{2}\log x + e^{-x}\frac{x^2}{4} + x^2 e^{-x}\log x - x^2 e^{-x}$$

or $$\text{P.I} = \frac{x^2e^{-x}}{2}\log x - \frac{3}{4}x^2 e^{-x}$$

Hence, complete solution is y = C.F + P.I

or $$y = (C_1 + C_2 x)e^{-x} + \frac{x^2e^{-x}}{2}\log x - \frac{3}{4}x^2 e^{-x}$$

Example 19. Using variation of parameters method, solve

$$x^2\frac{d^2y}{dx^2} + 2x\frac{dy}{dx} - 12y = x^3\log x$$

(U.P.T.U. 2004)

Solution. On changing the independent variable by substituting $x = e^z$ or $z = \log_e x$ and $\frac{d}{dz} \equiv D$, the differential equation becomes

$$\{D(D-1) + 2D - 12\}\,y = z\,e^{3z}$$

or $$(D^2 + D - 12)\,y = z\,e^{3z}$$

The auxiliary equation is $m^2 + m - 12 = 0$

$\Rightarrow$ $$m = 3, -4$$

$\therefore$ $$\text{C.F} = C_1 e^{3z} + C_2 e^{-4z}$$

or $$\text{C.F} = C_1 x^3 + C_2\frac{1}{x^4} \qquad (1)$$

Here $y_1 = x^3$, $y_2 = \frac{1}{x^4}$

$$\text{P.I.} = uy_1 + vy_2 \qquad (2)$$

where $$u = \int\frac{-X\,y_2}{y_1y_2' - y_1'y_2}\,dx$$

$$= -\int\frac{x\log x\cdot x^{-4}}{x^3(-4x^{-5}) - 3x^2(x^{-4})}\,dx = -\int\frac{x^{-3}\log x}{-7x^{-2}}\,dx$$

$$= \frac{1}{7}\int\frac{\log x}{x}\,dx$$

$$= \frac{1}{7}\,\frac{(\log x)^2}{2}$$

$$= \frac{1}{14}\,(\log x)^2$$

and $$v = \int \frac{X\,y_1}{y_1 y_2' - y_1' y_2}\,dx$$

$$= \int \frac{x \log x \cdot x^3}{-7\,x^{-2}}\,dx = \frac{-1}{7}\int x^6 \log x\,dx$$

$$= -\frac{1}{7}\left[\log x \cdot \frac{x^7}{7} - \int \frac{1}{x}\,\frac{x^7}{7}\,dx\right]$$

$$= -\frac{1}{7}\left[\frac{x^7 \log x}{7} - \frac{1}{7}\left(\frac{x^7}{7}\right)\right]$$

$$= \frac{x^7}{49}\left(\frac{1}{7} - \log x\right)$$

putting the values of u and v in (1) we get

$$\text{P.I} = \frac{1}{14}\,(\log x)^2\,x^3 + \frac{1}{x^4}\,\frac{x^7}{49}\left(\frac{1}{7} - \log x\right)$$

$$= \frac{x^3}{14}\,(\log x)^2 + \frac{x^3}{49}\left(\frac{1}{7} - \log x\right)$$

Therefore, the required solution is y = C.F + P.I

or $$y = C_1\,x^3 + \frac{C_2}{x^4} + \frac{x^3}{14}\,(\log x)^2 + \frac{x^3}{343} - \frac{x^3}{49}\log x$$

or $$y = \left(C_1 + \frac{1}{343}\right)x^3 + \frac{C_2}{x^4} + \frac{x^3}{98}\log x\,(7\log x - 2)$$

SECOND ORDER LINEAR DIFFERENTIAL EQUATIONS WITH VARIABLE COEFFICIENTS (Solution by Changing dependent and independent variables)

INTRODUCTION

The general form of linear differential equation of the second order may be written as

$$\frac{d^2y}{dx^2} + P\frac{dy}{dx} + Qy = R \tag{1}$$

where P, Q and R are functions of x only. There is no general method for the solution of this type of equations. Some particular methods used to solve these equations are, change of independent variables, Variation of parameters and removal of first order derivatives etc. As this kind of differential equations are of great significance in physics, especially in connection with vibrations in mechanics and theory of electric circuit. In addition many profound and beautiful ideas in pure mathematics have grown out to the study of these equations.

Method I: Complete solution is terms of known integral belonging to the complementary function (i.e. part of C.F. is known or one solution is known).

Let u be a part of complementary function of equation (1) and v is remaining solution of differential equation (1)

Then the complete solution of equation (1) is

$$y = u\,v \tag{2}$$

$$\Rightarrow \quad \frac{dy}{dx} = v\frac{du}{dx} + u\frac{dv}{dx} \text{ and } \frac{d^2y}{dx^2} = v\frac{d^2u}{dx^2} + 2\frac{du}{dx}\frac{dv}{dx} + u\frac{d^2v}{dx^2}$$

Putting these values in equation (1) then, we get

$$v\frac{d^2u}{dx^2} + 2\frac{du}{dx}\frac{dv}{dx} + u\frac{d^2v}{dx^2} + P\left(v\frac{du}{dx} + u\frac{dv}{dx}\right) + Quv = R$$

or $$v\left[\frac{d^2u}{dx^2} + P\frac{du}{dx} + Qu\right] + u\left[\frac{d^2v}{dx^2} + P\frac{dv}{dx}\right] + 2\frac{du}{dx}\frac{dv}{dx} = R \tag{3}$$

Since u is a part of C.F. i.e. solution of (1)

$$\therefore \frac{d^2u}{dx^2} + P\frac{du}{dx} + Qu = 0$$

Hence equation (3) becomes

$$u\left(\frac{d^2v}{dx^2}+P\frac{dv}{dx}\right)+2\frac{du}{dx}\frac{dv}{dx}=R$$

or $$\frac{d^2v}{dx^2}+\left(P+\frac{2}{u}\frac{du}{dx}\right)\frac{dv}{dx}=\frac{R}{u} \quad (4)$$

Let $$\frac{dv}{dx}=z, \text{ so that } \frac{d^2v}{dx^2}=\frac{dz}{dx}$$

Equation (4) becomes

$$\frac{dz}{dx}+\left(P+\frac{2}{u}\frac{du}{dx}\right)z=\frac{R}{u},$$

which is linear in z, Hence z can be determined

We obtain v, by integration the relation $\frac{dv}{dx}=z$

$$\Rightarrow \quad v=\int z\,dx+C_1$$

Therefore, the solution of (1) is $y=u\left[\int z\,dx+C_1\right]$

i.e. $$y=uv$$

Remark. Solving by the above method, u determined by inspection of the following rules

(1) If $P+Qx=0$, then $u=x$

(2) If $1+P+Q=0$, then $u=e^x$

(3) If $1-P+Q=0$, then $u=e^{-x}$

(4) If $1+\frac{P}{a}+\frac{Q}{a^2}=0$, then $u=e^{ax}$

(5) If $2+2Px+Qx^2=0$, then $u=x^2$

(6) If $m(m-1)+Pmx+Qx^2=0$, then $u=x^m$

Example 20. Solve $y''-4xy'+(4x^2-2)y=0$ given that $y=e^{x^2}$ is an integral induced in the complementary function.

(U.P.T.U. 2004)

Solution. The given equation may be written as

$$\frac{d^2y}{dx^2} - 4x\frac{dy}{dx} + (4x^2 - 2)\,y = 0$$

Here $P = -4x$, $Q = 4x^2 - 2$, $R = 0$

and $u = e^{x^2}$, so that $\frac{du}{dx} = 2\,xe^{x^2}$

Let $y = uv \Rightarrow y = e^{x^2}\,v$ (1)

we know that

$$\frac{d^2v}{dx^2} + \left(P + \frac{2}{u}\frac{du}{dx}\right)\frac{dv}{dx} = 0, \text{ As } R = 0$$

$$\Rightarrow \quad \frac{d^2v}{dx^2} + \left(\frac{2}{e^{x^2}}\,2xe^{x^2} - 4x\right)\frac{dv}{dx} = 0$$

$$\Rightarrow \quad \frac{d^2v}{dx^2} = 0$$

$$\Rightarrow \quad \frac{dv}{dx} = C$$

$$\Rightarrow \quad v = C_1x + C_2$$

Hence the complete solution is $y = e^{x^2}\,v$

or $\quad y = e^{x^2}\,(C_1x + C_2)$

Example 21. By the method of variation of parameters, solve the differential equation **(U.P.T.U. 2004)**

$$\frac{d^2y}{dx^2} + (1 - \cot x)\frac{dy}{dx} - y\cot x = \sin^2 x$$

Solution. Here $P = 1 - \cot x$, $Q = -\cot x$

Therefore $1 - P + Q = 1 - (1 - \cot x) - \cot x = 0$

That is $y = e^{-x}$ is a part of the C.F. putting $y = ve^{-x}$

$$\frac{dy}{dx} = -ve^{-x} + e^{-x}\frac{dv}{dx}$$

and $$\frac{d^2y}{dx^2} = e^{-x}\frac{d^2v}{dx^2} - 2e^{-x}\frac{dv}{dx} + ve^{-x}$$

on putting these values is the given differential equation, we have

$$\frac{d^2v}{dx^2} - (1 + \cot x)\frac{dv}{dx} = 0$$

or $$\frac{dp}{dx} - (1 + \cot x)\, p = 0 \qquad \text{where } p = \frac{dv}{dx}$$

$\Rightarrow$ $$\frac{dp}{p} = (1 + \cot x)\, dx$$

on integrating we get

$$\log p = x + \log \sin x + \log C_1$$

$\Rightarrow$ $$p = C_1 e^x \sin x$$

Substituting for p

$$\frac{dv}{dx} = C_1 e^x \sin x$$

or $$dv = C_1 e^x \sin x\, dx$$

Integrating

$$v = C_1 \int e^x \sin x\, dx + C_2$$

$$= C_1 \frac{1}{2} e^x (\sin x - \cos x) + C_2$$

Therefore, solution of the given differential equation i.e. C.F. is given by

$$y = ve^{-x} = C_1 \frac{1}{2} (\sin x - \cos x) + C_2 e^{-x}$$

Let y = Au + Bv be the complete solution of the given differential equation where A and B are the functions of x, i.e.

$$y = A(\sin x - \cos x) + Be^{-x} \qquad (1)$$

Differentiating on both sides

$$\frac{dy}{dx} = A(\cos x + \sin x) - Be^{-x} + \frac{dA}{dx}(\sin x - \cos x) + \frac{dB}{dx} e^{-x}$$

Let us choose A and B such that

$$\frac{dA}{dx}(\sin x - \cos x) + \frac{dB}{dx}e^{-x} = 0 \qquad (2)$$

$$\Rightarrow \qquad \frac{dy}{dx} = A(\cos x + \sin x) - Be^{-x}$$

and $\dfrac{d^2y}{dx^2} = \dfrac{dA}{dx}(\cos x + \sin x) - \dfrac{dB}{dx}e^{-x} + A(-\sin x + \cos x) + Be^{-x}$

putting these values of $\dfrac{d^2y}{dx^2}, \dfrac{dy}{dx}$ and y is the given equation, we get

$$\frac{dA}{dx}(\cos x + \sin x) - \frac{dB}{dx}e^{-x} = \sin^2 x \qquad (3)$$

on solving equation (2) and (3), we get

$$\frac{dA}{dx} = \frac{1}{2}\sin x$$

Integrating,

$$A = -\frac{1}{2}\cos x + C_1$$

and $\dfrac{dB}{dx} = \dfrac{1}{2}e^x(\sin x \cos x - \sin^2 x)$

$$= \frac{e^x}{4}(\sin 2x + \cos 2x - 1)$$

on integration, we have

$$B = \frac{e^x}{20}(3\sin 2x - \cos 2x) - \frac{e^x}{4} + C_2$$

putting the values of A and B in equation (1) we get

$$y = \left(-\frac{1}{2}\cos x + C_1\right)(\sin x - \cos x)\left(\frac{e^x}{20}(3\sin 2x - \cos 2x) - \frac{e^x}{4} + C_2\right)e^{-x}$$

$$= -\frac{1}{2}\cos x \sin x + \frac{1}{2}\cos^2 x + C_1 \sin x - C_1 \cos x + \frac{3}{20}\sin 2x - \frac{1}{20}\cos 2x - \frac{1}{4} + C_2 e^{-x}$$

or $$y = C_1 (\sin x - \cos x) + C_2 e^{-x} - \frac{1}{10} (\sin 2x - 2\cos 2x)$$

Method II. Normal form (Removal of first derivative)

Let $$\frac{d^2y}{dx^2} + P\frac{dy}{dx} + Qy = R \qquad (1)$$

putting y = uv, we get

$$v\left(\frac{d^2u}{dx^2} + P\frac{du}{dx} + Qu\right) + u\left(\frac{d^2v}{dx^2} + P\frac{dv}{dx}\right) + 2\frac{du}{dx}\frac{dv}{dx} = R$$

$$\Rightarrow \quad \frac{d^2v}{dx^2} + \left(\frac{2}{u}\frac{du}{dx} + P\right)\frac{dv}{dx} + v\left(\frac{1}{u}\frac{d^2u}{dx^2} + \frac{P}{u}\frac{du}{dx} + Q\right) = \frac{R}{u} \qquad (2)$$

But the first order derivative must be remove

so $$\frac{2}{u}\frac{du}{dx} + P = 0 \Rightarrow \frac{du}{u} = -\frac{1}{2}Pdx$$

$$\Rightarrow \log u = -\int\frac{P}{2}dx$$

$$\Rightarrow u = e^{-\int\frac{P}{2}dx}$$

Since $$\frac{du}{dx} = -\frac{Pu}{2} \Rightarrow \frac{d^2u}{dx^2} = -\frac{1}{2}\left[P\frac{du}{dx} + u\frac{dP}{dx}\right]$$

$$\Rightarrow \frac{d^2u}{dx^2} = -\frac{1}{2}\left[P\left(-\frac{Pu}{2}\right) + u\frac{dP}{dx}\right] = \frac{P^2u}{4} - \frac{u}{2}\frac{dp}{dx}$$

From (2) $$\frac{d^2v}{dx^2} + v\left(\frac{P^2}{4} - \frac{1}{2}\frac{dP}{dx} - \frac{P^2}{2} + Q\right) = \frac{R}{u}$$

$$\Rightarrow \frac{d^2v}{dx^2} + v\left(Q - \frac{1}{2}\frac{dP}{dx} - \frac{P^2}{4}\right) = \frac{R}{u}$$

$$\Rightarrow \frac{d^2v}{dx^2} + Iv = \frac{R}{u} \qquad (3)$$

This equation is called normal form of equation (1)

where $\quad I = Q - \frac{1}{2}\frac{dP}{dx} - \frac{P^2}{4}$

Example 22. Solve

$$\frac{d^2y}{dx^2} - 4x\frac{dy}{dx} + (4x^2 - 1)\,y = -3\,e^{x^2}\,\sin 2x$$

[I.A.S. 2000, U.P.T.U. (C.O.) 2004]

Solution. Here $P = -4x$, $Q = 4x^2 - 1$, $R = -3e^{x^2}\,\sin 2x$

so $\quad I = Q - \frac{1}{2}\frac{dP}{dx} - \frac{1}{4}P^2 = 4x^2 - 1 - \frac{1}{2}(-4) - \frac{1}{4}(-4x)^2$

$= 4x^2 - 1 + 2 - 4x^2 = 1$

$$u = e^{-\frac{1}{2}\int P dx} = e^{-\frac{1}{2}\int(-4x)\,dx}$$

$$= e^{2\int x\,dx} = e^{x^2}$$

Then substituting these values in the equation

$$\frac{d^2v}{dx^2} + Iv = \frac{R}{u}, \text{ We have}$$

$$\frac{d^2v}{dx^2} + v = \frac{-3e^{x^2}\,\sin 2x}{e^{x^2}} = -3\sin 2x$$

its $C.F = C_1\cos x + C_2\sin x$

and $P.I = -\frac{1}{D^2+1}\,2\sin 2x = -3\,\frac{1}{-2^2+1}\sin 2x$

$= \sin 2x$

Thus $\quad v = C_1\cos x + C_2\sin x + \sin 2x$

Therefore required solution is $y = uv$

or $\quad y = e^{x^2}\,(C_1\cos x + C_2\sin x + \sin 2x)$

Example 23. Solve

$$\frac{d^2y}{dx^2} - 4x\frac{dy}{dx} + (4x^2 - 3)\,y = e^{x^2}$$

by removing first derivative

Solution. Here $P = -4x$, $Q = 4x^2 - 3$, $R = e^{x^2}$

$$I = Q - \frac{1}{2}\frac{dP}{dx} - \frac{P^2}{4} = 4x^2 - 3 - \frac{1}{2}(-4) - \frac{1}{4}(-4x)^2$$

$$= 4x^2 - 3 + 2 - 4x^2 = -1$$

$$w = e^{-\frac{1}{2}\int P dx}$$

$$= e^{-\frac{1}{2}\int (-4x)\, dx} = e^{2\int x\, dx} = e^{x^2}$$

Then substituting these values in the equation

$$\frac{d^2v}{dx^2} + Iv = \frac{R}{u}, \text{ we get}$$

$$\frac{d^2v}{dx^2} - v = \frac{e^{x^2}}{e^{x^2}}$$

or $$\frac{d^2v}{dx^2} - v = 1$$

its $\quad C.F = C_1 e^{x} + C_2 e^{-x}$

and $$P.I = \frac{1}{D^2 - 1} 1 = -(1 - D^2)^{-1} 1$$

$$= -(1 + D^2 + D^4 + ...) 1$$

$$= -1$$

Thus $v = C_1 e^{x} + C_2 e^{-x} - 1$

Hence the general solution of the given equation is

$$y = uv$$

or $$y = e^{x^2} (C_1 e^{x} + C_2 e^{-x} - 1)$$

Method III. Change of independent variable

consider $\dfrac{d^2y}{dx^2} + P\dfrac{dy}{dx} + Qy = R$ (1)

Let us change the independent variable x to z and $z = f(x)$.

Then $\frac{dy}{dx} = \frac{dy}{dz}\frac{dz}{dx}$ (2)

$$\frac{d^2y}{dx^2} = \frac{d}{dx}\left(\frac{dy}{dz}\frac{dz}{dx}\right) = \frac{dy}{dz}\frac{d^2z}{dx^2} + \left(\frac{dz}{dx}\right)^2 \frac{d^2y}{dx^2} \quad (3)$$

Putting the values of $\frac{dy}{dx}$ and $\frac{d^2y}{dx^2}$ in (1) we get

$$\frac{d^2y}{dz^2}\left(\frac{dz}{dx}\right)^2 + \frac{dy}{dz}\frac{d^2z}{dx^2} + P\frac{dy}{dz}\frac{dz}{dx} + Qy = R$$

or
$$\frac{d^2y}{dz^2}\left(\frac{dz}{dx}\right)^2 + \left(P\frac{dz}{dx} + \frac{d^2z}{dx^2}\right)\frac{dy}{dz} + Qy = R$$

or
$$\frac{d^2y}{dz^2} + \frac{\left(P\frac{dz}{dx} + \frac{d^2z}{dx^2}\right)}{\left(\frac{dz}{dx}\right)^2}\frac{dy}{dz} + \frac{Q}{\left(\frac{dz}{dx}\right)^2}y = \frac{R}{\left(\frac{dz}{dx}\right)^2}$$

$\Rightarrow$
$$\frac{d^2y}{dz^2} + P_1\frac{dy}{dz} + Q_1 y = R_1 \quad (4)$$

where
$$P_1 = \frac{\left(P\frac{dz}{dx} + \frac{d^2z}{dx^2}\right)}{\left(\frac{dz}{dx}\right)^2}, Q_1 = \frac{Q}{\left(\frac{dz}{dx}\right)^2} \text{ and } R_1 = \frac{R}{\left(\frac{dz}{dx}\right)^2}$$

Equation (4) is solved either by taking $P_1 = 0$ or Q_1 = a constant

Example 24. Solve by changing the independent variable

$$x\frac{d^2y}{dx^2} - \frac{dy}{dx} + 4x^3y = x^5$$ **(U.P.T.U. 2002, 2003)**

Solution. Given equation is

$$\frac{d^2y}{dx^2} - \frac{1}{x}\frac{dy}{dx} + 4x^2y = x^4 \quad (1)$$

Here $P = -\frac{1}{x}$, $Q = 4x^2$ and $R = x^4$

On changing the independent variable x to z, the equation (1) transformed as

$$\frac{d^2y}{dz^2} + P_1\frac{dy}{dz} + Q_1 y = R_1 \qquad (2)$$

where $$Q_1 = \frac{Q}{\left(\frac{dz}{dx}\right)^2} = \frac{4x^2}{\left(\frac{dz}{dx}\right)^2} = \text{constant} = 1 \text{ say}$$

or $$\left(\frac{dz}{dx}\right)^2 = 4x^2$$

$\Rightarrow$ $$\frac{dz}{dx} = 2x$$

$\Rightarrow$ $$z = x^2$$

$\Rightarrow$ $$\frac{d^2z}{dx^2} = 2$$

$$P_1 = \frac{P\frac{dz}{dx} + \frac{d^2z}{dx^2}}{\left(\frac{dz}{dx}\right)^2} = \frac{\left\{2 + \left(\frac{-1}{x}\right)2x\right\}}{4x^2} = 0$$

$$R_1 = \frac{R}{\left(\frac{dz}{dx}\right)^2} = \frac{x^4}{4x^2} = \frac{x^2}{4} = \frac{z}{4}$$

on putting the values of P_1, Q_1 and R_1 in (2), we get

$$\frac{d^2y}{dz^2} + y = \frac{z}{4}$$

or $$(D^2 + 1)\, y = \frac{z}{4}$$

its A.E. is $m^2 + 1 = 0 \quad \Rightarrow \quad m = \pm i$

$\therefore$ $$\text{C.F} = C_1 \cos z + C_2 \sin z$$

or $$\text{C.F} = C_1 \cos x^2 + C_2 \sin x^2$$

and $$\text{P.I} = \frac{1}{D^2 + 1}\,\frac{z}{4} = \frac{1}{4}(1 + D^2)^{-1} z$$

$$= \frac{1}{4}(1 - D^2 + \text{..........})\, z$$

$$= \frac{z}{4}$$

$$= \frac{x^2}{4}$$

Hence the complete solution is y = C.F + P.I

or $$y = C_1 \cos x^2 + C_2 \sin x^2 + \frac{x^2}{4}$$

Example 25. Solve the following differential equation by changing the independent variable $x \frac{d^2y}{dx^2} + (4x^2 - 1)\frac{dy}{dx} + 4x^3 y = 2x^3$ **(U.P.T.U. 2006)**

Solution. The given differential equation may be written as

$$\frac{d^2y}{dx^2} + \left(4x - \frac{1}{x}\right)\frac{dy}{dx} + 4x^2 y = 2x^2 \qquad (1)$$

Here $P = 4x - \frac{1}{x}$, $Q = 4x^2$, $R = 2x^2$

on changing the independent variable x to z, the equation (1) is transformed as

$$\frac{d^2y}{dz^2} + P_1\frac{dy}{dz} + Q_1 y = R_1 \qquad (2)$$

where $$Q_1 = \frac{Q}{\left(\frac{dz}{dx}\right)^2} = \frac{4x^2}{\left(\frac{dz}{dx}\right)^2} - 1 \text{ (constant) say}$$

$$\Rightarrow \quad \frac{dz}{dx} = 2x$$

$$\Rightarrow \quad z = x^2$$

$$\Rightarrow \quad \frac{d^2z}{dx^2} = 2$$

$$P_1 = \frac{P\frac{dz}{dx} + \frac{d^2z}{dx^2}}{\left(\frac{dz}{dx}\right)^2} = \frac{8x^2 - 2 + 2}{4x^2} = 2$$

and $$R_1 = \frac{R}{\left(\frac{dz}{dx}\right)^2} = \frac{2x^2}{4x^2} = \frac{1}{2}$$

Putting the values of P_1, Q_1 & R_1 in (2), we get

$$\frac{d^2y}{dz^2} + 2\frac{dy}{dz} + y = \frac{1}{2}$$

its Auxiliary equation is $m^2 + 2m + 1 = 0$

$\Rightarrow \qquad (m + 1)^2 = 0$

$\Rightarrow \qquad m = -1, -1$

$\therefore$ C.F. $= (C_1 + C_2 z)\, e^{-z}$

$= (C_1 + C_2 x^2)\, e^{-x^2}$

and $$P.I = \frac{1}{D^2 + 2D + 1}\left(\frac{1}{2}\right)$$

$$= \frac{1}{2}\,\frac{1}{D^2 + 2D + 1}\, e^{oz}$$

$$= \frac{1}{2}\,\frac{1}{(o)^2 + 2.0 + 1}\, 1 = \frac{1}{2}$$

$\therefore$ Complete solution is y = C.F + P.I

$$= (C_1 + C_2 x^2)\, e^{-x^2} + \frac{1}{2}$$

Example 26. Solve $\frac{d^2y}{dx^2} + \cot x\,\frac{dy}{dx} + 4y\,\text{cosex}^2 x = 0$

Solution. Here P = cot x, Q = 4 cosec²x and R = 0 on changing the independent variable x to z, the given differential equation transformed to

$$\frac{d^2y}{dz^2} + P_1\frac{dy}{dz} + Q_1 y = 0$$

where $P_1 = \dfrac{P\dfrac{dz}{dx} + \dfrac{d^2z}{dx^2}}{\left(\dfrac{dz}{dx}\right)^2}, Q_1 = \dfrac{Q}{\left(\dfrac{dz}{dx}\right)^2}$

Case I. Let us take $P_1 = 0$

$$\frac{P\dfrac{dz}{dx} + \dfrac{d^2z}{dx^2}}{\left(\dfrac{dz}{dx}\right)^2} = 0 \text{ or } P\frac{dz}{dx} + \frac{d^2z}{dx^2} = 0$$

$$\Rightarrow \quad \frac{d^2z}{dx^2} + \cot x \frac{dz}{dx} = 0 \tag{2}$$

put $\dfrac{dz}{dx} = v, \dfrac{d^2z}{dx^2} = \dfrac{dv}{dx}$

Using these, (2) becomes $\dfrac{dv}{dx} + (\cot x)\, v = 0$

$\Rightarrow \quad \dfrac{dv}{v} = -\cot x\, dx$

$\Rightarrow \quad \log v = -\log \sin x + \log C = \log C \operatorname{cosec} x$

$\Rightarrow \quad v = C \operatorname{cosec} x$

$\dfrac{dz}{dx} = c \operatorname{cosec} x$

or $\quad dz = (C \operatorname{cosec} x)\, dx$

$\Rightarrow \quad z = c \log \tan \dfrac{x}{2}$

Case II. Now, let us take Q_1 = constant

$Q_1 = \dfrac{Q}{\left(\dfrac{dz}{dx}\right)^2} = \dfrac{4 \operatorname{cosec}^2 x}{c^2 \operatorname{cosec}^2 x} = \dfrac{4}{c^2}$ which is constant

Hence the equation (1) reduce to

$$\frac{d^2y}{dz^2} + 0\frac{dy}{dz} + \frac{4}{c^2}y = 0$$

or $$\frac{d^2y}{dz^2} + \frac{4}{c^2}y = 0 \qquad \because P_1 = 0, Q_1 = \frac{4}{c^2}$$

$$\Rightarrow \quad \left(D^2 + \frac{4}{c^2}\right)y = 0$$

its auxiliary equation is $m^2 + \frac{4}{c^2} = 0 \Rightarrow m = \pm i\frac{2}{c}$

$$\therefore \quad C.F = c_1 \cos\frac{2z}{c} + c_2 \sin\frac{2z}{c}$$

$$\Rightarrow \quad y = c_1 \cos\left(2\log\tan\frac{x}{2}\right) + C_2 \sin\left(2\log\tan\frac{x}{2}\right) \because z = c\log\tan\frac{x}{2}$$

SIMULTANEOUS LINEAR EQUATIONS WITH CONSTANT COEFFICIENTS:

In Several applied mathematics problems, there are more than one dependent variables, each of which is a function of one independent variable, usually say time t. The formulation of such problems leads to a system of simultaneous linear differential equation with constant coefficients. Such a system can be solved by the method of elimination. Laplace transform method, using matrices and short cut operator methods.

Example 27. Solve $\frac{dx}{dt} + y = \sin t, \frac{dy}{dt} + x = \cos t$

$x(o) = 2, y(o) = 0$ **(U.P.T.U. 2004)**

Solution. We have

$$\frac{dx}{dt} + y = \sin t \qquad (1)$$

$$\frac{dy}{dt} + x = \cos t \qquad (2)$$

Differentiating (1) w.r.t. 't' we have

$$\frac{d^2x}{dt^2} + \frac{dy}{dt} = \cos t \qquad (3)$$

Using (2) in (3) we get

$$\frac{d^2x}{dt^2} - x = 0 \quad \Rightarrow \quad (D^2 - 1)\,x = 0$$

its auxiliary equation is

$m^2 - 1 = 0 \quad \Rightarrow \quad m = \pm 1$

$$\therefore \quad x = C_1 e^t + C_2 e^{-t} \tag{4}$$

$$\Rightarrow \quad \frac{dx}{dt} = C_1 e^t - C_2 e^{-t}$$

putting this value of $\frac{dx}{dt}$ in (1) we get

$$y = \sin t - C_1 e^t + C_2 e^{-t} \tag{5}$$

Using given conditions

$$\left.\begin{array}{l}\text{from (iv) } C_1 + C_2 = 2 \\ \text{from (v) } -C_1 + C_2 = 0\end{array}\right\} \Rightarrow C_1 = C_2 = 1$$

putting these values of C_1 and C_2 in (4) & (5) we get

$$x = e^t + e^{-t}$$

and $\quad y = \sin t - e^t + e^{-t}$

is the required solution

Example 28. Solve $\frac{dx}{dt} + 4x + 3y = t$

$$\frac{dy}{dt} + 2x + 5y = e^t$$ **(U.P.T.U. 2006)**

Solution. The given equation can be written as

$$(D + 4)\,x + 3y = t \tag{1}$$

$$2x + (D + 5)\,y = e^t \tag{2}$$

operating (D +5) on equation (1) and multiplied equation (2) by 3, we get

$$(D + 5)(D + 4)\,x + 3\,(D + 5)\,y = (D + 5)\,t \tag{3}$$

$$6x + 3\,(D + 5)\,y = 3e^t \tag{4}$$

Subtracting (4) from (3) we get

$(D^2 + 9D + 20 - 6)\, x = 1 + 5t - 3e^t$

$\Rightarrow \quad (D^2 + 9D + 14)\, x = 5t - 3e^t + 1$

Here auxiliary equation is $m^2 + 9m + 14 = 0$

$\Rightarrow \quad (m + 7)(m + 2) = 0$

$\Rightarrow \quad m = -7, -2$

$\therefore \quad C.F = C_1 e^{-7t} + C_2 e^{-2t}$

and $\quad P.I = \dfrac{1}{(D^2 + 9D + 14)} (5t - 3e^t + 1)$

$$= \frac{5}{14}\left(1 + \frac{D^2 + 9D}{14}\right)^{-1} t - \frac{3e^t}{(1)^2 + 9(1) + 14} + \frac{1}{(0)^2 + 9(0) + 14} e^{ot}$$

$$= \frac{5}{14}\left(1 - \frac{9D}{14}\right) t - \frac{3e^t}{24} + \frac{1}{14}$$

$$= \frac{5}{14}\left(t - \frac{9}{14}\right) - \frac{e^t}{8} + \frac{1}{14}$$

$\Rightarrow \quad P.I = \dfrac{5t}{14} - \dfrac{e^t}{8} - \dfrac{31}{196}$

$\therefore \quad x = C_1 e^{-7t} + C_2 e^{-2t} + \dfrac{5t}{14} - \dfrac{e^t}{8} - \dfrac{31}{196}$

Now $(D + 4)\, x = -7C_1 e^{-7t} - 2C_2 e^{-2t} + \dfrac{5}{14} - \dfrac{e^t}{8} + 4C_1 e^{-7t} + 4C_2 e^{-2t} + \dfrac{10}{7} t - \dfrac{e^t}{2} - \dfrac{31}{49}$

$\Rightarrow \quad (D + 4)\, x = -3C_1 e^{-7t} + 2\, C_2 e^{-2t} - \dfrac{5e^t}{8} + \dfrac{10}{7} t - \dfrac{27}{98}$

Using this value in equation (1) we get

$3y = t + 3C_1 e^{-7t} - 2C_2 e^{-2t} + \dfrac{5}{8} e^t - \dfrac{10}{7} t + \dfrac{27}{98}$

$\Rightarrow \quad y = -\dfrac{1}{7} t + C_1 e^{-7t} - \dfrac{2}{3} C_2 e^{-2t} + \dfrac{5}{24} e^t + \dfrac{9}{98}$

Thus the required solution is

$$x = C_1 e^{-7t} + C_2 e^{-2t} + \frac{5t}{14} - \frac{e^t}{8} - \frac{31}{196}$$

and $y = -\frac{1}{7}t + C_1 e^{-7t} - \frac{2}{3} C_2 e^{-2t} + \frac{5}{24} e^t + \frac{9}{98}$

Example 29. The equation of motion of a particle are given by $\frac{dx}{dt} + wy = 0$, $\frac{dy}{dt} - wx = 0$. Find the path of the particle and show that it is a circle.

(U.P.T.U. 2009)

Solution. Writing D for $\frac{d}{dt}$, the equations are

$$Dx + wy = 0 \quad (1)$$

and $$-wx + Dy = 0 \quad (2)$$

Differentiating (1) w.r.t. 't' we have

$D^2x + wDy = 0 \Rightarrow \quad D^2w + w\,(wx) = 0 \quad \Rightarrow \quad (D^2 + w^2)\,x = 0$ using (2)

$\Rightarrow \quad x = C_1 \cos wt + C_2 \sin wt$

Putting this value of x is (1) we have $y = -\frac{1}{w}\frac{d}{dt}(C_1 \cos wt + C_2 \sin wt)$

we get $y\,(t) = C_1 \cos wt + C_2 \sin wt$ (3)

and $x\,(t) = C_2 \cos wt - C_1 \sin wt$ (4)

Squaring (3) and (4) their adding, we get

$$x^2 + y^2 = C_1^2 + C_2^2$$

or $\quad x^2 + y^2 = R^2$

which is a circle

Applications to Engineering Problems

INTRODUCTION

Differential equations have many numerical applications in Physics, Chemistry, electrical engineering, mechanical engineering, biological sciences, social sciences etc. In this section, we discuss some applications.

Simple Harmonic Motion

A particle moving in a straight line, is said to execute simple harmonic motion, if its acceleration is always directed towards a fixed point in line and is proportional to the distance of the particle from the fixed point.

Since the acceleration is always directed towards a fixed point, the differential equation of the motion of the particle is given by

$$\frac{d^2x}{dt^2} = -w^2x \qquad (1)$$

$w^2x \leftarrow$

A' ——— O ← x → P ——— A

← - - - - - - - - - - a - - - - - - - - - - →

where x is the displacement of the particle from a fixed point o at any time t.

The solution of (1) is

$$x = C_1 \cos wt + C_2 \sin wt \qquad (2)$$

If the particle starts from rest at a point A, where

OA = a i.e. (x = a, when t = 0) then, from (2), we get

$$C_1 = a$$

Differentiating (2) with respect to t, we get

$$v = \frac{dx}{dt} = w(-C_1 \sin wt + C_2 \cos wt) \qquad (3)$$

Since $\frac{dx}{dt} = 0$, at t = 0, from (3), we get

$$O = C_2$$

Hence, the displacement of the particle is

$$x = a \cos wt \text{ (a is amplitude)} \qquad (4)$$

such that

$$\text{Velocity} = v = \frac{dx}{dt} = -aw \sin wt$$

$$= -w\sqrt{a^2 - x^2} \qquad (5)$$

Equation (5) gives the velocity of the particle at any time t, when its displacement from a fixed point O is x. Particle time (time for one complete oscillation) is denoted by T and is given by $T = \frac{2\pi}{w}$. The number of complete oscillations per second is called the frequency of motion and we have $n = \frac{1}{T} = \frac{w}{2\pi}$

In the figure O is the fixed point

we have OA = a

The acceleration is directed towards O. The particle moves towards O from A. The acceleration gradually decreases and vanishes at O. At O particle acquired maximum acceleration. Under retardation the particle further moves towards A' and comes to rest at A' such that

$$OA' = OA$$

The point O is called mean position.

Example 30. A point moves in a straight line towards a centre of force $\mu/(\text{distance})^3$, starting from rest at a distance a from the centre of force. Show that the time of reaching a point distance b from the centre of force is $\frac{a}{\sqrt{\mu}}\sqrt{a^2 - b^2}$ and that its velocity is $\frac{\sqrt{\mu}}{ab}\sqrt{a^2 - b^2}$

(U.P.T.U. 2001)

Solution. Let O is the centre of force and let a point moves from P towards the centre of force O.

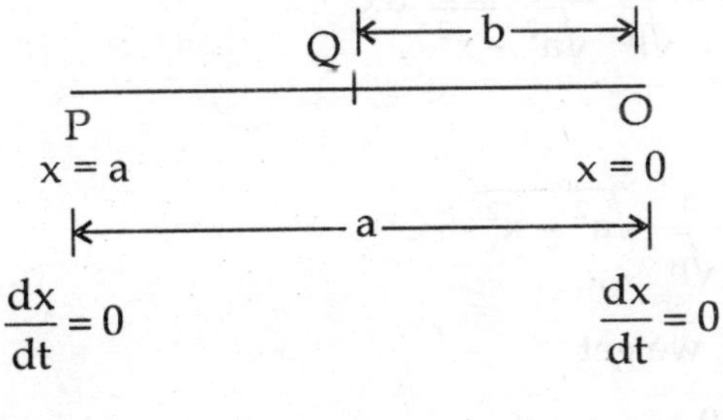

Q ←b→ O

P, x = a; O, x = 0

←a→

$\frac{dx}{dt} = 0$ $\qquad$ $\frac{dx}{dt} = 0$

The equation of motion is

$$\frac{d^2x}{dt^2} = \frac{-\mu}{x^3} \tag{1}$$

$$\Rightarrow \qquad 2\frac{dx}{dt}\frac{d^2x}{dt^2} = -2\frac{\mu}{x^3}\frac{dx}{dt}$$

on integrating, we get

$$\left(\frac{dx}{dt}\right)^2 = -2\mu \frac{1}{-2x^2} + C = \frac{\mu}{x^2} + C$$

$$\Rightarrow \qquad \frac{dx}{dt} = \pm\sqrt{\frac{\mu}{x^2} + C} \qquad (2)$$

At P, $\frac{dx}{dt} = 0$ and $x = a$

$$\Rightarrow \qquad o = \sqrt{\frac{\mu}{a^2} + C} \quad \Rightarrow \quad C = \frac{-\mu}{a^2}$$

From (2), $\frac{dx}{dt} = \pm\sqrt{\frac{\mu}{x^2} - \frac{\mu}{a^2}} = \pm\sqrt{\mu}\,\frac{\sqrt{a^2 - x^2}}{ax}$ (3)

The velocity at x = b is

$$v = \pm\sqrt{\mu}\,\frac{\sqrt{a^2 - b^2}}{ab}$$

or $\qquad v = -\sqrt{\mu}\,\frac{\sqrt{a^2 - b^2}}{ab}$ (As the point P is moving towards O)

From (3) $\frac{dx}{dt} = -\sqrt{\mu}\,\frac{\sqrt{a^2 - x^2}}{ax}$

$$\Rightarrow \qquad dt = -\frac{1}{\sqrt{\mu}}\,\frac{xa}{\sqrt{a^2 - x^2}}\,dx$$

on integration, we get

$$t = \frac{a}{\sqrt{\mu}}\sqrt{a^2 - x^2} + C \qquad (4)$$

At P, t = 0, x = a, in (4), we get

$$C = 0$$

Putting this value of C in (4), we have

$$t = \frac{a}{\sqrt{\mu}}\sqrt{a^2 - x^2}$$

At x = b

$$t = \frac{a}{\sqrt{\mu}} \sqrt{(a^2 - b^2)}$$

Vertical Motion In Resisting Medium

Example 31. A particle falls under gravity in a resisting medium whose resistance varies with velocity. Find the relation between distance and velocity if initially the particle starts from rest.

(Bihar P.C.S. 1997, U.P.T.U. 2003)

Solution. Let m be the mass of a particle falls from rest from a fixed point O. Let P be the position of a particle such that OP = x. The forces acting on the particle at P are:

(1) The weight mg of a particle acting vertically downwards.

(2) The resistance m kv acting vertically upwards.

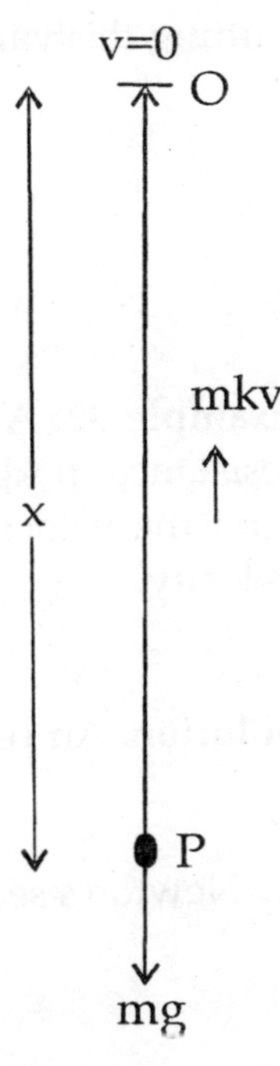

Now by Newton's second law of motion the equation of the motion of the body

$$\frac{md^2x}{dt^2} = mg - mkv$$

or $$\frac{d^2x}{dt^2} = g - kv$$

or $$v\frac{dv}{dx} = g - kv$$

$$\because \frac{d^2x}{dt^2} = \frac{dv}{dt} = \frac{dv}{dx}\frac{dx}{dt} = v\frac{dv}{dx}$$

or $$\frac{vdv}{g - kv} = dx$$

$\Rightarrow$ $$\frac{1}{k}\left(\frac{-g + kv + g}{g - kv}\right) dv = dx$$

$\Rightarrow$ $$-\frac{dv}{k} + \frac{g}{k\,(g - kv)} dv = dx$$

Integrating, we get

$$-\frac{v}{k} + \frac{g}{k}\left(-\frac{1}{k}\right) \log (g - kv) = x + C$$

or
$$-\frac{v}{k} - \frac{g}{k^2} \log (g - kv) = x + C \quad (1)$$

Initially, at point O, x = 0, v = 0

$$\Rightarrow \quad -\frac{g}{k^2} \log g = C$$

putting this value of C in (1) we have

$$-\frac{v}{k} - \frac{g}{k^2} \log (g - kv) = x - \frac{g}{k^2} \log g$$

$$\Rightarrow \quad -\frac{v}{k} - \frac{g}{k^2} \log \frac{g - kv}{g} = x$$

Example 32. A 4 kg object falls from rest of time t = 0 in a medium offering a resistance in kg numerically equal to twice its instantaneous velocity in m/sec. Find the velocity and distance travelled at any time t > 0 and also the limiting velocity.

(U.P.T.U. 2007)

Solution. Air resistance = 2v

∴ Upthrust = 2 × 4v = 8v As m = 4kg

By Newton's second law of motion the equation of motion of body

$$4 \frac{d^2x}{dt^2} = 4g - 8v$$

$$\Rightarrow \quad \frac{d^2x}{dt^2} = g - 2\frac{dx}{dt}$$

$$\Rightarrow \quad \frac{d^2x}{dt^2} + 2\frac{dx}{dt} = g \quad (1)$$

Let $\frac{dx}{dt} = p \quad \Rightarrow \quad \frac{d^2x}{dt^2} = \frac{dp}{dt}$

∴ From (1), we get

$$\frac{dp}{dt} + 2p = g \qquad (2)$$

which is linear in p.

its I.F $= e^{\int 2\,dt} = e^{2t}$

So, the solution of equation (2) is

$$p.e.^{2t} = \int g\, e^{2t}\, dt + C = \frac{g}{2}\, e^{2t} + C$$

$$\Rightarrow \qquad p = \frac{g}{2} + Ce^{-2t}$$

$$\Rightarrow \qquad \frac{dx}{dt} = \frac{g}{2} + c\, e^{-2t} \qquad (3)$$

At $t = 0, \frac{dx}{dt} = 0$ which gives $C = -\frac{g}{2}$

From (3) $\frac{dx}{dt} = \frac{g}{2}\,(1 - e^{-2t})$

$$\Rightarrow \qquad \text{velocity} = \frac{g}{2}\,(1 - e^{-2t})$$

Again integrating above equation, we get

$$x = \frac{gt}{2} + \frac{g}{4}\, e^{-2t} + C_1 \qquad (4)$$

At $t = 0, x = 0 \quad \Rightarrow \quad o = \frac{g}{4} + C_1 \quad \Rightarrow \quad C_1 = -g/4$

From (4), $x = \frac{gt}{2} + \frac{g}{4}\,(e^{-2t} - 1)$

$$\Rightarrow \qquad \text{distance} = \frac{gt}{2} + \frac{g}{4}\,(e^{-2t} - 1)$$

and Limiting velocity $= \left(\lim_{t \to \infty} \frac{dx}{dt}\right) = \lim_{t \to \infty} \frac{g}{2}\,(1 - e^{-2t})$

$$= \frac{g}{2}$$

Example 33. A mass M suspended from the end of a helical spring is subjected to a periodic force f = F sin wt in the direction of its length. The force f is measured positive vertically downwards and at zero time M is at rest. If the spring stiffness is S, prove that the displacement of M at time t from the commencement of motion is given by $x = \frac{F}{M(p^2 - w^2)}\left(\sin wt - \frac{w}{p}\sin pt\right)$, where $p^2 = \frac{S}{M}$ and daming effects are neglected.

(U.P.T.U. 2000)

Solution. Let x be the displacement from the equilibrium position at any time t then the equation of the motion is

$$M\frac{d^2x}{dt^2} = -Sx + F\sin wt$$

or $$\frac{d^2x}{dt^2} + \frac{S}{M}x = \frac{F}{M}\sin wt$$

or $$\frac{d^2x}{dt^2} + p^2x = \frac{F}{M}\sin wt \qquad (1)$$

$$\left(\text{As } \frac{S}{M} = p^2\right)$$

The A.E. is $m^2 + p^2 = 0 \Rightarrow \quad m = \pm ip$

C.F. = $C_1 \cos pt + C_2 \sin pt$

and P.I. = $\frac{1}{D^2 + p^2}\left(\frac{F}{M}\sin wt\right)$

$$= \frac{F}{M}\frac{1}{-w^2 + p^2}\sin wt$$

$$\therefore x = C_1\cos pt + C_2 \sin pt + \frac{F}{M}\frac{1}{(p^2 - w^2)}\sin wt \qquad (2)$$

Initially, at t = 0, x = 0 $\therefore C_1 = 0$

Differentiating equation (2) w.r.t. 't' we get

$$\frac{dx}{dt} = -p\,C_1 \sin pt + p\,C_2\cos pt + \frac{F}{M}\frac{w}{p^2 - w^2}\cos wt$$

At $t = 0, \frac{dx}{dt} = 0$

$$\therefore \qquad p\,C_2 + \frac{F}{M}\,\frac{w}{p^2 - w^2} = 0$$

or
$$C_2 = -\frac{w}{p}\,\frac{F}{M\,(p^2 - w^2)}$$

From (2), we have

$$x = -\frac{w}{p}\,\frac{F}{M\,(p^2 - w^2)}\sin pt + \frac{F}{M}\,\frac{1}{p^2 - w^2}\sin wt$$

or
$$x = \frac{F}{M\,(p^2 - w^2)}\left(\sin wt - \frac{w}{p}\sin pt\right)$$

Problems Related to Electric Circuit

There are some formulae which are useful to solve such type of problems

(1) $i = \frac{dq}{dt}$

(2) Voltage drop across resistance R is $V_R = Ri$

(3) Voltage drop across inductance L is $V_L = L\,\frac{di}{dt}$

(4) Voltage drop across capacitance C is $V_C = \frac{q}{C}$

Electro-Mechanical Analogy

The following correspondences between the electrical and mechanical quantities should be kept in mind

Mechanical system	Series Circuit	Parallel circuit
Displacement	Current i	Voltage E
Force or Couple	Voltage E	Current i
Mass m or M.I.	Inductance L	Capacitance C
Damping force	Resistance R	Conductance 1/R
Spring modulus	Elastance 1/C	Susceptance 1/L

Example 34. An uncharged condenser of capacity C is charged by applying an e.m.f. $E \sin \frac{t}{\sqrt{LC}}$ through leads of self-inductance L and negligible resistance. Prove that at time t, the charge on one of the plates is

$$\frac{EC}{2}\left[\sin \frac{t}{\sqrt{LC}} - \frac{t}{\sqrt{LC}} \cos \frac{t}{\sqrt{LC}}\right]$$

(U.P.T.U. 2003)

Solution. If q be the charge on the condenser, the differential equation of the circuit is

$$L \frac{d^2q}{dt^2} + \frac{q}{C} = Eo$$

$$L \frac{d^2q}{dt^2} + \frac{q}{C} = E \sin \frac{t}{\sqrt{LC}} \quad \text{As } E_o = E \sin \frac{t}{\sqrt{LC}}$$

or
$$\frac{d^2q}{dt^2} + \frac{1}{LC} q = \frac{E}{L} \sin \frac{t}{\sqrt{LC}} \quad (1)$$

Here auxiliary equation is $m^2 + \frac{1}{LC} = 0$

$\Rightarrow$
$$m = \pm \frac{i}{\sqrt{LC}}$$

$$\therefore \text{C.F} = C_1 \cos \frac{1}{\sqrt{LC}} t + C_2 \sin \frac{1}{\sqrt{LC}} t$$

$$P.I = \frac{1}{\left(D^2 + \frac{1}{LC}\right)} \frac{E}{L} \sin \frac{t}{\sqrt{LC}} \text{ (Case of failure)}$$

$$= \frac{E}{L}\left(-\frac{t}{2\frac{1}{\sqrt{LC}}} \cos \frac{t}{\sqrt{LC}}\right) \qquad \because \frac{1}{D^2 + a^2} \sin ax = \frac{-x}{2a} \cos ax$$

$$= \frac{E}{L}\left(\frac{-t\sqrt{LC}}{2} \cos \frac{t}{\sqrt{LC}}\right)$$

$$= -\frac{Et}{2}\sqrt{\left(\frac{C}{L}\right)} \cos \frac{t}{\sqrt{LC}}$$

Therefore, the solution of the equation is

$$q = C_1 \cos \frac{t}{\sqrt{LC}} + C_2 \sin \frac{t}{\sqrt{LC}} - \frac{Et}{2}\sqrt{\left(\frac{C}{L}\right)} \cos \frac{t}{\sqrt{LC}} \tag{2}$$

At $t = 0, q = 0 \therefore C_1 = 0$

Differentiating equation (2) w.r.t. "t" we get

$$\frac{dq}{dt} = -\frac{C_1}{\sqrt{LC}} \sin \frac{t}{\sqrt{LC}} + \frac{C_2}{\sqrt{LC}} \cos \frac{t}{\sqrt{LC}} + \frac{Et}{2}\sqrt{\left(\frac{C}{L}\right)} \sin\left(\frac{t}{\sqrt{LC}}\right)\frac{1}{\sqrt{LC}} - \frac{E}{2}\sqrt{\frac{C}{L}} \cos \frac{t}{\sqrt{LC}}$$

Initially $\frac{dq}{dt} = 0$, when $t = 0$

$$\therefore \quad \frac{C_2}{\sqrt{LC}} - \frac{E}{2}\sqrt{\frac{C}{L}} = 0 \quad \Rightarrow \quad C_2 = \frac{EC}{2}$$

From equation (2)we get

$$q = \frac{EC}{2} \sin \frac{t}{\sqrt{LC}} - \frac{Et}{2}\sqrt{\left(\frac{C}{L}\right)} \cos \frac{t}{\sqrt{LC}}$$

or

$$q = \frac{EC}{2}\left(\sin \frac{t}{\sqrt{LC}} - \frac{t}{\sqrt{LC}} \cos \frac{t}{\sqrt{LC}}\right)$$

Example 35. The equation of electromotive force in terms of current i for an electrical circuit having resistant R and a condenser of capacity C, in serires is $E = Ri + \int \frac{i}{C} dt$. Find the current i at any time t, when E = Eo sin wt

(U.P.T.U. 2006)

Solution. The given equation is

$$Ri + \int \frac{i}{C} dt = Eo \sin wt \quad \text{as } E = Eo \sin wt$$

Differentiating w.r.t. 't', we get

$$R \frac{di}{dt} + \frac{i}{C} = Eo\, w \cos wt$$

$$\Rightarrow \qquad \frac{di}{dt} + \frac{i}{RC} = \frac{Eo\, w}{R} \cos wt \qquad (1)$$

which is a linear differential equation

its
$$I.F = e^{\int \frac{1}{RC} dt}$$

$$= e^{\frac{t}{RC}}$$

The solution of (1) is

$$i.e^{\frac{t}{RC}} = \int \frac{E_o w}{R} \cos wt.\, e^{t/RC}\, dt + C_1$$

$$= \frac{E_o w}{R} \frac{e^{t/RC}}{\frac{1}{R^2C^2} + w^2} \left[\frac{1}{RC} \cos wt + w \sin wt\right] + C_1$$

$$\because \qquad \int e^{ax} \cos bx\, dx = \frac{e^{ax}}{a^2 + b^2} (a \cos bx + b \sin bx)$$

or
$$i = \frac{wE_o\, RC^2}{1 + w^2R^2C^2} \left[\frac{1}{RC} \cos wt + w \sin wt\right] + ke^{-t/RC}$$

Example 36. The damped LCR circuit is governed by the equation

$$L \frac{d^2Q}{dt^2} + R \frac{dQ}{dt} + \frac{1}{C} Q = 0$$

where L, C, R are positive constants. Find the conditions under which the cirucuit is overdamped, underdamped and critically damped. Find also the critical resistance.

(U.P.T.U. 2005)

Solution. The given equation is

$$L\frac{d^2Q}{dt^2} + R\frac{dQ}{dt} + \frac{1}{C}Q = 0$$

or

$$\frac{d^2Q}{dt^2} + 2k\frac{dQ}{dt} + w^2Q = 0 \qquad (1)$$

Where $2k = \frac{R}{L}$ and $w^2 = \frac{1}{LC}$

Here auxiliary equation is

$$m^2 + 2km + w^2 = 0$$

$$\Rightarrow \qquad m = -k \pm \sqrt{k^2 - w^2} \qquad (2)$$

Case I. when k<w i.e. $\frac{R}{2L} < \frac{1}{\sqrt{LC}}$, the roots of A.E. given by (2) are imaginary.

The general solution of (1) is

$$Q = e^{-kt}\left(C_1 \cos\sqrt{(w^2 - k^2)}\, t + C_2 \sin\sqrt{(w^2 - k^2)}\, t\right)$$

where C_1 and C_2 being arbitrary constants.

Time period $= \frac{2\pi}{\sqrt{w^2 - k^2}}$ which is greater than $\frac{2\pi}{w}$

Thus the effect of damping increases the period of oscillation and motion ultimately dies away. In this condition when $\frac{R}{2L} < \frac{1}{\sqrt{LC}}$ the circuit is under damped.

Case II. When k = w, then roots of A.E. (2) are equal, each being equal to –k. The general solution of (1) is

$$Q = (C_1 + C_2 t)\, e^{-kt}$$

In this case charge Q is always positive and decreases to zero as $t \to \infty$. In this case circuit is called critically damped and the resistance R is called critical resistance.

Thus $\quad k = w \Rightarrow \dfrac{R}{2L} = \dfrac{1}{\sqrt{LC}}$

$$\Rightarrow \quad R = 2\sqrt{\frac{L}{C}}$$

which is required critical resistance.

Case III. when k > w, the roots of A.E. are real and unequal.

Also, the roots

$m = -k + \sqrt{k^2 - w^2}$ and $m = -k - \sqrt{k^2 - w^2}$

are both negative. The general solution of (1) is

$$Q = C_1 e^{\left\{-k + \sqrt{k^2 - w^2}\right\}t} + C_2 e^{\left\{-k - \sqrt{k^2 - w^2}\right\}t}$$

In this case also change Q is positive and decreases to zero as $t \to \infty$, since exponential terms having negative powers approach to zero. In this case the circuit is called overdamped.

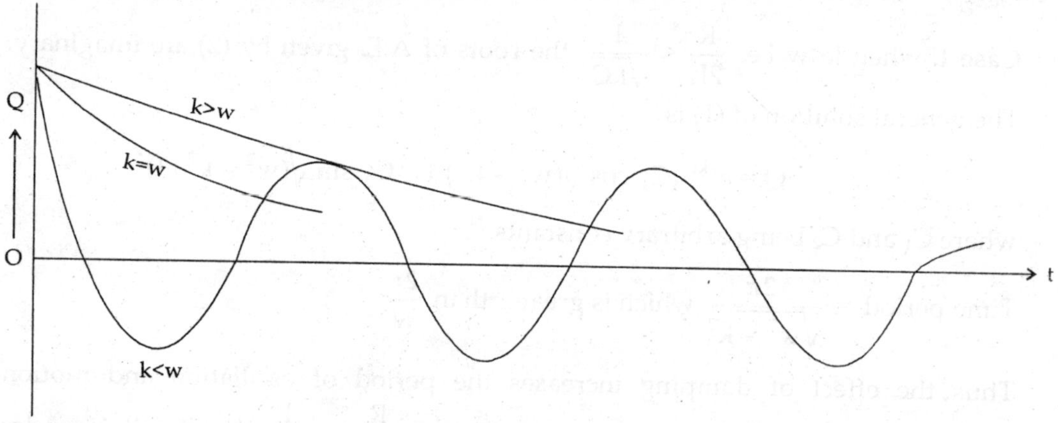

Example 37. The voltage V and the current i at a distance x from the sending end of the transmission line satisfying the equations.

$$-\frac{dv}{dx} = Ri, -\frac{di}{dx} = GV$$

where R and G are constants. If V = Vo at the sending end (x = 0) and V = 0 at receiving end (x = l), show that

$$V = Vo\left\{\frac{\sinh n\,(l - x)}{\sinh nl}\right\}$$

(U.P.T.U. 2006)

Solution. We have $-\frac{dV}{dx} = Ri$ (1)

and $-\frac{di}{dx} = GV$ (2)

when $x = 0$, $V = V_o$, when $x = l$, $V = 0$

From (1) and (2), we have

$$-\frac{d}{dx}\left(-\frac{dV}{dx}\frac{1}{R}\right) = GV$$

$\Rightarrow$ $$\frac{d^2V}{dx^2} = RGV$$

$\Rightarrow$ $$\frac{d^2V}{dx^2} - (RG)V = 0$$

or $$(D^2 - RG)V = 0, \; D \equiv \frac{d}{dx} \quad (3)$$

Here auxiliary equation is $m^2 - RG = 0$

$\Rightarrow$ $$m = \pm n, \; n^2 = RG$$

The solution of (3) is $V = C_1 e^{nx} + C_2 e^{-nx}$ (4)

where C_1 and C_2 are arbitrary constants.

putting $x = 0$ and $V = V_o$ is (4), we get

$$V_o = C_1 + C_2 \quad (5)$$

Again putting $x = l$ and $V = 0$ is (4), we get

$$o = C_1 e^{nl} + C_2 e^{-nl} \quad (5)$$

Solving equations (5) and (6), we have

$$C_1 = \frac{V_o}{1 - e^{2nl}}, C_2 = \frac{-V_o e^{2nl}}{1 - e^{2nl}}$$

Substituting the values of C_1 and C_2 in (4), we get

$$V = \frac{V_o}{1 - e^{2nl}} e^{nx} - \frac{V_o e^{2nl}}{1 - e^{2nl}} e^{-nx}$$

$$= \frac{V_o \left(e^{nx} - e^{2nl - nx}\right)}{1 - e^{2nl}}$$

or $$V = \frac{V_o \left\{e^{(nl - nx)} - e^{-(nl - nx)}\right\}}{e^{nl} - e^{-nl}} = V_o \left\{\frac{\sin hn(l - x)}{\sin h\, nl}\right\}$$

Example 38. An inductance of 2 henries and a resistance of 20 ohms are connected in series with an emf E votts. If the current is zero when t = 0, find the current at the end of 0.01 sec if E = 100 volts, using the following differential equation.

(U.P.T.U. 2008)

$$L\frac{di}{dt} + iR = E$$

Solution. we have $L\frac{di}{dt} + iR = E$

or $$\frac{di}{dt} + \frac{R}{L}i = \frac{E}{L} \qquad (1)$$

Equation (1) is linear differential equation of first order.

$$I.F = e^{\int \frac{R}{L} dt} = e^{(R/L)t}$$

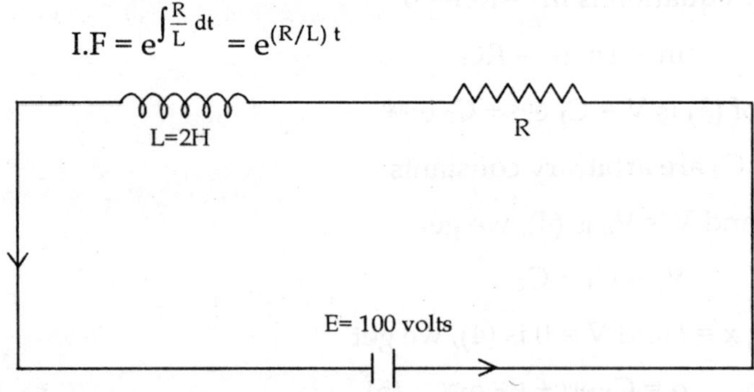

∴ Solution of (1) is

$$i.e^{(R/L)t} = \int \frac{E}{L} e^{(R/L)t}\, dt + C$$

where C is an arbitrary constant

or $$i\, e^{(R/L)t} = \frac{E}{R} e^{(R/L)t} + C$$

$$\Rightarrow \qquad i = \frac{E}{R} + C\,e^{-(R/L)\,t} \qquad (2)$$

Initially i = 0, when t = 0 ∴ From (2), we have

$$C = -\frac{E}{R}$$

∴ From (2), we have $i = \frac{E}{R}\left[1 - e^{-(R/L)\,t}\right]$ (3)

on putting E = 100 volts, R = 20 ohms and L = 2 henries in (3) we have

$$i = \frac{100}{5}\left[1 - e^{-\frac{20}{2}t}\right] = 5\,(1 - e^{-10\,t})$$

At t = 0.01 sec, i = 5 (1 – $e^{-0.1}$)

= 0.475 amp (approximately)

Example 39. In an LCR circuit, the charge q on a plate of a condenser is given by $L\frac{d^2q}{dt^2} + R\frac{dq}{dt} + \frac{q}{C} = E\sin pt$. The circuit is tuned to resonance so that $p^2 = \frac{1}{LC}$, if initially the current i and the charge q be zero, show that for small values of $\frac{R}{L}$, the current in the circuit at time t is given by $\frac{Et}{2L}\sin pt$

(U.P.T.U. 2004) (C.O.)

Solution. The given differential equation is

$$L\frac{d^2q}{dt^2} + R\frac{dq}{dt} + \frac{q}{C} = E\sin pt \qquad (1)$$

Here A.E. is $Lm^2 + Rm + \frac{1}{C} = 0$

$$\Rightarrow \qquad m = \frac{-R + \sqrt{R^2 - (4L/C)}}{2L} = -\frac{R}{2L} \pm \frac{1}{2}\sqrt{\frac{R^2}{L^2} - \frac{4}{CL}}$$

$$\Rightarrow \qquad m = -\frac{R}{2L} \pm \frac{1}{2}\sqrt{-\frac{4}{CL}} \quad \left(\text{neglected } \frac{R^2}{L^2} \text{ as } \frac{R}{LC} \text{ is small}\right)$$

$$\Rightarrow \qquad m = -\frac{R}{2L} \pm \frac{i}{\sqrt{CL}} = -\frac{R}{2L} \pm ip \quad \left(\text{since } p^2 = \frac{1}{LC} \text{ given}\right)$$

$$\therefore \qquad C.F = e^{-Rt/2L}(C_1 \cos pt + C_2 \sin pt)$$

But $e^{-Rt/2L} = 1 - \frac{Rt}{2L} + \frac{1}{L^2}\frac{R^2t^2}{4L^2} - \ldots\ldots\ldots\ldots$

$$= 1 - \frac{Rt}{2L} \quad \text{neglecting } \frac{R^2}{L^2} \text{ etc}$$

$$\therefore \qquad C.F. = \left(1 - \frac{Rt}{2L}\right)(C_1 \cos pt + C_2 \sin pt)$$

where C_1 and C_2 are arbitrary constants

$$P.I = \frac{1}{LD^2 + RD + \frac{1}{C}} E \sin pt \text{ where } D \equiv \frac{d}{dt}$$

$$= E \frac{1}{L(-p^2) + RD + \frac{1}{C}} \sin pt$$

$$= E \frac{1}{RD} \sin pt \qquad \text{since } p^2 = \frac{1}{LC}$$

$$= \frac{E}{R} \int \sin pt \, dt = \frac{-E}{pR} \cos pt$$

Hence the general solution of (1) is given by

$$q = \left(1 - \frac{Rt}{2L}\right)(C_1 \cos pt + C_2 \sin pt) - \frac{E}{pR} \cos pt \qquad (2)$$

Differentiating (2) w.r.t. 't' we have

$$i = \frac{dq}{dt} = \left(1 - \frac{Rt}{2L}\right)(-pC_1 \sin pt + pC_2 \cos pt) - \frac{R}{2L}(C_1 \cos pt + C_2 \sin pt) + \frac{E}{R} \sin pt \qquad (3)$$

Initially given that $t = 0$, $q = 0$

$\therefore$ (2) gives

$$o = C_1 - \frac{E}{PR} \quad \Rightarrow \quad C_1 = \frac{E}{PR}$$

and (3) gives

$$o = pC_2 - \frac{RC_1}{2L} \Rightarrow \qquad C_2 = \frac{RC_1}{2pL} = \frac{E}{2Lp^2}$$

Now putting values of C_1 and C_2 in (3), the current i in the circuit at any time t is given by

$$i = \left(1 - \frac{Rt}{2L}\right)\left(-\frac{E}{R}\sin pt + \frac{E}{2pL}\cos pt\right) - \frac{R}{2L}\left(\frac{E}{pR}\cos pt + \frac{E}{2Lp^2}\sin pt\right) + \frac{E}{R}\sin pt$$

$$= \frac{Et}{2L}\sin pt - \frac{ERt}{4pL^2}\cos pt - \frac{ER}{4L^2p^2}\sin pt$$

$= \frac{Et}{2L}\sin pt$ since $\frac{R}{L}$ is small, also $\frac{R}{L^2} = \frac{1}{R}\left(\frac{R}{L}\right)^2$, so neglecting second and third terms

BEAM

A bar whose length is much greater than its cross-section and its thickness is called a beam

Cantilever: If one end of a beam is fixed and the other end is loaded, it is callled a cantilever.

Bending of Beam: Let a beam be fixed at one end and the other end is loaded. Then the upper surface is elongated and therefore under tension and the lower surface is shortended so under compression.

Bending Moment: Whenever a beam is loaded it deflects from its original position. If M is the bending moment of the forces acting on it, then

$$M = \frac{EI}{R} \tag{1}$$

where E = Modulus of elasticity of the beam

I = Moment of inertia of the cross-section of beam about neutral axis

R = Radius of curvature of the curved beam

$$R = \frac{\left[1 + \left(\frac{dy}{dx}\right)^2\right]^{3/2}}{d^2y/dx^2} = \frac{1}{d^2y/dx^2} \text{ neglecting } \frac{dy}{dx}$$

Thus equation (1) becomes $M = EI\ \dfrac{d^2y}{dx^2}$

Example 40. A beam of length l is clamped horizontally at its end x = 0and is free at the end x = l. A point load W is applied at the end x = l, in addition of a uniform load w per unit length from x = 0 to $x = \frac{l}{2}$. Find the deflection at any point.

(U.P.T.U. 2002)

Solution. Let OA be a beam, clamped (i.e. fixed) at one end **O** and free at end A. Let B be its mid point. The weight $\frac{wl}{2}$ of the beam OB acts at C (the mid point of OB). The weight w acts at A.

Let R be the force acting at O. The directions of all the force acting on the beam are as shown in figure.

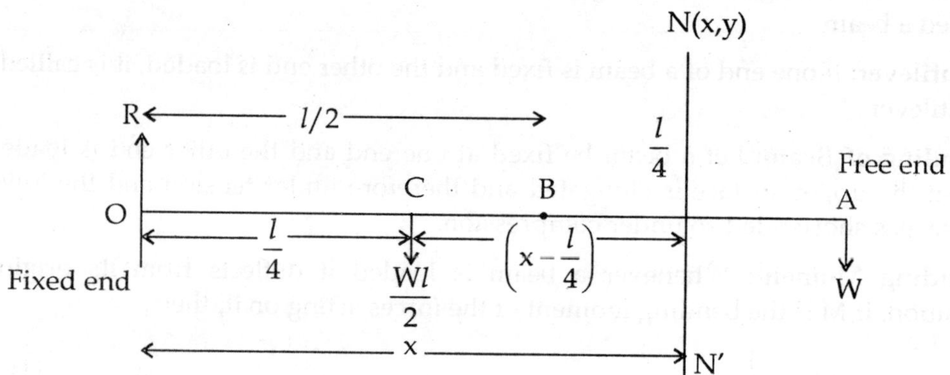

∴ From balance equation, we have

$$R = W + \frac{wl}{2} \qquad (1)$$

Now we choose a random axis NN', if (x, y) are the co-ordinates of N, then taking moments about N, we get

$$EI\ \frac{d^2y}{dx^2} = -Rx + \frac{Wl}{2}\left(x - \frac{l}{4}\right)$$

or

$$EI\ \frac{d^2y}{dx^2} = -\left(W + \frac{wl}{2}\right)x + \frac{wl}{2}x - \frac{wl^2}{8} \text{ using (1) for R}$$

or $$EI\frac{d^2y}{dx^2} = -\frac{Wl^2}{8} - Wx$$

Integrating both sides w.r.t. x, we get

$$E.I.\frac{dy}{dx} = -\frac{wl^2}{8}x - \frac{1}{2}Wx^2 + C_1 \qquad (2)$$

Applying boundary conditions at the fixed (i.e. clamped) end O i.e., at $x = 0$, $dy/dx = 0$, we get from (2), $C_1 = 0$

$\therefore$ (2) becomes, $EI\dfrac{dy}{dx} = -\dfrac{wl^2}{8}x - \dfrac{1}{2}Wx^2$

Again integrating $EIy = -\dfrac{wl^2x^2}{16} - \dfrac{1}{6}Wx^3 + C_2$ (3)

Again boundary conditions, at $x = 0$, $y = 0$ gives $C_2 = 0$

$\therefore$ (3) becomes,

$$EIy = \frac{-wl^2x^2}{16} - \frac{Wx^3}{6}$$

or $$y = -\frac{1}{EI}\left(\frac{wl^2x^2}{16} + \frac{Wx^3}{6}\right)$$

which gives the deflection at any point.

Example 41. The deflection of a strut of length l with one end ($x = 0$) built in and the other end supported and subjected to end thrust P, satisfies the equation

$$\frac{d^2y}{dx^2} + a^2y = \frac{a^2R}{P}(l - x)$$

Prove that the deflection curve $y = \dfrac{R}{P}\left(\dfrac{\sin ax}{a} - l\cos ax + l - x\right)$

where $a\,l = \tan al$

(U.P.T.U. 2001)

Solution. we have

$$\frac{d^2y}{dx^2} + a^2y = \frac{a^2R}{P}(l - x) \qquad (1)$$

its auxiliary equation is $m^2 + a^2 = 0 \quad \Rightarrow \quad m = \pm ai$

$\therefore$ C.F. = $C_1 \cos ax + C_2 \sin ax$

where C_1 and C_2 are arbitrary constants.

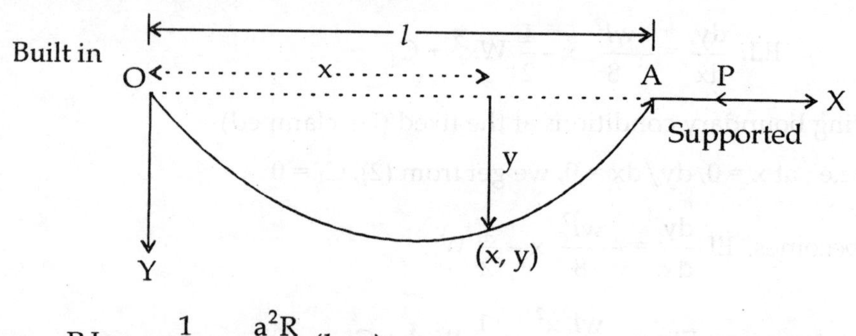

$$P.I = \frac{1}{D^2 + a^2} \frac{a^2R}{P} (l - x)$$

$$= \frac{R}{P}\left(1 + \frac{D^2}{a^2}\right)(l - x)$$

$$= \frac{R}{P}\left(1 - \frac{D^2}{a^2}\right)(l - x)$$

$$= \frac{R}{P}(l - x)$$

$\therefore$ The general solution is

$$y = C_1 \cos ax + C_2 \sin ax + \frac{R}{P}(l - x) \qquad (2)$$

Differentiating (2) w.r.t. x, we get

$$\frac{dy}{dx} = -C_1 \, a \sin ax + C_2 \, a \cos ax - \frac{R}{P} \qquad (3)$$

The end O of the strut is buit in, so at x = 0, y = dy/dx = 0

$\therefore$ (2) gives

$$0 = C_1 + \frac{Rl}{P} \quad \Rightarrow \quad C_1 = -\frac{Rl}{P}$$

and (3) gives $o = C_2 \, a - \frac{R}{P} \quad \Rightarrow \quad C_2 = \frac{R}{aP}$

Putting for C_1 and C_2 is (2)

$$y = -\frac{Rl}{P}\cos ax + \frac{R}{aP}\sin ax + \frac{R}{P}(l - x)$$

$$\Rightarrow \quad y = \frac{R}{P}\left(\frac{\sin ax}{a} - l\cos ax + l - x\right) \qquad (4)$$

Also, the end A of the strut is supported, so at $x = l$, $y = 0$, so that (4) becomes

$$O = \frac{R}{P}\left(\frac{\sin al}{a} - l\cos al + l - l\right)$$

or $$\frac{\sin al}{a} = l\cos al$$

or $$al = \tan al$$

Hence the required equation of deplection curve is given by (4) where $al = \tan al$.

EXERCISE

Solve the following differential equations

1. (i) $x^2\frac{d^2y}{dx^2} + x\frac{dy}{dx} + y = \sin(\log x^2)$

Ans. $y = C_1\cos(\log x) + C_2\sin(\log x) - \frac{1}{3}\sin(\log x^2)$

(ii) $(1+x)^2\frac{d^2y}{dx^2} + (1+x)\frac{dy}{dx} + y = \sin 2\{\log(1+x)\}$ **(I.A.S. 2003)**

Ans. $y = C_1\cos\{\log(1+x) + C_2\sin\{\log(1+x)\} - \frac{1}{3}\sin 2\{\log(1+x)\}$

2. $x^2\frac{d^3y}{dx^3} + 3x\frac{d^2y}{dx^2} + \frac{dy}{dx} = x^2\log x$

Ans. $y = C_1 + C_2\log x + C_3(\log x)^2 + \frac{x^3}{27}(\log x - 1)$

3. $x^4\frac{d^4y}{dx^4} + 6x^3\frac{d^3y}{dx^3} + 4x^2\frac{d^2y}{dx^2} - 2x\frac{dy}{dx} - 4y = 2\cos(\log x)$

Ans. $y = C_1x^2 + C_2x^{-2} + C_3\cos\log x + C_4\sin\log x - \frac{1}{5}\log x\sin\log x$

4. $x^2 \frac{d^2y}{dx^2} - x \frac{dy}{dx} - 3y = x^2 \log x$ **(I.A.S. 2001)**

Ans. $y = C_1 x^3 + C_2 x^{-1} - \frac{1}{3} x^3 \left(\log x + \frac{2}{3} \right)$

5. $\frac{d^2y}{dx^2} + y = \text{cosec } x$ by the method of variation of Parameters

Ans. $y = C_1 \cos x + C_2 \sin x - x \cos x + \sin x. \log \sin x$

6. $\frac{d^2y}{dx^2} + (1 - \cot x) \frac{dy}{dx} - y \cot x = \sin^2 x$ by variation of parameters.

(U.P.T.U. Special Exam 2001)

Ans. $y = C_1 (\sin x - \cos x) + C_2 e^{-x} - \frac{1}{10} \sin 2x + \frac{1}{5} \cos 2x$

7. $x^2 \frac{d^2y}{dx^2} - (x^2 + 2x) \frac{dy}{dx} + (x + 2) y = x^3 e^x$ of which $y = x$ is a solution.

Ans. $y = x (C_1 + C_2 e^x + x e^x)$

8. $(x \sin x + \cos x) \frac{d^2y}{dx^2} - x \cos x \frac{dy}{dx} + y \cos x = 0$ of which $y = x$ is a solution.

Ans. $y = -C_1 \cos x + C_2 x$

9. $\frac{d^2y}{dx^2} - 2 \tan x \frac{dy}{dx} + 5y = e^x . \sec x$ by reducing normal form.

Ans. $y = \sec x \left(C_1 \cos\sqrt{6}\, x + C_2 \sin\sqrt{6}\, x + \frac{1}{7} e^x \right)$

10. $(1 + x^2)^2 \frac{d^2y}{dx^2} + 2x (1 + x^2) \frac{dy}{dx} + 4y = 0$ by changing independent variable.

Ans. $y = C_1 \cos (2 \tan^{-1} x) + C_2 \sin (2 \tan^{-1} x)$

11. $x \frac{d^2y}{dx^2} - \frac{dy}{dx} + 4 x^3 y = 8x^3 \sin x^2$ by changing independent variable.

Ans. $y = C_1 \cos x^2 + C_2 \sin x^2 - x^2 \cos x^2$

12. $\frac{d^2y}{dx^2} + (3 \sin x - \cot x) \frac{dy}{dx} + 2y \sin^2 x = e^{-\cos x} . \sin^2 x$ by changing independent variable.

Ans. $y = C_1 e^{\cos x} + C_2 e^{2\cos x} + \frac{1}{6} e^{-\cos x}$

13. $\frac{d^2x}{dt^2} + \frac{dy}{dt} + 3x = e^{-t}, \; \frac{d^2y}{dt^2} - 4\frac{dx}{dt} + 3y = \sin 2t$

(U.P.T.U. 2007)

Ans. $x = -\frac{C_1}{2}\sin 3t + \frac{C_2}{2}\cos 3t + \frac{C_3}{2}\sin t - \frac{C_4}{2}\cos t + \frac{1}{5}e^{-t} + \frac{1}{5}\sin 2t + A$ where $A = -\frac{C_5}{4}$

$y = C_1 \cos 3t + C_2 \sin 3t + C_3 \cos t + C_4 \sin t + \frac{1}{5}\left(-e^{-t} + \frac{1}{3}\sin 2t\right)$

14. $\frac{dx}{dt} + \frac{dy}{dt} - 2y = 2\cos t - 7\sin t$

$\frac{dx}{dt} - \frac{dy}{dt} + 2x = 4\cos t - 3\sin t$

(U.P.T.U. 2001)

Ans. $x = -\left(1 - \sqrt{2}\right) C_1 e^{\sqrt{2}\,t} - (1 + \sqrt{2}) C_2 e^{-\sqrt{2}\,t} + 3\cos t + C_3$

$y = C_1 e^{\sqrt{2}\,t} + C_2 e^{-\sqrt{2}\,t} + 2\sin t$

15. $\frac{d^2x}{dt^2} - 3x - 4y = 0, \; \frac{d^2y}{dt^2} + x + y = 0$

(U.P.T.U. 2005)

Ans. $x = -[(2C_1 + 2C_2 + 2C_2 t)e^t + 2C_3 - 2C_4 + 2C_4 t)e^{-t}]$

$y = (C_1 + C_2 t)e^t + (C_3 + C_4 t)e^{-t}$

16. Solve $\frac{dx}{dt} + 5x - 2y = t$

$\frac{dy}{dt} + 2x + y = 0$

Also show that $x = y = 0$ when $t = 0$ for some definite values of constants

(U.P.T.U. 2008)

Ans. $x = C_1 e^{-3t} + C_2 te^{-3t} - \frac{1}{2}C_2 e^{-3t} - \frac{1}{2}C_1 e^{-3t} + \frac{t}{9} + \frac{1}{27}$, $y = (C_1 + C_2 t) e^{-3t}$

$-\frac{2t}{9} + \frac{4}{27}$

17. A particle moving in a straight line with S.H.M. has velocities v_1 and v_2 when its distances from the centre are x_1 and x_2 respectively. Show that the period of motion is $2\pi\sqrt{\dfrac{x_1^2 - x_2^2}{v_2^2 - v_1^2}}$ and its amplitude is $\sqrt{\dfrac{\left(v_1^2x_2^2 - v_2^2x_1^2\right)}{\left(v_1^2 - v_2^2\right)}}$

(Bihar P.C.S. 2005)

18. A particle is performing a simple harmonic motion of period T about a centre O and its passes through a point P, where OP = b with velocity v in the direction OP. Prove that the time which elapses before it return to P is $\dfrac{T}{\pi}\tan^{-1}\left(\dfrac{vT}{2\pi b}\right)$.

(I.A.S. 2007)

19. A particle of mass m is projected vertically under gravity, the resistance of the air being mk times the velocity. Show that the greatest height attained by the particle is $\dfrac{V^2}{g}[\lambda - \log(1+\lambda)]$ where V is terminal velocity of the particle and λV is the initial velocity.

(U.P.P.C.S. 2004)

20. If u and V are the velocity of projection and the terminal velocity respectively of a particle rising vertically against a resistance varying as the square of the velocity. Prove that the time taken by the particle to reach the highest point is $\dfrac{V}{g}\tan^{-1}\left(\dfrac{u}{V}\right)$.

(I.A.S. 2006)

21. In the LCR circuit, the charge q on a plate of a condenser is given by $L\dfrac{d^2q}{dt^2} + R\dfrac{dq}{dt} + \dfrac{q}{C} = E \sin pt$. The circuit is turned to resonance so that $p^2 = \dfrac{1}{LC}$. If initially the current i and the charge q be zero, show that for small value of $\dfrac{R}{L}$, the current in the circuit at time t is given by $\left(\dfrac{Et}{2L}\right)\sin pt$.

(U.P.T.U. 2004)

Objective Type of Questions

Choose a correct answer from the four answers given in each of the following questions.

1. The solution of the differential equation

$$x^2 \frac{d^2y}{dx^2} + x \frac{dy}{dx} + y = \sin(\log x^2) \text{ is}$$

(a) $C_1 \cos(\log x) + C_2 \sin(\log x) + \frac{1}{3} \sin(\log x^2)$

(b) $C_1 \cos(\log x) + C_2 \sin(\log x) - \frac{1}{3} \sin(\log x^2)$

(c) $C_1 \cos(\log x) + C_2 \sin(\log x) - \frac{1}{3} \sin(\log x)$

(d) $C_1 \cos(\log x^2) + C_2 \sin(\log x^2) - \frac{1}{3} \sin(\log x)$

Ans. (b)

2. A particular integral of the differential equation

$$x^2 \frac{d^3y}{dx^3} + 3x \frac{d^2y}{dx^2} + \frac{dy}{dx} = x^2 \log x$$

(a) $\frac{x^2}{27}(\log x - 1)$ (b) $\frac{x^2}{27}(\log x + 1)$

(c) $\frac{x^3}{27}(\log x - 1)$ (d) $\frac{x^3}{27}(\log x + 1)$

Ans. (c)

3. A particular integral of $x^2 \frac{d^2y}{dx^2} + 4x \frac{dy}{dx} + 2y = e^x$ is

(a) $\frac{1}{x^2} e^x$ (b) $\frac{1}{x} e^x$

(c) $\frac{1}{x^2} e^{2x}$ (d) $\frac{1}{x^2} e^{3x}$

Ans. (a)

4. The C.F. of the differential equation $x^2 \frac{d^2y}{dx^2} + x\frac{dy}{dx} - y = x^3e^x$

(a) $C_1 x^2 + \frac{C_2}{x^2}$ (b) $C_1 x^2 + \frac{C_2}{x}$

(c) $C_1 x + \frac{C_2}{x^2}$ (d) $C_1 x + \frac{C_2}{x}$

Ans. (d)

5. On putting $x = e^z$, the transformed differential equation of $x^2 \frac{d^2y}{dx^2} + x\frac{dy}{dx} + y = x$ is

(a) $\frac{d^2y}{dz^2} = e^z$ (b) $\frac{d^2y}{dz^2} + y = e^z$

(c) $\frac{d^2y}{dz^2} + y = e^{2z}$ (d) $\frac{d^2y}{dz^2} + y = z$

Ans. (b)

6. The equation of motion of a particle are given by simultaneous differential equations $\frac{dx}{dt} + wy = 0$, $\frac{dy}{dt} - wx = 0$, Then the path of the particle is

(a) Straight line (b) Circle

(c) Ellipse (d) Parabola

Ans. (b)

7. A Particular integral of $\frac{d^2y}{dx} + P\frac{dy}{dx} + Qy = 0$ is $y = e^{mx}$ if

(a) $m^2 + Pm + Q = 0$ (b) $m^2 - Pm + Q = 0$

(c) $m + Pm^2 + Q = 0$ (d) $m^2 + Pm - Q = 0$

Ans. (a)

8. $y = e^x$ is a part of C.F. of differential equation $\frac{d^2y}{dx^2} + P\frac{dy}{dx} + Qy = 0$ if

(a) $1 + P + Q = 0$ (b) $1 - P + Q = 0$

(c) $P + Qx = 0$ (d) $P - Qx = 0$

Ans. (a)

9. In a differential equation

$$x^2 \frac{d^2y}{dx^2} - (x^2 + 2x) \frac{dy}{dx} + (x + 2)\, y = x^3 e^x, \; y = x \text{ is a part of C.F. because}$$

(a) $P - Qx = 0$ (b) $P + Qx = 0$

(c) $1 + P + Q = 0$ (d) $1 - P - Q = 0$

Ans. (b)

10. The solution for $\frac{d^2y}{dx} - 4x \frac{dy}{dx} + (4x^2 - 2)\, y = 0$, given that $y = e^{x^2}$ is an integral included in the complementary function is

(a) $y = (C_1 x + C_2)$ (b) $e^x (C_1 x + C_2)$

(c) $e^{x^2} C_1 x$ (d) $e^{x^2} (C_1 x + C_2)$

Ans. (d)

11. A resistance of 100 ohms, an inductance of 0.5 henry are connected in series with a battery of 20 volts. The current in the circuit is

(a) $i = \frac{1}{5} (1 - e^{-20\, t})$ (b) $i = \frac{1}{5} (1 - e^{20\, t})$

(c) $i = \frac{1}{5} (1 - e^{-200\, t})$ (d) $i = \frac{1}{5} (1 - e^{200\, t})$

Ans. (c)

12. The solution of the differential equation $L \frac{di}{dt} + Ri = E_o \sin wt$ is

(a) $i = \frac{E_o L}{\sqrt{R^2 + L^2 w^2}} \sin\left(wt + \tan^{-1} \frac{wL}{R}\right)$

(b) $i = \frac{E_o L}{\sqrt{R^2 + w^2 L^2}} \sin\left(wt - \tan^{-1} \frac{wL}{R}\right)$

(c) $i = \frac{E_o L}{\sqrt{R^2 + w^2 L^2}} \sin\left(wt + \tan^{-1} \frac{R}{wL}\right)$

(d) $i = \frac{E_o L}{\sqrt{R^2 + w^2 L^2}} \sin\left(wt - \tan^{-1} \frac{R}{wL}\right)$

Ans. (b)

13. A particle executes S.H.M. Such that in two of its positions, the velocities are u, v and the corresponding accelerations α, β. The distance between the position is

(a) $\dfrac{\alpha+\beta}{u^2-u^2}$ (b) $\dfrac{\alpha^2+\beta^2}{u^2-v^2}$

(c) $\dfrac{u^2-v^2}{\alpha+\beta}$ (d) $\dfrac{u^2-v^2}{\alpha-\beta}$

Ans. (c)

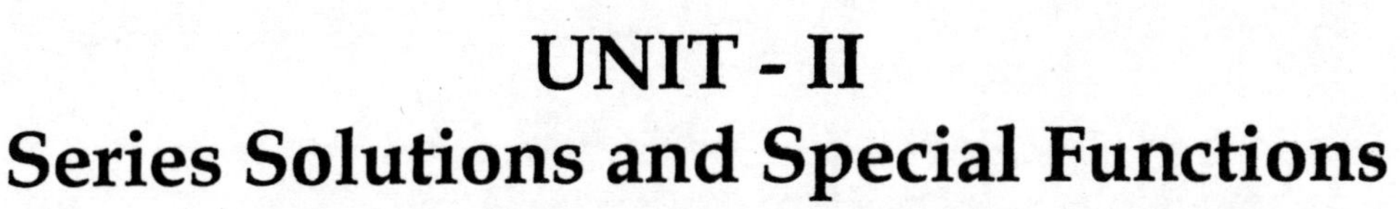

UNIT - II
Series Solutions and Special Functions

Chapter 5

Series Solutions and Special Functions

INTRODUCTION

In this chapter, we shall describe methods of solving variable coefficient differential equations.

Consider the second order homogeneous linear differential equation

$$a_0(x)\frac{d^2y}{dx^2}+a_1(x)\frac{dy}{dx}+a_2(x)y=0, \qquad a_0(x)\neq 0 \tag{i}$$

where $a_0(x)$, $a_1(x)$ and $a_2(x)$ are polynomials.

The equation (i) is called a variable coefficient second order homogeneous equation. A number of problems of physics and engineering involve differential equations of the form (i) in which $a_0(x)$, $a_1(x)$ and $a_2(x)$ are polynomials. Solution of such equations can be obtained in terms of infinite series. In this chapter, we shall present methods for determining the solution of (i) in terms of infinite series. The method can be classified into two categories, power series method and generalized power series method (Frobenius method). In the following sections, we shall discuss the applications of these methods in solving special functions. The study of those solutions (and of other "higher" functions not discussed in calculus) is called the theory of special functions. The series solution of certain differential equations give rise to special functions such as Bessel's function, Legendre's Polynomial, Lagurre's polynomial, Hermite's polynomial, chebyshev polynomial, strum - Lioville problem based on the orthogonality of function is also included which shows that Bessel's Legendre's and other equations can be considered from a common point of view. These special functions, have many applications in mathematical physics and engineering.

ORDINARY AND SINGULAR POINTS

Consider the second order linear homogeneous equation

$$a_0(x)\frac{d^2y}{dx^2}+a_1(x)\frac{dy}{dx}+a_2(x)y=0, \qquad a_0(x)\neq 0 \tag{i}$$

where $a_0(x)$, $a_1(x)$ and $a_2(x)$ are polynomials. Dividing the equation by $a_0(x)$, we get

$$\frac{d^2y}{dx^2}+\frac{a_1(x)}{a_0(x)}\frac{dy}{dx}+\frac{a_2(x)}{a_0(x)}y=0 \qquad \text{(ii)}$$

Taking $p(x)=\frac{a_1(x)}{a_0(x)}$, $q(x)=\frac{a_2(x)}{a_0(x)}$. The equation (ii) Can be written as

$$\frac{d^2y}{dx^2}+p(x)\frac{dy}{dx}+q(y)y=0$$

Equation is called the standard form or normal form of (i). It is also called the canonical form of equation (i). Consider the interval I. let $x_0 \in I$ and $a_1(x)$, $a_2(x)$ be analytic (differentiable) at the point x_0. We define the following.

Ordinary Point

A point $x_0 \in I$ is said to be ordinary point (a regular point) of equation (i) if $a_0(x_0) \neq 0$.

Example $x = 0$ is an ordinary point of the equation

$$(1+x^2)\frac{d^2y}{dx^2}+x\frac{dy}{dx}-y=0$$

Singular point A point $x_0 \in I$ is said to be a singular point of the equation (i) if $a_0(x_0) = 0$

Example $x = 0$ is a singular point of the equation

$$x^3\frac{d^2y}{dx^2}+3x\frac{dy}{dx}+6y=0$$

REGULAR AND IRREGULAR SINGULAR POINTS

Consider the normal form of equation (i) i.e.

$$\frac{d^2y}{dx^2}+p(x)\frac{dy}{dx}+q(x)y=0$$

where $p(x)=\frac{a_1(x)}{a_0(x)}$, $q(x)=\frac{a_2(x)}{a_0(x)}$

Let $x_0 \in I$ be a singular point of (i) and $a_1(x)$, $a_2(x)$ be analytic (differentiable) at $x = x_0$.

write above equation as

$$\frac{d^2y}{dx^2}+\frac{p_1(x)}{(x-x_0)}\frac{dy}{dx}+\frac{q_1(x)}{(x-x_0)^2}y=0 \qquad \text{(iii)}$$

where $p_1(x) = (x - x_0)\, p(x)$ and $q_1(x) = (x - x_0)^2\, q(x)$ we define the following.

Regular singular point

A singular point $x_0 \in I$ of (i) is said to be a regular singular point if the function $p_1(x) = (x - x_0)\, p(x) = \dfrac{(x-x_0)a_1(x)}{a_0(x)}$ and $q_1(x) = (x-x_0)^2 q(x) = \dfrac{(x-x_0)^2 a_2(x)}{a_0(x)}$ have removable discontinuities at x_0 and become analytic when these discontinuities are removed.

Irregular singular point

A singular point of the equation (i) is said to be an irregular singular point of the equation (i) if and only if x_0 is not a regular singular point of (i).

EXAMPLES

Example 1. Find the regular and singular points of

(i) $\quad x^3\dfrac{d^2y}{dx^2} + 3x\dfrac{dy}{dx} + 6y = 0$

(ii) $\quad \left(1-x^2\right)\dfrac{d^2y}{dx^2} - 2x\dfrac{dy}{dx} + 2y = 0$

(iii) $\quad x^2\dfrac{d^2y}{dx^2} + ax\dfrac{dy}{dx} + by = 0$ (a, b are contants)

Solution

(i) we have $a_0(x) = x^3$

$a_0(x) = 0 \quad \Rightarrow \quad x^3 = 0$

or $x = 0$

x_0 is the singular point of the given equation i.e.,

$$x^3\frac{d^2y}{dx^2} + 3x\frac{dy}{dx} + 6y = 0$$

All other points are regular (ordinary) points.

(ii) Here we have $a_0(x) = 1 - x^2$

$a_0 = 0 \Rightarrow 1 - x^2 = 0$

or $x = \pm 1$

Hence, $x = -1$ and $x = 1$ are singular points of the equation

$$\left(1-x^2\right)\frac{d^2y}{dx^2} - 2x\frac{dy}{dx} + 2y = 0$$

while all other points are regular (ordinary) points.

(iii) Here we have $a_0(x) = x^2$

setting $a_0(x) = x^2 = 0$, we get $x = 0$

$\therefore$ $x = 0$ is the singular point of the equation while all other point are ordinary points

SERIES SOLUTION OF THE DIFFERENTIAL EQUATION WHEN X =0 IS AN ORDINARY POINT

There are following steps for solution of equation

Step I: Assume that $y = \sum_{n=0}^{\infty} a_n x^n$ or $y = a_0 + a_1 x + a_2 x^2 + \ldots\ldots\ldots\ldots$ (i)

Step II: Substitute $y, \frac{dy}{dx}$ and $\frac{d^2y}{dx^2}$ obtained by equation (i) in equation

$$\frac{d^2y}{dx^2} + p(x)\frac{dy}{dx} + q(x)y = 0$$

Step III. Equate to zero the coefficients of various powers of x and find a_2, a_3.......... in terms of a_0 and a_1.

Step IV. Substitute the value of a_2, a_3...........in relation (i) which will be the required series solution.

EXAMPLE 2

Solve the following differential equation in series

$$\left(1-x^2\right)\frac{d^2y}{dx^2} - x\frac{dy}{dx} + 4y = 0 \qquad \text{(U.P.T.U. 2006)}$$

Solution The given equation can be written as

$$\frac{d^2y}{dx^2} - \frac{x}{\left(1-x^2\right)}\frac{dy}{dx} + \frac{4y}{\left(1-x^2\right)} = 0 \qquad \text{(i)}$$

Here $p(x) = -\frac{x}{1-x^2}, \; q(x) = \frac{4}{1-x^2}$

p(x) and q(x) both exist at x =0, so x = 0 is an ordinary point of the equation

$$\text{Let } y = \sum_{n=0}^{\infty} a_n x^n \qquad \text{(ii)}$$

$$\therefore \frac{dy}{dx} = \sum_{n=0}^{\infty} a_n \cdot n x^{n-1} \text{ and } \frac{d^2y}{dx^2} = \sum_{n=0}^{\infty} a_n \cdot n(n-1)x^{n-2}$$

putting these values in given equation, we get

$$(1-x^2)\sum_{n=0}^{\infty} a_n \cdot n(n-1)x^{n-2} - x\sum_{n=0}^{\infty} a_n \cdot nx^{n-1} + 4\sum_{n=0}^{\infty} a_n \cdot x^n = 0$$

$$\Rightarrow \sum_{n=0}^{\infty} a_n \cdot n(n-1)x^{n-2} - \sum_{n=0}^{\infty} a_n.(n^2-4)x^n = 0$$

Equating to zero the coefficients of various powers of x.

∴ Coefficient of x^0

$a_2. 2.1 + 4 a_0 = 0 \Rightarrow a_2 = -2a_0$

Now equating the coefficient of x^n, we get

$$a_{n+2} (n+2) (n+1) - (n-2) (n+2) a_n = 0$$

$$\Rightarrow \quad a_{n+2} = \frac{n-2}{n+1} a_n$$

putting n = 1, 2, 3, ... etc we get

$$a_3 = -\frac{1}{2} a_1$$

$$a_4 = -\frac{0}{2} a_2 a_2 = 0, a_5 = \frac{1}{4} a_3 = \frac{1}{4}\left(-\frac{1}{2} a_1\right) = \frac{-a_1}{8}$$

$$a_6 = \frac{2}{5} a_4 = \frac{2}{5}.0 = 0 \text{.......... and so on.}$$

Substituting these values in equation (ii) we get

$$y = a_0 + a_1 x - 2a_0 x^2 + \left(-\frac{1}{2} a_1\right)x^3 + 0x^4 + \left(\frac{-a_1}{8}\right)x^5 + \text{..........}$$

$$= a_0 + a_1 x - 2a_0 x^2 - \frac{a_1}{2} x^3 - \frac{a_4}{8} x^5 + \text{..........}$$

$$\therefore y = a_0 (1 - 2x^2) + a_1 x\left(1 - \frac{x^2}{2} - \frac{x^4}{8} + \text{............}\right)$$

EXAMPLE 3. Find the power series solution of the following differential equation about x = 0

$$(1-x^2)\frac{d^2y}{dx^2} - 2x\frac{dy}{dx} + 2y = 0 \qquad \textbf{(U.P.T.U. 2004)}$$

Solution We have

$$(1-x^2)\frac{d^2y}{dx^2} - 2x\frac{dy}{dx} + 2y = 0 \qquad \text{(i)}$$

This can be written as

$$\frac{d^2y}{dx^2}-\frac{2x}{1-x^2}+\frac{2}{(1-x^2)}y=0$$

Here x =0 is an ordinary point because p(x) and q(x) both exist at x =0

Assume the power series solution as

$$y=\sum_{n=0}^{\infty} a_n x^n \quad \text{(ii)}$$

$$\therefore \frac{dy}{dx}=\sum_{n=0}^{\infty} a_n \cdot nx^{n-1},\ \frac{d^2y}{dx^2}=\sum_{n=0}^{\infty} a_n \cdot n(n-1)x^{n-2}$$

substituting these values in equation (i) we get

$$(1-x^2)\sum_{n=0}^{\infty} a_n \cdot n(n-1)x^{n-2}-2x\sum_{n=0}^{\infty} a_n \cdot nx^{n-1}+2\sum_{n=0}^{\infty} a_n \cdot x^n=0$$

$$\Rightarrow \sum_{n=0}^{\infty} a_n \cdot n(n-1)x^{n-2}-\sum_{n=0}^{\infty} a_n.(n-1)(n+2)x^n=0$$

$\therefore$ Coefficient of x^0

$a_2 . 2 - a_0 (0-1)(0+2) = 0 \Rightarrow 2a_2 + 2a_0 = 0$

$\Rightarrow \qquad a_2 = -a_0$

Coefficient of x^n

$$a_{n+2}(n+2)(n+1) - a_n(n-1)(n+2) = 0$$

$$\Rightarrow a_{n+2}=\frac{(n-1)(n+2)}{(n+2)(n+1)}a_n$$

$$\Rightarrow a_{n+2}=\frac{n-1}{n+1}a_n$$

putting n = 1, 2, 3,........etc. we get

$$a_3=\frac{0}{2}a_1 \Rightarrow a_3=0$$

$$a_4=\frac{1}{3}a_2 \Rightarrow a_4=-\frac{1}{3}a_0$$

$$a_5 = 0 = a_7 = \ldots\ldots\ldots\ldots \text{etc}$$

$$a_6 = \frac{3}{5}a_4 = \frac{3}{5}\left(-\frac{1}{3}a_0\right) \Rightarrow a_6 = -\frac{a_0}{5} \text{ and so on}$$

substituting these values in (ii), we get the required series solution

$$y = a_0 + a_1 x - a_0 x^2 - \frac{1}{3}a_0 x^4 - \frac{a_0}{5}x^6 - \ldots\ldots\ldots\ldots$$

$$\text{Therefore } y = a_1 x + a_0\left(1 - x^2 - \frac{x^4}{3} - \frac{x^6}{5} \ldots\ldots\ldots\ldots\right)$$

FROBENIUS METHOD (SERIES SOLTUTION WHEN x =0 IS NOT AN ORDIANRY POINT)

If x =0 is not an ordinary point (is regular singular point) then we use the following steps for solution.

Step I. Assume a solution of the form

$$y = \sum_{r=0}^{\infty} a_r x^{m+r}$$

Step II. Find $\frac{dy}{dx}, \frac{d^2y}{dx^2}$ and substitute these values in the given equation.

Step III Equate to zero the coefficients of lowest power of x in m gives a quadratic equation in m known as the identical equation of given differential equation.

Thus we get two values of m. The series solution of given equation depends on nature of the roots of the identical equation. Let $y_1(x)$ and $y_2(x)$ be two nontrivial linearly independent solution of given differential equation.

Then the general solution is

$y(x) = Ay_1(x) + By_2(x)$, where A and B are arbitrary constants

There are following cases to solve the equation

Case I: The roots i.e. m_1, and m_2 are distinct ($m_1 \neq m_2$) and not differing by an integer 1, 2, 3,.........

The complete solution is $y = A\,(y)_{m_1} + B(y)_{m_2}$.

Case II: ($m_1 \neq m_2$) and differ by an integer.

There are two cases for solving the equation when $m_1 \neq m_2$ and differ by an integer.

(a) If $m_1 < m_2$ and making some coefficients indeterminate at $m = m_1$, then solution is $y = (y)_{m_1}$.

Similarly if $m_1 > m_2$ then the solution is $y = (y)_{m_2}$.

Remark : The solution at m_1 or m_2 has already contain all those terms which present in the solution at other root, so we obtained solution only m_1 or m_2

(b) Let $m_1 > m_2$ if some of the coefficients of y become infinite when $m = m_2$ then we substitute $a_0 = b_0(m - m_2)$ where $b_0 \neq 0$ in the summation complete solution.

$$y = A(y)_{m_1} + B\left(\frac{\partial y}{\partial m}\right)_{m_2}$$

Remark: The value of y at $m = m_2$ after replacing a_0 by $b_0(m - m_2)$ gives only the multiple of $(y)_{m_1}$ and it not be an independent solution.

Case III. When the roots are equal i.e.

$(m_1 = m_2 = k)$

Then the complete solution is

$$y = A(y)_k + B\left(\frac{\partial y}{\partial m}\right)_k$$

$$\text{As } m_1 = m_2$$

$$= k$$

Case I.

EXAMPLE 4: Solve the following equation in power series about x =0.

$$2x^2\frac{d^2y}{dx^2} + x\frac{dy}{dx} - (x+1)y = 0 \qquad \text{(U.P.T.U. 2005)}$$

Solution The given equation can be written as

$$\frac{d^2y}{dx^2} + \frac{1}{2x}\frac{dy}{dx} - \left(\frac{x+1}{2x^2}\right)y = 0$$

$$p(x) = \frac{1}{2x} \text{ and } Q(x) = -\left(\frac{x+1}{2x^2}\right)$$

Hence x P(x) and $x^2Q(x)$ both exist at the point x =0. So x =0 is a regular point, we assume the solution in the form

$$y = \sum_{r=0}^{\infty} a_r x^{m+r} \qquad \text{(i)}$$

$$\therefore \frac{dy}{dx} = \sum_{r=0}^{\infty} a_r (m+r)x^{m+r-1}$$

and $\frac{d^2y}{dx^2}=\sum_{r=0}^{\infty}a_r(m+r)(m+r-1)x^{m+r-2}$

substituting the values of $y, \frac{dy}{dx}$ and $\frac{d^2y}{dx^2}$ in given equation we get

$$2x^2\sum_{r=0}^{\infty}a_r(m+r)(m+r-1)x^{m+r-2}+x\sum_{r=0}^{\infty}a_r(m+r)x^{m+r-1}-(x+1)\sum_{r=0}^{\infty}a_rx^{m+r}=0$$

$$\text{or }\sum_{r=0}^{\infty}a_r\left[2(m+r)(m+r-1)+(m+r)-1\right]x^{m+r}-\sum_{r=0}^{\infty}a_rx^{m+r-1}=0$$

Equating to zero the coefficients of lowest power of x, we get

$\Rightarrow$ $a_0[2m(m-1)+m-1]=0$

$\Rightarrow$ $2m(m-1)+(m-1)=0$

$\Rightarrow$ $(m-1)(2m+1)=0$

$\Rightarrow$ $m=1, -\frac{1}{2}$

Coefficient of next lowest power of x i.e. x^{m+1}

$$a_1[2(m+1)m+(m+1)-1]-a_0=0$$

$$\Rightarrow a_1=\frac{a_0}{2m^2+2m+m}=\frac{a_0}{m(2m+3)}$$

Now equating to zero the coefficient of x^{m+r} by putting r = r-1 in second summation, we get

$a_r\{2(m+r)(m+r-1)+(m+r-1)\}-a_{r-1}=0$

$$\Rightarrow a_r=\frac{a_{r-1}}{(m+r-1)(2m+2r+1)}$$

Now putting r = 1, 2, 3, 4,................, we get

$$a_1=\frac{a_0}{m(2m+3)}$$

$$a_2=\frac{a_1}{(m+1)(2m+5)}, a_3\frac{a_2}{(m+2)(2m+7)}, a_4=\frac{a_3}{(m+3)(2m+9)}$$

For m=1

$$a_1=\frac{a_0}{5}, a_2=\frac{a_1}{14}=\frac{a_0}{70}, a_3=\frac{a_2}{27}=\frac{a_0}{1890}, \ldots\ldots\ldots\ldots$$

Here $y_1 = a_0 x^{1+0} a_1 x^{1+1} + a_2 x^{1+2} + a_3 x^{1+3} + \dots\dots\dots$ (from (i))

$$\Rightarrow y_1 = a_o x + \frac{a_o}{5}x^2 + \frac{a_o}{70}x^3 + \frac{a_o}{1890}x^4 + \dots\dots\dots$$

$$\Rightarrow y_1 = a_0 x\left[1 + \frac{x}{5} + \frac{x^2}{70} + \frac{x^3}{1890} + \dots\dots\dots\right] \quad \text{(ii)}$$

For $m = -\frac{1}{2}$

$$a_1 = \frac{a_0}{-\frac{1}{2}(-1+3)} = \frac{a_0}{-\frac{1}{2}2} = -a_0$$

$$a_2 = -\frac{a_0}{2},\ a_3 = \frac{a_2}{9} = -\frac{a_0}{18}, \dots\dots\dots \text{ (from (i))}$$

or $y_2 = a_0 x^{-\frac{1}{2}} + (-a_0)x^{-\frac{1}{2}+1} - \frac{a_0}{2}x^{-\frac{1}{2}+2} - \frac{a_0}{18}x^{-\frac{1}{2}+3} \dots\dots\dots$

or $y_2 = a_0 x^{-\frac{1}{2}}\left[1 - x - \frac{x^2}{2} - \frac{x^3}{18} \dots\dots\dots\right]$

Hence the required solution is

$$y_1 = Aa_0 x\left(1 + \frac{x}{5} + \frac{x^2}{70} + \frac{x^3}{1890} + \dots\dots\dots\right) + Ba_0 x^{-\frac{1}{2}}\left(1 - x - \frac{x^2}{2} - \frac{x^3}{18} - \dots\dots\dots\right)$$

$$\text{or } y_1 = c_1 x\left(1 + \frac{x}{5} + \frac{x^2}{70} + \frac{x^3}{1890} + \dots\dots\dots\right) + c_2 x^{-\frac{1}{2}}\left(1 - x - \frac{x^2}{2} - \frac{x^3}{18} - \dots\dots\dots\right)$$

Case II (a) Non logarithm case

EXAMPLE 5 Solve in series the differential equation

$$x\frac{d^2y}{dx^2} + 2\frac{dy}{dx} + xy = 0 \quad \text{(U.P.T.U. 2003)}$$

Solution: Here $p(x) = \frac{2}{x}$, $Q(x) = 1$

At $x = 0$, $x\,P(x)$ and $x^2\,Q(x)$ both exist $\therefore$ $x = 0$ is a regular singular point

Let us suppose

$$y = \sum_{r=0}^{\infty} a_r r^{m+r} \quad \text{(i)}$$

$$\therefore \frac{dy}{dx} = \sum_{r=0}^{\infty} a_r (m+r) x^{m+r-1}$$

$$\frac{d^2y}{dx^2} = \sum_{r=0}^{\infty} a_r (m+r)(m+r-1) x^{m+r-2}$$

substituting these values in the given equation we get

$$x\sum_{r=0}^{\infty} a_r (m+r)(m+r-1) x^{m+r-2} + 2\sum_{r=0}^{\infty} a_r (m+r) x^{m+r-1} + x\sum_{r=0}^{\infty} a_r x^{m+r} = 0$$

$$\Rightarrow \sum_{r=0}^{\infty} a_r (m+r)(m+r+1) x^{m+r-1} + \sum_{r=0}^{\infty} a_r x^{m+r+1} = 0$$

Now, equate to zero the lowest power of x i.e. x^{m-1}, we get

$a_0(m + 0)(m + 1) = 0 \Rightarrow m = 0, -1$

Here $m_1 - m_2 = 0 - (-1) = 1$

and $m_1 > m_2$

So, obtained the solution for $m = m_2$ i.e. $m_2 = -1$.

next equate to zero the coefficient of x^m, then we get

$a_1(m+1)(m+2) = 0 \Rightarrow a_1 = \frac{0}{0}$ indeterminate form

and coefficient of x^{m+1}

$a_2(m + 2)(m + 3) + a_0 = 0$

$$\Rightarrow a_2 = -\frac{a_0}{(m+2)(m+3)} \Rightarrow a_2 = \frac{a_0}{(-1+2)(-1+3)}$$

$$= -\frac{a_0}{1.2} \text{(put } m = -1)$$

$$= -\frac{a_0}{\lfloor 2}$$

Now equating to zero the coefficient of x^{m+r}, we get

$a_{r+1}(m + r + 1)(m + r + 2) + a_{r-1} = 0$

$$\Rightarrow a_{r+1} = -\frac{a_{r-1}}{(m+r+1)(m+r+2)}$$

Putting $m_1 = m_2 = -1$, we get

$$a_{r+1} = -\frac{a_{r+1}}{r(r+1)}$$

putting r = 2, 3, 4, ...

$$a_3 = -\frac{a_1}{2.3} = -\frac{a_1}{\lfloor 3}$$

$$a_4 = -\frac{a_2}{3.4} = -\frac{1}{3.4}\left(-\frac{a_0}{1.2}\right) = \frac{1}{4.3.2.1}a_0 \Rightarrow a_4 = \frac{a_0}{\lfloor 4}$$

$$a_5 = -\frac{a_3}{4.5} = -\frac{1}{4.5}\left(-\frac{a_1}{2.3}\right) = \frac{1}{5.4.3.2.1}a_1 = \frac{1}{\lfloor 5}a_1$$

$$a_6 = -\frac{a_4}{5.6} = -\frac{1}{5.6}\left(\frac{a_0}{4.3.2.1}\right) = \frac{-1}{6.5.4.3.2.1}a_0 \Rightarrow a_6 = -\frac{a_0}{\lfloor 6}$$

and so on.

Hence the required solution is

$$y = (y)_{m_2} = \sum_{r=0}^{\infty} a_r x^{-1+r} = a_0 x^{-1} + a_1 x^{-1+1} + a_2 x^{-1+2} + a_3 x^{-1+3} + \ldots\ldots \qquad \text{(From (i))}$$

$$y = a_0 x^{-1} + a_1 + \left(-\frac{a_0}{\lfloor 2}\right)x + \left(-\frac{a_1}{\lfloor 3}\right)x^2 + \left(\frac{a_0}{\lfloor 4}\right)x^3 + \left(\frac{a_1}{\lfloor 5}\right)x^4 + \ldots\ldots$$

$$y = x^{-1}\left[a_0\left(1 - \frac{x^2}{\lfloor 2} + \frac{x^4}{\lfloor 4} - \ldots\ldots\right) + a_1\left(x - \frac{x^3}{\lfloor 3} + \frac{x^5}{\lfloor 5} - \ldots\ldots\right)\right]$$

$y = x^{-1}\,[a_0 \cos x + a_1 \sin x]$

Case II (b) Logarithmic case

EXAMPLE 6

Obtain the general solution of the equation

$$x^2\frac{d^2y}{dx^2} + 5x\frac{dy}{dx} + x^2 y = 0 \qquad \text{(U.P.T.U 2002)}$$

Solution Let us assume the series solution of the given equation as

$$y = \sum_{r=0}^{\infty} a_r x^{m+r} \qquad \text{(i)}$$

Substituting this in the given equation we get

$$\sum_{r=0}^{\infty}(m+r)(m+r-1)a_r x^{m+r} + 5\sum_{r=0}^{\infty}(m+r)a_r x^{m+r} + \sum_{r=0}^{\infty} a_r x^{m+r+2} = 0$$

$$\text{or } \sum_{r=0}^{\infty}(m+r)(m+r+4)a_r x^{m+r} + \sum_{r=0}^{\infty} a_r x^{m+r+2} = 0 \qquad \text{(ii)}$$

Equating to zero the coefficient of the lowest degree term x^m, we get the identical equation as $m(m+4)\, a_0 = 0$, which gives $m = 0, -4$ as $a_0 \neq 0$

Equating to zero the coefficient of x^{m+1}, we get $(m+1)(m+5)\, a_1 = 0$, $\Rightarrow a_1 = 0$ as $m \neq -1$ and $m \neq -5$. Equating to zero the coefficient of x^{m+r}, we get the recurrence relation as

$(m+r)(m+r+4)\, a_r + a_{r-2} = 0$

$$\Rightarrow a_r = -\frac{a_{r-2}}{(m+r)(m+r+4)}$$

putting r =2, 3, 4,..........

$$a_2 = -\frac{a_0}{(m+2)(m+6)}, a_4 = \frac{-a_2}{(m+4)(m+8)} = \frac{(-1)^2 a_0}{(m+2)(m+4)(m+6)(m+8)}$$

and so on

Therefore (i) becomes

$$y = a_0 x^m \left[1 - \frac{x^2}{(m+2)(m+6)} + \frac{x^4}{(m+2)(m+4)(m+6)(m+8)} - \frac{x^6}{(m+2)(m+6)(m+6)(m+8)(m+10)} + \cdots\cdots\right] \quad \text{(iii)}$$

putting m =0, we get one independent solution as

$$y_1 = a_0 \left[1 - \frac{x^2}{2.6} + \frac{x^4}{2.4.6.8} - \frac{x^6}{2.4.6.8.10} + \cdots\cdots\cdots\cdots\right] \quad \text{(iv)}$$

Next putting m=–4 in (iii) the denominator of each term from third onwards vanishes, To overcome this problem we take

$a_0 = b_0(m+4)$ in (iii) and taking limit as $m \to -4$, we get

$$y = \lim_{m \to -4} b_0 (m+4) x^m \left\{1 - \frac{x^2}{(m+2)(m+6)} + \frac{x^4}{(m+2)(m+4)(m+6)(m+8)} - \cdots\cdots\cdots\cdots\right\}$$

$$= b_0 x^{-4} \left[0 - 0 + \frac{x^4}{(-2)(2)(4)} - \frac{x^6}{(-2)(2)(4)(6)} + \cdots\cdots\cdots\right]$$

$$= \frac{b_0}{(-2)(2)(4)} \left[1 - \frac{x^2}{2.6} + \cdots\cdots\cdots\right]$$

which is nothing but a multiple of y_1 hence not an independent solution.

Here the second independent solution of the given differential equation is taken as

$$y_2 = \left[\frac{\partial}{\partial m} \{ b_0 (m+4) y_m \}_{m=-4} \right]$$

Now

$$y = b_0 (m+4) x^m - \frac{b_0 (m+4)}{(m+2)(m+6)} x^{m+2} + \frac{b_0 (m+4)}{(m+2)(m+4)(m+6)(m+8)} x^{m+4} - \dots\dots\dots$$

$$y = b_0 x^m \left[(m+4) - \frac{(m+4)}{(m+2)(m+6)} x^2 + \frac{1}{(m+2)(m+6)(m+8)} x^4 - \dots\dots\dots\dots\dots\dots \right]$$

Differentiating partially the above equation w.r.t. m, we get

$$\frac{\partial y}{\partial m} = y \log x + b_0 x^m \left[1 + \frac{m^2 + 8m + 20}{(m^2 + 8m + 12)^2} - \frac{(3m^2 + 32m + 76)}{(m^3 + 16m^2 + 75m + 96)^2} x^4 + \dots\dots\dots\dots\dots\dots \right]$$

$$\therefore y_2 = \left(\frac{\partial y}{\partial m} \right)_{m=-4} = (y)_{m=-4} \log x + b_0 x^{-4} \left(1 + \frac{x^2}{4} - \frac{x^4}{4} + \dots\dots\dots\dots \right)$$

$$= b_0 x^{-4} \log x \left[-\frac{x^4}{16} - \frac{x^6}{16} + \dots\dots\dots\dots \right] + b_0 x^{-4} \left(1 + \frac{x^2}{4} - \frac{x^4}{4} + \dots\dots\dots \right)$$

Hence the complete solution is given by

$y = A(y_1) + B(y_2)$

$$= Aa_0 \left[1 - \frac{x^2}{12} - \frac{x^4}{384} - \dots\dots\dots\dots \right] + Bb_0 x^{-4} \log x \left(-\frac{x^4}{16} - \frac{x^6}{16} - \dots\dots\dots \right) + Bb_0 x^{-4} + \left(1 - \frac{x^2}{4} - \frac{x^4}{4} + \dots\dots\dots \right)$$

$$\therefore y = c_1 \left(1 - \frac{x^2}{12} + \frac{x^4}{384} - \dots\dots\dots\dots \right) + c_2 x^{-4} \left(1 + \frac{x^2}{4} - \frac{x^4}{4} + \dots\dots\dots \right) - c_2 \log x \left(\frac{1}{16} - \frac{x^2}{16} + \dots\dots\dots \right)$$

where $c_1 = A\, a_0$ and $c_2 = B\, b_0$

Case III

EXAMPLE 7 Find the series solution of the equation

$$x \frac{d^2 y}{dx^2} + \frac{dy}{dx} + x^2 y = 0 \qquad \text{(U.P.T.U. 2007)}$$

Solution we have

$$x \frac{d^2 y}{dx^2} + \frac{dy}{dx} + x^2 y = 0 \qquad \text{(i)}$$

Here x P(x) and x^2 Q(x) exist at the point x = 0. So x =0 is a regular singular point.

Let $y = \sum_{r=0}^{\infty} a_r x^{m+r}$

$$\therefore \frac{dy}{dx} = \sum_{r=0}^{\infty} a_r (m+r) x^{m+r-1}$$

$$\frac{d^2y}{dx^2} = \sum_{r=0}^{\infty} a_r (m+r)(m+r-1) x^{m+r-2}$$

Substituting these values in equation (i) we get

$$x\sum_{r=0}^{\infty} a_r (m+r)(m+r-1)x^{m+r-2} + \sum_{r=0}^{\infty} a_r (m+r)x^{m+r-1} + x^2\sum_{r=0}^{\infty} a_r x^{m+r} = 0$$

$$\Rightarrow \sum_{r=0}^{\infty} a_r (m+r)^2 x^{m+r-1} + \sum_{r=0}^{\infty} a_r x^{m+r-2} = 0$$

Equating to zero the coefficient of lowest power of x i.e. x^{m-1}, we get the identical equation as

$a_0 m^2 = 0 \quad \Rightarrow \quad m = 0, 0$ as $a_0 \neq 0$

Equating the coefficient of x^m, we get

$a_1 (m+1)^2 = 0 \quad \Rightarrow \quad a_1 = 0,$ as $m \neq -1$

equating the coefficient of x^{m+1}, we get

$a_2 (m+2)^2 = 0 \quad \Rightarrow \quad a_2 = 0,$ as $m \neq -2$

Equating the coefficient of x^{m+2}

$$a_3 (m+3)^2 + a_0 = 0 \quad \Rightarrow \quad a_3 = \frac{-a_0}{(m+3)^2}$$

Next equating the coefficient of x^{m+r+2}, we get $a_{r+3} (m+r+3)^2 + a_r = 0$

$$a_{r+3} = -\frac{a_r}{(m+r+3)^2}$$

putting r =1,2,3………………….., we get

$$a_4 = -\frac{a_1}{(m+4)^2} \Rightarrow a_4 = 0, a_5 = -\frac{a_2}{(m+5)^2} \Rightarrow a_5 = 0$$

$$a_6 = \frac{-a_3}{(m+6)^2} = \frac{a_0}{(m+3)^2(m+6)^2}$$

$$a_7 = 0, a_8 = 0,\ a_9 = \frac{-a_6}{(m+9)^2} \Rightarrow a_9 = -\frac{a_0}{(m+3)^2(m+6)^2(m+9)^2}$$

and so on.

From equation (ii), we have

$y_1 = (y)_{m=0} = a_0 + a_1x + a_2x^2 + a_3x^3 + a_4x^4 + \ldots\ldots\ldots\ldots\ldots\ldots$

$$\Rightarrow y_1 = (y)_{m=o} = a_0 - \frac{a_0}{(3)^2}x^3 + 0 + 0 + \frac{a_0}{(3)^2(6)^2}x^6 + 0 + 0 - \frac{a_0}{(3)^2(6)^2(9)^2}x^9 + \ldots\ldots\ldots$$

$$\Rightarrow y_1 = a_0\left[1 - \frac{x^3}{3^2} + \frac{x^6}{3^2.6^2} - \frac{x^9}{3^2.6^2.9^2} + \ldots\ldots\ldots\right] \quad \text{(iii)}$$

Again

$$y = a_0x^m - \frac{a_0}{(m+3)^2}x^{m+3} + \frac{a_0}{(m+3)^2(m+6)^2}x^{m+6} - \ldots\ldots\ldots\ldots \quad \text{(From (ii))}$$

or $$y = a_0x^m\left[1 - \frac{x^3}{(m+3)^2} + \frac{x^6}{(m+3)^2(m+6)^2} - \ldots\ldots\ldots\ldots\right]$$

$$\frac{\partial y}{\partial m} = a_0x^m \log x\left[1 - \frac{x^3}{(m+3)^2} + \frac{x^6}{(m+3)^2(m+6)^2} - \ldots\ldots\ldots\ldots\right]$$

$$+ a_0x^m\left[0 + \frac{2x^3}{(m+3)^3} - \frac{4m^3 + 54m^2 + 234m + 324}{(m+3)^4(m+6)^4}x^6 + \ldots\ldots\ldots\ldots\right]$$

$$y_2 = \left(\frac{\partial y}{\partial m}\right)_{m=0} = \log x(y)_{m=0} + a_0\left[\frac{2}{3^3}x^3 - \frac{324}{3^4.6^4}x^6 + \ldots\ldots\ldots\ldots\right]$$

$$y_2 = a_0 \log x\left[1 - \frac{x^3}{3^2} + \frac{x^6}{3^2.6^2} - \frac{x^9}{3^2.6^2.9^2} + \ldots\ldots\ldots\ldots\right] + a_0\left[\frac{2}{3^3}x^3 - \frac{324}{3^4.6^4}x^6 + \ldots\ldots\ldots\ldots\right]$$

Hence the complete solution is

$$y = Ay_1 + B\left(\frac{\partial y}{\partial m}\right)_{m=0}$$

or $$y = (Aa_0 + Ba_0 \log x)\left(1 - \frac{x^3}{9} + \frac{x^6}{108} - \ldots\ldots\ldots\right) + \frac{2}{3^3}Ba_0x^3\left(1 - \frac{x^3}{24} + \ldots\ldots\ldots\right)$$

or $$y = (c_1 + c_2 \log x)\left(1 - \frac{x^3}{9} + \frac{x^6}{108} - \ldots\ldots\ldots\right) + \frac{2c_2}{9}x^3\left(1 - \frac{x^3}{24} + \ldots\ldots\ldots\right)$$

where $c = Aa_0$, $c_2 = B\,a_0$

BESSEL'S FUNCTIONS

INTRODUCTION

In many problems of mathematical physics and engineering we come across differential equation of order two, whose solution give rise to special functions. Bessel's equation has a number of applications in engineering are involve in (i) the oscillatory motion of a hanging chain (ii) Euler's theory of a circular membrane (iii) the studies of planetary motion (iv) The propagation of waves (v) The Elasticity (vi) The fluid motion (vii) The potential theory (viii) cylindrical and spherical waves (ix) Theory of plane waves. Bessel's function are also known as cylindrical and spherical function.

BESSEL'S EQUATION

The differential equation of the form

$x^2 y'' + xy' + (x^2 - n^2) y = 0$

or $y''+\left(\frac{1}{x}\right)y'+\left(1-\frac{n^2}{x^2}\right)y=0$

is called Bessel's equation of order n, n being a non-negative constant. The point x =0 is a regular singular point of the equation and Frobenius series solution exists. Solution of equation (i) are called Bessel's function of order n, one of the solutions is denoted by $J_n(x)$ and is called Bessel's function of the first kind. The another solution denoted by $J_{-n}(x)$ and is called Bessel's function of the second kind. We now solve (i) in series by using the method of Frobenius.

SOLUTION OF BESSEL'S EQUATION

$$x^2 y'' + xy' + (x^2 - n^2) y = 0 \qquad \text{(i)}$$

Let the series solution of (i) be

$$y - \sum_{r=0}^{\infty} a_r x^{m+r} \qquad \text{(ii)}$$

so that $y'=\sum_{r=0}^{\infty} a_r (m+r)x^{m+r-1}$

and $y''=\sum_{r=0}^{\infty} a_r (m+r)(m+r-1)x^{m+r-2}$

substituting for y, y', y'' in (i) we have

$$x^2\sum_{r=0}^{\infty} a_r (m+r)(m+r-1)x^{m+r-2}+x\sum_{r=0}^{\infty} a_r (m+r)x^{m+r-1}+\left(x^2-n^2\right)\sum_{r=0}^{\infty} a_r x^{m+r}=0$$

$$\Rightarrow \sum_{r=0}^{\infty} a_r (m+r)(m+r-1)x^{m+r} + \sum_{r=0}^{\infty} a_r (m+r)x^{m+r} + \sum_{r=0}^{\infty} a_r x^{m+r+2} - n^2 \sum_{r=0}^{\infty} a_r x^{m+r} = 0$$

$$\Rightarrow \sum_{r=0}^{\infty} a_r \left[(m+r)(m+r-1)(m+r) - n^2\right] x^{m+r} + \sum_{r=0}^{\infty} a_r x^{m+r+2} = 0$$

$$\Rightarrow \sum_{r=0}^{\infty} a_r \left[(m+r)^2 - n^2\right] x^{m+r} + \sum_{r=0}^{\infty} a_r x^{m+r+2} = 0 \qquad \text{(iii)}$$

Equating the coefficient of lowest degree term of x^m in the identity (iii) to zero, by putting r =0 in the first summation we get the indicial equation

$a_0[(m+0)^2 - n^2] = 0$ (r =0)

$\Rightarrow$ $m^2 = n^2$ i.e. $m = n, -n$; $a_0 \neq 0$

Equating the coefficient of the next lowest degree term x^{m-1} in the identity (iii), we put r =1 in the first summation

$a_1[(m+1)^2 - n^2] = 0$ i.e. $a_1 = 0$, since $(m+1)^2 - n^2 \neq 0$

Equating the coefficient of x^{m+r+2} in (iii) to zero, to find relation in successive coefficients we get

$a_{r+2}[(m+r+2)^2 - n^2] + a_r = 0$

$$\Rightarrow a_{r+2} = -\frac{1}{(m+r+2)^2 - n^2}.a_r$$

Therefore, $a_3 = a_5 = a_7 = ... = 0$, since $a_1 = 0$

If r = 0, $a_2 = -\dfrac{1}{(m+2)^2 - n^2} a_0$

If r =2, $a_4 = -\dfrac{1}{(m+4)^2 - n^2} a_2 = \dfrac{1}{\left[(m+2)^2 - n^2\right]\left[(m+4)^2 - n^2\right]} a_0$ and so on.

on substituting the values of the coefficients $a_1, a_2, a_3, a_4, ...$in (ii) we get

$$y = a_0 x^m - \frac{a_0}{(m+2)^2 - n^2} x^{m+2} + \frac{a_0}{\left[(m+2)^2 - n^2\right]\left[(m+4)^2 - n^2\right]} x^{m+4} + ...$$

$$y = a_0 x^m \left[1 - \frac{1}{(m+2)^2 - n^2} x^2 + \frac{1}{\left[(m+2)^2 - n^2\right]\left[(m+4)^2 - n^2\right]} x^4 - ....\right]$$

For m = n

$$y = a_0x^n\left[1-\frac{1}{4(n+1)}x^2+\frac{1}{4^2\lfloor 2.(n+1)(n+2)}x^4-\ldots\ldots\ldots\ldots\right] \quad \text{(iv)}$$

where a_0 is an arbitrary constant.

For m = -n

$$y = a_0x^{-n}\left[1-\frac{1}{4(-n+1)}x^2+\frac{1}{4^2\lfloor 2.(-n+1)(-n+2)}x^4-\ldots\ldots\ldots\ldots\right]$$

$$\text{or } y = a_0x^{-n}\left[1-\frac{1}{4(1-n)}+\frac{1}{4^2\lfloor 2.(1-n)(2-n)}x^4-\ldots\ldots\ldots\ldots\right] \quad \text{(v)}$$

The particular solution of (i), obtained from (iv) above by taking the arbitrary constant $a_0 = \dfrac{1}{\{2^n\,\overline{|(n+1)}\}}$ is called the Bessel function of the first kind of order n.

It will be denoted by $J_n(x)$.

Thus we have

$$J_n(x) = \frac{x^n}{2^n\,\overline{|(n+1)}}\left[1-\frac{x^2}{4(n+1)}+\frac{1}{4.8}\frac{x^4}{(n+1)(n+2)}-\ldots\ldots\ldots\ldots\right] \quad \text{(vi)}$$

$$\text{or } J_n(x) = \sum_{r=0}^{\infty}(-1)^r\frac{1}{\lfloor r\,\overline{|(n+r+1)}}\left(\frac{x}{2}\right)^{n+2r} \quad \text{(vii)}$$

$$\text{if } n=0,\ J_0(x) = \sum_{r=0}^{\infty}\frac{(-1)^r}{\{\lfloor r\}^2}\left(\frac{x}{2}\right)^{2r}$$

$$\Rightarrow J_0(x) = 1-\frac{x^2}{2^2}+\frac{x^4}{2^2.4^2}-\frac{x^6}{2^2.4^2.6^2}+\ldots\ldots\ldots\ldots$$

$$\text{if } n=1,\ J_1(x) = \frac{x}{2}-\frac{x^3}{2^2.4}+\frac{x^5}{2^2.4^2.6}-\ldots\ldots\ldots\ldots$$

Draw the graph of these two functions. Both the functions are oscillatory with a varying period and a decreasing amplitude

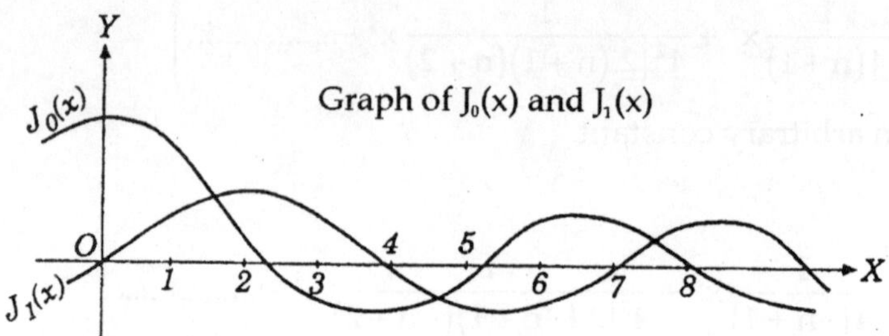

Replacing n by –n in (vii), we get

$$J_{-n}(x) = \sum_{r=0}^{\infty} \frac{(-1)^r}{\lfloor r \, \Gamma(-n+r+1)} \left(\frac{x}{2}\right)^{-n+2r}$$

Case I. If n is not integer or zero, then the complete solution of (i) is

$y = A\,J_n(x) + B\,J_{-n}(x)$

Case II. if n =0, then $y_1 = y_2$ and the complete solution of (i) is the Bessels function of order zero.

Case III. If n is positive integer, then y_2 is not the solution of (i). And y_1 fails to give a solution for negative value of n. Let us find out the general solution when n is an integer.

Recurrence formulae for $J_n(x)$

(i) $xJ'_n = nJ_n - xJ_{n+1}$

Proof. We have $J_n(x) = \sum_{r=0}^{\infty} \frac{(-1)^r}{\lfloor r \, \Gamma(n+r+1)} \left(\frac{x}{2}\right)^{n+2r}$ where n is a positive integer.

Differentiating w.r.t. x, we get

$$J'_n(x) = \sum_{r=0}^{\infty} \frac{(-1)^r (n+2r)}{\lfloor r \, \Gamma(n+r+1)} \left(\frac{x}{2}\right)^{n+2r-1} \cdot \frac{1}{2}$$

$$\therefore xJ'_n(x) = \sum_{r=0}^{\infty} \frac{(-1)^r (n+2r)}{\lfloor r \, \Gamma(n+r+1)} \left(\frac{x}{2}\right)^{n+2r}$$

$$\text{or } xJ'_n(x) = \sum_{r=0}^{\infty} \frac{(-1)^r n}{\lfloor r \, \Gamma(n+r+1)} \left(\frac{x}{2}\right)^{n+2r} + \sum_{r=0}^{\infty} \frac{(-1)^r 2r}{\lfloor r \, \Gamma(n+r+1)} \cdot \frac{x}{2} \left(\frac{x}{2}\right)^{n+2r-1}$$

$$= nJ_n(x) + x\sum_{r=1}^{\infty}(-1)^r \frac{1}{\underline{|(r-1)}\,\underline{|(n+r+1)}}\left(\frac{x}{2}\right)^{n+2r-1}$$

$$= nJ_n(x) + x\sum_{s=0}^{\infty}\frac{(-1)^{s+1}}{\underline{|s}\,\underline{|(n+s+2)}}\left(\frac{x}{2}\right)^{n+2s-1} \quad \{\text{by putting } r-1=s\}$$

$$= nJ_n(x) - x\sum_{s=0}^{\infty}\frac{(-1)^{s}}{\underline{|s}\,\underline{|(n+1+s+1)}}\left(\frac{x}{2}\right)^{(n+1)+2s}$$

or $\quad xJ'_n(x) = nJ_n(x) - x\,J_{n+1}(x)$

$\Rightarrow \quad xJ'_n = nJ_n - x\,J_{n+1}$

(ii) $xJ'_n = -n\,J_n + x\,J_{n-1}$ (U.P.T.U 2004, 2006)

Proof. we have

$$J_n = \sum_{r=0}^{\infty}\frac{(-1)^r}{\underline{|r}\,\underline{|(n+r+1)}}\left(\frac{x}{2}\right)^{n+2r}$$

$$\therefore J_n^1 = \sum_{r=0}^{\infty}\frac{(-1)^r(n+2r)}{\underline{|r}\,\underline{|(n+r+1)}}\cdot\frac{1}{2}\left(\frac{x}{2}\right)^{n+2r-1}$$

Therefore,

$$xJ'_n = \sum_{r=0}^{\infty}\frac{(-1)^r(n+2r)}{\underline{|r}\,\underline{|(n+r+1)}}\left(\frac{x}{2}\right)^{n+2r}$$

$$= \sum_{r=0}^{\infty}\frac{(-1)^r(2n+2r-n)}{\underline{|r}\,\underline{|(n+r+1)}}\cdot\left(\frac{x}{2}\right)^{n+2r}$$

$$= \sum_{r=0}^{\infty}\frac{(-1)^r\,2(n+r)}{\underline{|r}\,\underline{|(n+r+1)}}\frac{x}{2}\left(\frac{x}{2}\right)^{n+2r-1} - n\sum_{r=0}^{\infty}\frac{(-1)^r}{\underline{|r}\,\underline{|(n+r+1)}}\left(\frac{x}{2}\right)^{n+2r}$$

$$= x\sum_{r=0}^{\infty}\frac{(-1)^r(n+r)}{\underline{|r}(n+r)\underline{|(n+r)}}\left(\frac{x}{2}\right)^{n+2r-1} - nJ_n$$

$$= x\sum_{r=0}^{\infty}\frac{(-1)^r}{\underline{|r}\,\underline{|\{(n-1)+r+1\}}}\left(\frac{x}{2}\right)^{(n-1)+2r} - nJ_n$$

$= xJ_{n-1} - nJ_n$

or $\quad x\,J'_n = -nJ_n + x\,J_{n-1}$

(iii) $2J'_n = J_{n-1} - J_{n+1}$

Proof. By Recurrence formula (i), we have

$x J'_n = nJ_n - x J_{n+1}$ (1)

and by Recurrence formula (ii), we have

$x J'_n = -nJ_n + x J_{n-1}$ (2)

Adding (1) and (2), we get

$2x J_n^1 = x(J_{n-1} - J_{n+1})$

or $2 J_n^1 = J_{n-1} - J_{n+1}$

(iv) $2nJ_n = x(J_{n-1} + J_{n+1})$ **(I.A.S. 1980, 82, U.P.T.U. 2007)**

Proof By formula (i) we have $x J_n^1 = nJ_n - xJ_{n+1}$ (1)

and by formula (ii) we have $x J_n^1 = -nJ_n + xJ_{n-1}$ (2)

subtracting (2) from (1), we get

$0 = 2nJ_n - x(J_{n-1} + J_{n+1})$

or $2nJ_n = x(J_{n-1} + J_{n+1})$

(v) $\frac{d}{dx}(x^{-n}J_n) = -x^{-n}J_{n+1}$ (I.A.S. 1980, U.P.T.U. (CO) 2003)

Proof. By formula (i) we have

$x J_n' = nJ_n - x J_{n+1}$

Multiplying both sides of above by x^{-n-1}, we get

$x^{-n} J_n' = nx^{-n-1}J_n - x^{-n} J_{n+1}$

or $x^{-n} J_n' - nx^{-n-1}J_n = -x^{-n} J_{n+1}$

or $\frac{d}{dx}(x^{-n}J_n) = -x^{-n}J_{n+1}$

(vi) $\frac{d}{dx}(x^{n}J_n) = x^{n}J_{n-1}$ (I.A.S. 1970, U.P.T.U. 2004, 05)

Proof. By formula (ii) we have

$x J_n' = -nJ_n + x J_{n-1}$

Multiplying both sides by x^{n-1}, we get

$x^n J_n' = -nx^{n-1}J_n + x^n J_{n-1}$

or $x^n J_n' + nx^{n-1}J_n = x^n J_{n-1}$

or $\frac{d}{dx}(x^{n}J_n) = x^{n}J_{n-1}$

EXAMPLE 1 Prove that

(i) $J_{-1/2}(x) = \sqrt{\left(\frac{2}{\pi x}\right)} \cos x$

(ii) $J_{1/2}(x) = \sqrt{\left(\frac{2}{\pi x}\right)} \sin x$ (U.P.T.U. 2009)

(iii) $\left[J_{1/2}(x)\right]^2 + \left[J_{-1/2}(x)\right]^2 = \frac{2}{\pi x}$ (M.P.S.L.E.T 2000)

Solution (i) By definition of $J_n(x)$, we have

$$J_n(x) = \frac{x^n}{2^n \overline{|(n+1)}}\left[1 - \frac{x^2}{2(2n+2)} + \frac{x^4}{2.4(2n+2)(2n+4)} - \ldots\ldots\ldots\ldots\right]$$

$$\text{or } J_n(x) = \frac{x^n}{2^n \overline{|(n+1)}}\left[1 - \frac{x^2}{2.2(n+1)} + \frac{x^2}{2.4.2^2(n+1)(n+2)} - \ldots\ldots\ldots\ldots\right]$$

putting $n = -\frac{1}{2}$, we get

$$J_{-1/2}(x) = \frac{x^{-1/2}}{2^{-1/2}\overline{\left|\left(\frac{1}{2}\right)\right.}}\left[1 - \frac{x^2}{1.2} + \frac{x^4}{1.2.3.4} - \ldots\ldots\ldots\ldots\ldots\ldots\right]$$

$$\text{or } J_{-1/2}(x) = \sqrt{\left(\frac{2}{\pi x}\right)}\left[1 - \frac{x^2}{\underline{|2}} + \frac{x^4}{\underline{|4}} - \ldots\ldots\ldots\ldots\ldots\ldots\right]$$

$$J_{-1/2}(x) = \sqrt{\left(\frac{2}{\pi x}\right)} \cos x \quad \text{(A)}$$

(ii) putting $n = \frac{1}{2}$, we have

$$J_{1/2}(x) = \frac{x^{1/2}}{2^{1/2}\overline{\left|\left(\frac{3}{2}\right)\right.}}\left[1 - \frac{x^2}{2.3} + \frac{x^4}{2.3.4.5} - \ldots\ldots\ldots\ldots\ldots\ldots\right]$$

$$= \frac{x^{1/2}}{2^{1/2}\frac{1}{2}\overline{\left|\left(\frac{1}{2}\right)\right.}} \cdot \frac{1}{x}\left[x - \frac{x^3}{\underline{|3}} + \frac{x^5}{\underline{|5}} - \ldots\ldots\ldots\ldots\ldots\ldots\right]$$

or $J_{1/2}(x)=\sqrt{\left(\frac{2}{\pi x}\right)}\sin x$ (B)

(iii) squaring and adding the results (A) and (B), we have

$$\{J_{-1/2}(x)\}^2+\{J_{1/2}(x)\}^2=\frac{2}{\pi x}(\sin^2 x+\cos^2 x)$$

or $\left[J_{-1/2}(x)\right]^2+\left[J_{1/2}(x)\right]^2=\frac{2}{\pi x}$

EXAMPLE 2. Prove that

(i) $J_{-3/2}(x)=\sqrt{\left(\frac{2}{\pi x}\right)}.\left(-\frac{\cos x}{x}-\sin x\right)$

(ii) $J_{3/2}(x)=\sqrt{\left(\frac{2}{\pi x}\right)}.\left(\frac{\sin x}{x}-\cos x\right)$

Solution (i) we know that

$J_{-1/2}(x)=\sqrt{\left(\frac{2}{\pi x}\right)}.\cos x$

and $J_{1/2}(x)=\sqrt{\left(\frac{2}{\pi x}\right)}.\sin x$

By recurrence formula

$2n\,J_n(x) = x(J_{n-1} + J_{n+1})$

or $J_{n-1}+J_{n+1}=\frac{2n}{x}J_n(x)$ (3)

putting $n=-\frac{1}{2}$ in (3), we get

$$J_{-3/2}(x)+J_{1/2}(x)=-\left(\frac{2}{2x}\right)J_{-1/2}(x)$$

or $J_{-3/2}(x)=-\sqrt{\left(\frac{2}{\pi x}\right)}\sin x-\frac{1}{x}\sqrt{\left(\frac{2}{\pi x}\right)}\cos x$ by (1) and (2)

or $J_{-3/2}(x)=-\sqrt{\left(\frac{2}{\pi x}\right)}\left(-\frac{\cos x}{x}-\sin x\right)$

(ii) putting $n=\frac{1}{2}$ in (3), we have

$$J_{-1/2}(x) + J_{3/2} = \left(\frac{2}{\pi x}\right) J_{1/2}(x)$$

or $J_{3/2}(x) = -J_{-1/2}(x) + \frac{1}{x} J_{1/2}(x)$

$= -\sqrt{\left(\frac{2}{\pi x}\right)} \cos x + \frac{1}{x}\sqrt{\left(\frac{2}{\pi x}\right)}$ sin x by (1) and (2)

or $J_{3/2}(x) = \sqrt{\left(\frac{2}{\pi x}\right)}\left(\frac{\sin x}{x} - \cos x\right)$

EXAMPLE 3. Prove that

(i) $J_{n+3} + J_{n+5} = \frac{2}{x}(n+4) J_{n+4}$ (U.P.T.U. 2007)

(ii) $4J_n'' = J_{n-2} - 2J_n + J_{n+2}$

Solution (i) By recurrence formula (iv), we have

$2n J_n = x(J_{n-1} + J_{n+1})$

Now putting (n+4) for n in the above formula, we get

$2(n+4) J_{n+4} = x(J_{n+3} + J_{n+5})$

or $\frac{2}{x}(n+4) J_{n+4} = J_{n+3} + J_{n+5}$

or $J_{n+3} + J_{n+5} = \frac{2}{x}(n+4) J_{n+4}$

(ii) By recurrence formula (iii), we have

$2J_n' = J_{n-1} - J_{n+1}$ (1)

Differentiating it, we get

$2J_n'' = J_{n-1}' - J_{n+1}'$

or $4J_n'' = 2J_{n-1}' - 2J_{n+1}'$

Applying (1) for $2J_{n-1}'$ and $2J_{n+1}'$, we have

$4J_n'' = (J_{n-2} - J_n) - (J_n - J_{n+2})$

$\Rightarrow \quad 4J_n'' = J_{n-2} - 2J_n + J_{n+2}$

EXAMPLE 4 Prove that

(i) $\frac{d}{dx}\left(J_n^2 + J_{n+1}^2\right) = 2\left(\frac{n}{x}J_n^2 - \frac{n+1}{x}J_{n+1}^2\right)$ (U.P.T.U. 2005)

(ii) $J_0^2 + 2\left(J_1^2 + J_2^2 + J_3^2 + \ldots\ldots\ldots\right) = 1$ (U.P.T.U. 2003, 2008)

Solution (i) By recurrence formula (i), we have

$$xJ_n' = nJ_n - xJ_{n+1} \quad (1)$$

and by recurrence formula (ii) we have

$$xJ_n' = -nJ_n + xJ_{n-1} \quad (2)$$

Replacing n by (n +1) in (2) we get

$$xJ_{n+1}' = -(n+1)J_{n+1} + xJ_n$$

or $J_{n+1}' = -\frac{(n+1)}{x}J_{n+1} + J_n$ (3)

Now $\frac{d}{dx}\left(J_n^2 + J_{n+1}^2\right) = 2J_nJ_n' + 2J_{n+1}J_{n+1}'$

$= 2J_n\left(\frac{n}{x}J_n - J_{n+1}\right) + 2J_{n+1}\left(-\frac{n+1}{x}J_{n+1} + J_n\right)$ using (1) and (3)

$= 2\left(\frac{n}{x}J_n^2 - \frac{n+1}{x}J_{n+1}^2\right)$, on simplification

(ii) since we have

$$\frac{d}{dx} = \left(J_n^2 + J_{n+1}^2\right) = 2\left(\frac{n}{x}J_n^2 - \frac{n+1}{x}J_{n+1}^2\right) \quad (1)$$

Replacing n by 0,1,2,3,...........successively in (1) we get

$$\frac{d}{dx}\left(J_0^2 + J_1^2\right) = 2\left(0 - \frac{1}{x}J_1^2\right)$$

$$\frac{d}{dx}\left(J_1^2 + J_2^2\right) = 2\left(\frac{1}{x}J_1^2 - \frac{2}{x}J_2^2\right)$$

$$\frac{d}{dx}\left(J_2^2 + J_3^2\right) = 2\left(\frac{2}{x}J_2^2 - \frac{3}{x}J_3^2\right)$$

....

....

....

Adding these column wise and noting that $J_n \to 0$ as $n \to \infty$ we get

$$\frac{d}{dx}\left[J_0^2 + +2\left(J_1^2 + J_2^2 + ...\right)\right] = 0$$

Integrating, above we have

$$J_0^2(x) + 2\left[J_1^2(x) + J_2^2(x) + ...\right] = c \qquad (2)$$

Replacing x by 0 in (2) and noting that

$J_0(x) = 1$ & $J_n(0) = 0$ for $n \geq 1$, we get

$1 + 2(0 + 0 + 0 + ...) = c$

$\therefore c = 1$

Hence (2) becomes

$$J_0^2 + 2\left(J_1^2 + J_2^2 + ...\right) = 1$$

EXAMPLE 5 : Prove that following relations

$x^2 J_n''(x) = (n^2 - n - x^2)J_n(x) + xJ_{n+1}(x)$ (U.P.T.U. 2006, 07)

Solution The solution of $x^2\frac{d^2y}{dx^2} + x\frac{dy}{dx} + \left(x^2 - n^2\right)y = 0$ (1)

is $J_n(x)$

$$\Rightarrow x^2 J_n'' + xJ_n' + \left(x^2 - n^2\right)J_n = 0 \qquad (2)$$

By recurrence relation (i) we have

$$xJ_n' = nJ_n - xJ_{n+1} \qquad (3)$$

putting the value of xJ_n' from (3) in (2), we have

$$x^2 J_n''(x) = -nJ_n + xJ_{n+1} + \left(n^2 - x^2\right)J_n$$

or $x^2 J_n''(x) = (n^2 - n - x^2)J_n + xJ_{n+1}$

EXAMPLE 6 Prove that

$\int xJ_0^2 dx = \frac{1}{2}x\left[J_0^2(x) + J_1^2(x)\right] + c$ (U.P.T.U. 2003, 2004)

Solution

L.H.S = $\int xJ_0^2(x)dx$

$$= J_0^2 \cdot \frac{x^2}{2} - \int 2J_0 J_0' \cdot \frac{x^2}{2} dx + c$$

Integrating by parts, taking x as IInd functions

$$= \frac{x^2}{2} J_0^2 - \int x^2 J_0 (-J_1) dx + c \text{ As } J_0^1 = -J_1$$

$$= \frac{x^2}{2} J_0^2 + \int xJ_1 . xJ_0 dx + c \quad \text{from (vi) th recurrence formula}$$

$$\frac{d}{dx}(x^n J_n) = x^n J_{n-1} \text{ for n =1}$$

$$= \frac{x^2}{2} J_0^2 + \int xJ_1 \frac{d}{dx}(xJ_1) dx + c$$

$$= \frac{x^2}{2} J_0^2 + \frac{(xJ_1)^2}{2} + c \qquad \frac{d}{dx}(xJ_1) = xJ_0$$

$$\Rightarrow \int xJ_0^2 dx = \frac{x^2}{2}\left[J_0^2 + J_1^2\right] + c$$

EXAMPLE 7 prove that

$$J_2^1 = \left(1 - \frac{4}{x^2}\right) J_1 + \frac{2}{x} J_0$$ (U.P.T.U. 2001, 2008)

Solution From recurrence formula (ii), we have

$$xJ_n' = -nJ_n + xJ_{n-1} \quad \text{(i)}$$

putting n =2 in (i) we get

$$xJ_2' = -2J_2 + xJ_1$$

$$\Rightarrow J_2' = -\frac{2}{x} J_2 + J_1 \quad \text{(ii)}$$

From recurrence formula (i), we have

$$xJ_n' = nJ_n - xJ_{n+1} \quad \text{(iii)}$$

From (i) and (iii), we have

$$-nJ_n + xJ_{n-1} = nJ_1 - xJ_{n+1}$$

putting n = 1, we have

$$-J_1 + xJ_0 = J_1 - xJ_2$$

or $-\frac{1}{x}J_1 + J_0 = \frac{1}{x}J_1 - J_2$

$$\Rightarrow J_2 = \frac{2}{x}J_1 - J_0 \qquad \text{(iv)}$$

Putting J_2 from (iv) in (ii), we get

$$J_2^1 = -\frac{2}{x}\left(\frac{2}{x}J_1 - J_0\right) + J_1 = -\frac{4}{x^2}J_1 + \frac{2}{x}J_0 + J_1$$

or $J_2^1 = \left(1 - \frac{4}{x^2}\right)J_1 + \frac{2}{x}J_0$

EXAMPLE 8 Prove that

$$J_3(x) + 3J_0'(x) + 4J_0'''(x) = 0 \qquad \text{(U.P.T.U. 2001)}$$

Solution we know that $J_0' = -J_1$ by 1st recurrence formula $xJ_n' = nJ_n - xJ_{n+1}$ putting n =0 we have $J_0' = -J_1$

$$\Rightarrow J_0'' = -J_1' \qquad (1)$$

By (iii) rd recurrence relation

$$2J_n' = J_{n-1} - J_{n+1}$$

Putting n =1, we have

$$2J_1' = J_0 - J_2 \qquad (2)$$

Using (2) in (1), we get

$$J_0'' = -\frac{1}{2}(J_0 - J_2)$$

$$\Rightarrow J_0''' = -\frac{1}{2}(J_0' - J_2')$$

$= -\frac{1}{2}J_0' + \frac{1}{2}\cdot\frac{1}{2}(J_1 - J_3)$ $\because J_2^1 = \frac{1}{2}(J_1 - J_3)$ by putting n =1 in third recurrence formula

$$= -\frac{1}{2}J_0' + \frac{1}{4}J_1 - \frac{1}{4}J_3$$

$$= -\frac{1}{2}J_0' + \frac{1}{4}J_0' - \frac{1}{4}J_3$$

$$\Rightarrow J_0''' = -\frac{3}{4}J_0' - \frac{1}{4}J_3$$

$$\Rightarrow 4J_0''' + 3J_0' + J_3 = 0$$

Generating Function for $J_n(x)$

To show that when n is a positive integer, $J_n(x)$ is the coefficient of z^n in the expansion of $e^{x(z-1/z)/2}$ in ascending and descending power of z.

Proof. We know that

$$e^t = 1 + t + \frac{t^2}{\lfloor 2} + \frac{t^3}{\lfloor 3} + \ldots\ldots\ldots\ldots + \frac{t^n}{\lfloor n} + \ldots\ldots\ldots$$

$$\text{and } e^{-t} = 1 - t + \frac{t^2}{\lfloor 2} - \frac{t^3}{\lfloor 3} + \ldots\ldots\ldots\ldots + (-1)^n \frac{t^n}{\lfloor n} + \ldots\ldots\ldots$$

Now $e^{x(z-1/z)/2} = e^{xz/2} . e^{-x/2z}$

$$= \left[1 + \frac{x}{2}z + \left(\frac{x}{2}\right)^2 \frac{z^2}{\lfloor 2} + \ldots\ldots\ldots + \left(\frac{x}{2}\right)^n \frac{z^n}{\lfloor n} + \left(\frac{x}{2}\right)^{n+1} \frac{z^{n+1}}{\lfloor (n+1)} + \ldots\ldots\right]$$

$$X\left[1 - \frac{x}{2z} + \left(\frac{x}{2z}\right)^2 \frac{1}{\lfloor 2} - \ldots\ldots\ldots\right]$$

Coefficient of z^n in the product is

$$= \left[\left(\frac{x}{2}\right)^n \frac{1}{\lfloor n} - \left(\frac{x}{2}\right)^{n+2} \frac{1}{\lfloor (n+1)} + \left(\frac{x}{2}\right)^{n+4} \frac{1}{\lfloor (n+2) \lfloor 2} + \ldots\ldots\ldots\right]$$

$$= \sum_{r=0}^{\infty} \frac{(-1)^r}{\lfloor (r) \lfloor (n+r)} \left(\frac{x}{2}\right)^{n+2r}$$

$$= \sum_{r=0}^{\infty} \frac{(-1)^r}{\lfloor (r) \Gamma(n+r+1)} \left(\frac{x}{2}\right)^{n+2r} \qquad \because \lfloor (n+r) = \Gamma(n+r+1)$$

$= J_n(x)$

EXAMPLE 9

Prove that

$$\frac{d}{dx}[J_0(x)] = -J_1(x)$$

Solution we know that

$$J_n(x) = \sum_{r=0}^{\infty} \frac{(-1)^r}{\lfloor r \lfloor(n+r+1)} \left(\frac{x}{2}\right)^{n+2r}$$

$$\Rightarrow J_0(x) = \sum_{r=0}^{\infty} (-1)^r \frac{1}{(\lfloor r)^2} \left(\frac{x}{2}\right)^{2r}$$

$$= 1 - \left(\frac{x}{2}\right)^2 + \frac{1}{(\lfloor 2)^2}\left(\frac{x}{2}\right)^4 - \frac{1}{(\lfloor 3)^2}\left(\frac{x}{2}\right)^6 + \dots\dots\dots + (-1)^{n+1} \frac{1}{\{\lfloor(n+1)\}^2}\left(\frac{x}{2}\right)^{2n+2} + \dots\dots\dots$$

Differentiating it w.r.t. x, we get

$$\frac{d}{dx}\{J_0(x)\} = -\frac{x}{2} + \frac{1}{\lfloor 1 \lfloor 2}\left(\frac{x}{2}\right)^3 - \frac{1}{\lfloor 2 \lfloor 3}\left(\frac{x}{2}\right)^5 + \dots\dots\dots (-1)^{n+1} \frac{1}{\lfloor n \lfloor(n+1)}\left(\frac{x}{2}\right)^{2n+1} + \dots\dots\dots$$

$$= -\left[\left(\frac{x}{2}\right) - \frac{1}{\lfloor 1 \lfloor 2}\left(\frac{x}{2}\right)^3 + \dots\dots\dots\dots + (-1)^n \frac{1}{\lfloor n \lfloor(n+1)}\left(\frac{x}{2}\right)^{2n+1} + \dots\dots\dots\right]$$

$$= -\sum_{r=0}^{\infty} (-1)^r \frac{1}{\lfloor r \lfloor(r+1)}\left(\frac{x}{2}\right)^{2r+1}$$

$= - J_1(x)$

$$\therefore \frac{d}{dx}[J_0(x)] = -J_1(x)$$

EXAMPLE 10: Express $J_5(x)$ in terms of $J_0(x)$ and $J_1(x)$

Solution we know that

$$J_n(x) = \frac{x}{2n}[J_{n-1}(x) + J_{n+1}(x)]$$

i.e. $J_{n+1}(x) = \frac{2n}{x} J_n(x) - J_{n-1}(x)$

putting n =1, 2, 3, 4 successively, we have

$$J_2(x) = \frac{2}{x} J_1(x) - J_0(x) \qquad \text{(i)}$$

$$J_3(x) = \frac{4}{x} J_2(x) - J_1(x) \qquad \text{(ii)}$$

$$J_4(x) = \frac{6}{x} J_3(x) - J_2(x) \qquad \text{(iii)}$$

$$J_5(x)=\frac{8}{x}J_4(x)-J_3(x) \qquad \text{(iv)}$$

Substituting the value of $J_2(x)$ in (ii), we have

$$J_3(x)=\frac{4}{x}\left\{\frac{2}{x}J_1(x)-J_0(x)\right\}-J_1(x)$$

$$=\left(\frac{8}{x^2}-1\right)J_1(x)-\frac{4}{x}J_0(x) \qquad \text{(v)}$$

Now substituting the values of $J_3(x)$ from (v) and $J_2(x)$ from (i) in (iii) we get

$$J_4(x)=\left(\frac{48}{x^3}-\frac{8}{x}\right)J_1(x)+\left(1-\frac{24}{x^2}\right)J_0(x) \qquad \text{(vi)}$$

Finally putting the values of $J_4(x)$ from (vi) and $J_3(x)$ from (v) in (iv) we obtain.

$$J_5(x)=\left(\frac{384}{x^4}-\frac{72}{x^2}-1\right)J_1(x)+\left(\frac{12}{x}-\frac{192}{x^3}\right)J_0(x)$$

EXAMPLE 11 Show that Bessel's function $J_n(x)$ is an even function when n is even and is odd function when n is odd. Express $J_6(x)$ in terms of $J_0(x)$ and $J_1(x)$

(U.P.T.U. 2009)

Solution:

Hint

when n is even $J_n(-x) = J_n(x)$ since $(-1)^n = 1$

when n is odd $J_n(-x) = - J_n(x)$

since $(-1)^n = -1$

Since we know that $J_{n+1}(x)=\frac{2n}{x}J_n(x)-J_{n-1}(x)$

Putting n=1, 2, 3, 4, 5, 6 successively, we have

$$J_2(x)=\frac{2}{x}J_1(x)-J_0(x), J_3(x)=\frac{4}{x}J_2-J_1$$

$$J_4(x)=\frac{6}{x}J_3-J_2, J_5(x)=\frac{8}{x}J_4-J_3$$

$$J_6(x)=\frac{10}{x}J_5-J_4$$

$$=\left[\frac{3840}{x^5}-\frac{768}{x^3}-\frac{2}{x}\right]J_1+\left[\frac{144}{x^2}-\frac{1920}{x^4}-1\right]J_0$$

Legendre Functions

Introduction

In series solutions of differential equations, we applied the power series method for solving second order differential equations. In this title we will discuss an important problem of mathematical physics known as Legendre polynomials.

LEGENDRE'S EQUATION

The differential equation

$$\left(1-x^2\right)\frac{d^2y}{dx^2}-2x\frac{dy}{dx}+n(n+1)y=0 \qquad \text{(i)}$$

is called Legendre's differential equation, where n is a real constant. In most applications n takes positive integral values. This equation is important in applied mathematics, particularly in boundary value problem involving spherical configurations.

The equation (i) can also be written as

$$\frac{d}{dx}\left\{\left(1-x^2\right)\frac{dy}{dx}\right\}+n(n+1)y=0$$

Solution of Legendre's Equation in series

The Legendre's equation is

$$\left(1-x^2\right)\frac{d^2y}{dx^2}-2x\frac{dy}{dx}+n(n+1)y=0 \qquad \text{(i)}$$

Let us consider the solution of Legendre's equation in a series of descending powers of x as

$$y=x^k\left(a_0+a_1x^{-1}+a_2x^{-2}+\ldots\ldots\ldots\ldots\right)=\sum_{r=0}^{\infty}a_rx^{k-r} \qquad \text{(ii)}$$

$$\therefore \frac{dy}{dx}=\sum_{r=0}^{\infty}a_r(k-r)x^{k-r-1} \text{ and } \frac{d^2y}{dx^2}=\sum_{r=0}^{\infty}a_r(k-r)(k-r-1)x^{k-r-2}$$

Now putting the values of $\frac{dy}{dx}$ and $\frac{d^2y}{dx^2}$ in Legendre's equation i.e. in (i), we get

$$\left(1-x^2\right)\left[\sum_{r=0}^{\infty}a_r(k-r)(k-r-1)x^{k-r-2}\right]-2x\left[\sum_{r=0}^{\infty}a_r(k-r)x^{k-r-1}\right]+n(n+1)\sum_{r=0}^{\infty}a_rx^{k-r}=0$$

on simplifying, we get

$$\sum_{r=0}^{\infty}\left[(k-r)(k-r-1)x^{k-r-2}-\{(k-r)(k-r-1)+2(k-r)-n(n+1)\}x^{k-r}\right]a_r=0$$

or $\sum_{r=0}^{\infty}\left[(k-r)(k-r-1)x^{k-r-2}+\{n(n+1)-(k-r)(k-r+1)\}x^{k-r}\right]a_r=0$ (iii)

Equation (iii) is an identity and therefore, coefficient of various powers of x should vanish. First we equate to zero the coefficients of x^k (the highest power of x) by putting r =0 in (iii) i.e.

$[n(n+1) - k(k+1)]\, a_0 = 0$

where a_0 being the coefficient of the first term of the series, cannot be zero, hence

$n(n+1) - k(k+1) = 0$ or $n^2 + n - k^2 - k = 0$

or $(n^2 - k^2) + (n - k) = 0$

i.e. $k = n$ or $k = -(n+1)$ (iv)

Now equating to zero the coefficient of x^{k-1} by putting r =1 in (iii), we get

$[n(n+1) - (k-1)k]a_1 = 0$

or $[(k+n)(k-n-1)]a_1 = 0$ (v)

From (v), $a_1 = 0$, since form (iv) $(k+n)(k-n-1) \neq 0$

To find a relation between successive coefficients, we equate to zero the coefficient of x^{k-r-2}

$(k-r)(k-r-1)a_r + \{n(n+1) - (k-r-2)(k-r-1)\}\, a_{r+2} = 0$

(As IInd terms of above relation is obtained by putting (r+2) for r in IInd terms of (iii))

$$\therefore a_{r+2} = -\frac{(k-r)(k-r-1)}{n(n+1)-(k-r-2)(k-r-1)}a_r \qquad \text{(vi)}$$

Again $[n(n+1) - \{(k-r)-2\}\{(k-r)-1\} = n^2 + n - (k-r)^2 + 3(k-r) - 2$

$= (n^2 + n - 2) + 3(k-r) - (k-r)^2$

$= (n+2)(n-1) + 3(k-r) - (k-r)^2$

$= -\{(k-r)^2 - 3(k-r) - (n+2)(n-1)\}$

$= -[\{k-r+(n-1)\}\{k-r-(n+1)\}]$ (vii)

Thus, from (vi) and (vii), we get

$$a_{r+2} = \frac{(k-r)(k-r-1)}{\{k-r+(n-1)\}\{k-r-(n+2)\}}a_r \qquad \text{(viii)}$$

we know that $a_1 = 0$, hence from (viii) we get

$a_3 = a_5 = a_7 \ldots\ldots\ldots\ldots\ldots = 0$ (ix)

Now for k =n or -(n +1), the two values of k given by (iv), we have the following two cases:

Case 1: when k = n

From (viii), we get

$$a_{r+2} = \frac{(n-r)(n-r-1)}{(2n-r-1)(-r-2)} a_r$$

putting r = 0, 2, 4, ...we get $a_2 = -\frac{n(n-1)}{(2n-1).2} a_0$

$$a_4 = -\frac{(n-2)(n-3)}{(2n-3).4} a_2 = \frac{n(n-1)(n-2)(n-3)}{(2n-1)(2n-3).2.4} a_0$$

Similarly, we get $a_6 = \frac{n(n-1)(n-2)(n-3)(n-4)(n-5)}{(2n-1)(2n-3)(2n-5).2.4.6} a_0$ and so on.

Also, we have $a_1 = a_3 = a_5 = ...=0$ by (ix)

putting the values of a_0, a_1, a_2, a_3 etc, and k =n, the series (ii) reduces to

$$y = a_0 \left[x^n - \frac{n(n-1)}{(2n-1).2} x^{n-2} + \frac{n(n-1)(n-2)(n-3)}{(2n-1)(2n-3)2.4} x^{n-4} - .... \right] \tag{x}$$

Which is a solution of Legendre's equation (i) in series of descending powers of x.

Case II. when k = -(n +1)

From (viii), we get

$$a_{r+2} = \frac{(n+r+1)(n+r+2)}{(r+2)(2n+r+3)} a_r$$

putting r =0, 2, 4, ..., we get

$$a_2 = \frac{(n+1)(n+2)}{2.4(2n+3)} a_0$$

$$a_4 = \frac{(n+3)(n+4)}{4(2n+5)} a_2$$

$$= \frac{(n+1)(n+2)(n+3)(n+4)}{2.4(2n+3)(2n+5)} a_0$$

and so on by putting the values of a_2

Also from (ix), we have $a_1 = a_3 = a_5 = ...= 0$

putting the values of a_0, a_2, a_4,............ etc, and k= -(n+1), the series (ii) reduce to

$$y = a_0\left[x^{-n-1} + \frac{(n+1)(n+2)}{2(2n+3)}.x^{-n-3} + \frac{(n+1)(n+2)(n+3)(n+4)}{2.4(2n+3)(2n+5)}x^{-n-5} + \dots\dots\right] \quad \text{(xi)}$$

which is another solution of Legendre's equation (i) in a series of descending powers of x.

Legendre's Functions

The two independent solution found in Legendre's equation are called Legendre's function. The solution of the Legendre's equation in series of descending powers of x (for first case) is

$$y = a_0\left[x^n - \frac{n(n-1)}{2(2n-1)}.x^{n-2} + \frac{n(n-1)(n-2)(n-3)}{2.4(2n-1)(2n-3)}x^{n-4} + \dots\dots\right]$$

If n is a positive integer and $a_0 = \frac{1.3.5\dots\dots(2n-1)}{\lfloor n}$ then above solution is denoted by $P_n(x)$ and is called Legendre's function of the first kind.

$$\therefore P_n(x) = \frac{1.3.5\dots\dots(2n-1)}{\lfloor n}\left[x^n - \frac{n(n-1)}{2(2n-1)}x^{n-2} + \frac{n(n-1)(n-2)(n-3)}{2.4(2n-1)(2n-3)}x^{n-4} + \dots\dots\right.$$

$P_n(x)$ is a terminating series which gives Legendre's polynomials for different values of n.

we can write

$$P_n(x) = \sum_{r=0}^{\left[\frac{n}{2}\right]}(-1)^r \frac{\lfloor(2n-2r)}{2^n\lfloor r\lfloor(n-r)\lfloor(n-2r)}x^{n-2r}$$

$$\text{where } \left[\frac{n}{2}\right] = \begin{cases} \frac{n}{2}, \text{if n is even} \\ \frac{(n-1)}{2}, \text{if n is odd} \end{cases}$$

Again when n is positive integer and

$a_0 = \frac{\lfloor n}{1.3.5\dots\dots(2n+1)}$ then the second solution

$y = a_0\left[x^{-n-1} + \frac{(n+1)(n+2)}{2(2n+3)}x^{-n-3} + \dots\dots\right]$ is denoted by $Q_n(x)$ and is called the Legendre's function of the second kind.

$$\therefore Q_n(x) = \frac{\lfloor n}{1.3.5.............(2n+1)}\left[x^{-n-1} + \frac{(n+1)(n+2)}{2(2n+3)}x^{-n-3} + \frac{(n+1)(n+2)(n+3)(n+4)}{2.4.(2n+3)(2n+5)}x^{-n-5} + ...\right]$$

$Q_n(x)$ is an infinite or non terminating series as n is positive.

General solution of Legendre's equation

The most general solution of Legendre's equation is

$y = AP_n(x) + BQ_n(x)$

where A and B are arbitrary constants.

Generating function for Legendre Polynomials

To show that $P_n(x)$ is the coefficient of z^n in the expansion of $(1-2xz+z^2)^{-1/2}$ in ascending powers of z. where $|z| < 1$, $|x| \leq 1$ (U.P.T.U. 2005)

Proof. Since $|z| < 1$ and $|x| \leq 1$, we have

$[1-2xz+z^2]^{-1/2} = [1-z(2x-z)]^{-1/2}$

$$= 1 + \frac{1}{2}z(2x-z) + \frac{1.3}{2.4}z^2(2z-z)^2 + ... + \frac{1.3.............(2n-3)}{2.4..............(2n-2)}z^{n-1}(2x-z)^{n-1}$$

$$+ \frac{1.3.............(2n-1)}{2.4..............(2n)}z^n(2x-z)^n + ... \quad \text{(i)}$$

Now coefficient of z^n in $\frac{1.3.5............(2n-1)}{2.4.6.............(2n)}z^n(2x-z)^n$ is obviously

$$= \frac{1.3.5.............(2n-1)}{2.4.6..............(2n)}(2x)^n = \frac{1.3.5.............(2n-1)}{(2.1)(2.2)(2.3)..............(2n)}(2x)^n$$

$$= \frac{1.3.5.............(2n-1)}{2^n(1.2.3..........n)}2^n x^n = \frac{1.3.5.............(2n-1)}{\lfloor n}x^n \quad \text{(ii)}$$

Coefficient of z^n in $\frac{1.3.5...............(2n-3)}{2.4.6...........(2n-2)}z^{n-1}(2x-z)^{n-1}$ is

$$= \frac{1.3.5...............(2n-3)}{(2.1)(2.2)...........2(n-1)}\left\{-(n-1)(2x)^{n-2}\right\}$$

$$= -\frac{1.3.5...............(2n-3)}{2^{n-1}\{1.2.3...........(n-1)\}}\cdot\frac{2n-1}{n}\cdot\frac{n}{2n-1}\{(n-1)2^{n-2}x^{n-2}\}$$

As on multiplying and dividing by $\frac{2n-1}{n}$

$$= -\frac{1.3.5..........(2n-1)}{\lfloor n}\frac{n(n-1)}{2(2n-1)}x^{n-2} \qquad \text{(iii)}$$

coefficient of z^n in

$$= \frac{1.3.5.............(2n-5)}{2.4.6................(2n-4)}z^{n-2}(2x-z)^{n-2}$$

$$= \frac{1.3.5..........(2n-5)}{2^{n-2}\{1.2.3..........(n-2)\}}\frac{(n-2)(n-3)}{\lfloor 2}(2x)^{n-2}$$

$$= \frac{1.3.5..........(2n-3)(2n-1)}{1.2.3..........(n-1)n}\frac{n(n-1)(n-2)(n-3)}{1.2.2^2(2n-3)(2n-1)}x^{n-4}$$

As multiplying and dividing by $\dfrac{(2n-3)(2n-1)}{(n-1)n}$

$$= \frac{1.3.5..........(2n-1)}{\lfloor n} \times \frac{n(n-1)(n-2)(n-3)}{2.4(2n-1)(2n-3)}x^{n-4} \qquad \text{(iv)}$$

Proceeding further as above, we conclude form (ii) (iii), (iv)..................that the coefficient of z^n in the expansion of (i) is

$$\frac{1.3.5..........(2n-1)}{\lfloor n}\left[x^n - \frac{n(n-1)}{2(2n-1)}x^{n-2} + \frac{n(n-1)(n-2)(n-3)}{2.4(2n-1)(2n-3)}x^{n-4} - ....\right]$$

which is clearly equal to $P_n(x)$

Thus, we can say that in the expansion of $(1-2xz+z^2)^{-1/2}$, $P_1(x)$, $P_2(x)$, $P_3(x)$, ...etc. will be the coefficients of z, z^2, z^3,...........respectively or $P_n(x)$ is the coefficient of z^n. That is why it is known as generating function of Legendre polynomials.

Hence $(1 - 2xz + z^2)^{-1/2} = 1+z P_1(x) + z^2 P_2(x) + z^3 P_3(x) + ... + z^n P_n(x) +............$

or $\left(1-2xz+z^2\right)^{-1/2} = \sum_{n=0}^{\infty} z^n P_n(x)$

Corollary 1. To show that $P_n(1) = 1$

Proof. Since we know

$(1 - 2xz + z^2)^{-1/2} = 1+z P_1(x) + z^2 P_2(x) + z^3 P_3(x) + ... + z^n P_n(x) +............$

putting $x = 1$ on both sides of above, we have

$(1 - 2z + z^2)^{-1/2} = 1+z P_1(1) + z^2 P_2(1) + z^3 P_3(1) + ... + z^n P_n(x) +............$

$(1-z)^{-1} = 1+z P_1(x) + z^2 P_2(1) + z^3 P_3(1) + ... + z^n P_n(1) +............$

or $1+z+z^2+z^3+\ldots\ldots+z^n+\ldots\ldots=1+zP_1(1)+z^2P_2(1)+z^3P_3(1)+\ldots\ldots+z^nP_n(1)+\ldots\ldots$

Equating the coefficient of z^n on both sides, we get

$P_n(1) = 1$

Corollary 2. Show that $\sum_{n=0}^{\infty} P_n(x) = \frac{1}{\sqrt{2-2x}}$

Proof. $(1-2xz+z^2)^{-1/2} = \sum_{n=0}^{\infty} z^n P_n(x)$

$$\Rightarrow (1-2x+1)^{-1/2} = \sum_{n=0}^{\infty} P_n(x) \quad \text{for } z=1$$

$$\Rightarrow \frac{1}{\sqrt{2-2x}} = \sum_{n=0}^{\infty} P_n(x)$$

EXAMPLE 2. To show that $P_n(-x) = (-1)^n P_n(x)$ and $P_n(-1) = (-1)^n$

Solution. we know that by generating function formula

$$(1-2xz+z^2)^{-1/2} = \sum_{n=0}^{\infty} z^n P_n(x) \quad \text{(i)}$$

putting –x for x in both sides of (i), we get

$$(1+2xz+z^2)^{-1/2} = \sum_{n=0}^{\infty} z^n P_n(-x) \quad \text{(ii)}$$

Again putting –z for z in (i), we get

$$(1+2xz+z^2)^{-1/2} = \sum_{n=0}^{\infty} (-1)^n z^n P_n(x) \quad \text{(iii)}$$

From (ii) and (iii) we get

$$\sum_{n=0}^{\infty} z^n P_n(-x) = \sum_{n=0}^{\infty} (-1)^n z^n P_n(x) \quad \text{(iv)}$$

Comparing the coefficients of z^n from both sides of (iv), we get

$P_n(-x) = (-1)^n P_n(x)$

Putting x =1 in the above result, we have

$P_n(-1) = (-1)^n P_n(1)$

or $P_n(-1) = (-1)^n \because P_n(1) = 1$

Theorem.

Rodrigue's Formula To show that

$$P_n(x)=\frac{1}{2^n\lfloor n}\frac{d^n}{dx^n}(x^2-1)^n$$ (I.A.S. 1979, U.P.T.U. 2004 2007)

Proof. By the definition of Legendre polynomial,

$$P_n(x)=\sum_{r=0}^{\left[\frac{n}{2}\right]}(-1)^r\frac{\lfloor(2n-2r)}{2^n\lfloor r\lfloor(n-r)\lfloor(n-2r)}x^{n-2r} \quad \text{(i)}$$

$$\text{where } \left[\frac{n}{2}\right]=\begin{cases}\frac{n}{2}, \text{if n is even}\\ \frac{(n-1)}{2}, \text{if n is odd}\end{cases} \quad \text{(ii)}$$

$$\therefore (x^2-1)^n=\sum_{r=0}^{n}{}^nc_r(x^2)^{n-r}(-1)^r, \text{(by Binomial theorem)}$$

$$=\sum_{r=0}^{n}{}^nc_r(-1)^r x^{2n-2r}$$

$$\therefore \frac{1}{2^n\lfloor n}\frac{d^n}{dx^n}(x^2-1)^n=\frac{1}{2^n.\lfloor n}\sum_{r=0}^{n}{}^nc_r(-1)^r\frac{d^n}{dx^n}x^{2n-2r} \quad \text{(iii)}$$

But $\frac{d^n}{dx^n}x^m=0$ if $m<n$; $\frac{d^n}{dx^n}x^m=\frac{\lfloor m}{\lfloor(m-n)}x^{m-n}$, if $m\geq n$ (iv)

$$\therefore \frac{d^n}{dx^n}x^{2n-2r}=0, \text{ if } 2n-2r<n \text{ i.e. } r>\frac{n}{2} \quad \text{(v)}$$

Making use of (v) in (iii), we see that we must replace $\sum_{r=0}^{n}$ by $\sum_{r=0}^{\frac{n}{2}}$ if n is even and by $\sum_{r=0}^{\frac{(n-1)}{2}}$ if n is odd, i.e. we must replace $\sum_{r=0}^{n}$ by $\sum_{r=0}^{\left[\frac{n}{2}\right]}$. Hence (iii) reduces to

$$\frac{1}{2^n.\lfloor n}\frac{d^n}{dx^n}(x^2-1)^n=\frac{1}{2^n.\lfloor n}\sum_{r=0}^{\left[\frac{n}{2}\right]}{}^nc_r(-1)^r\frac{d^n}{dx^n}x^{2n-2r}$$

$$=\frac{1}{2^n.\lfloor n}\sum_{r=0}^{\left[\frac{n}{2}\right]}{}^nc_r(-1)^r\frac{\lfloor(2n-2r)}{\lfloor(2n-2r-n)}x^{2n-2r-n} \quad [\text{using (iv)}]$$

$$= \sum_{r=0}^{\left[\frac{n}{2}\right]} \frac{1}{2^n \lfloor n} \frac{\lfloor n (-1)^r}{\lfloor r \lfloor (n-r)} \frac{\lfloor (2n-2r)}{\lfloor (n-2r)} x^{n-2r}$$

$= P_n(x)$ using (i)

Thus we have

$$P_n(x) = \frac{1}{2^n \lfloor n} \frac{d^n}{dx^n} (x^2 - 1)^n$$

Orthogonal properties of the Legendre polynomials.

Theorem Prove that

(a) $\int_{-1}^{1} P_m(x) P_n(x) dx = 0$ if $m \neq n$ (I.A.S. 1969, 1981)

(b) $\int_{-1}^{1} [P_n(x)]^2 dx = \frac{2}{2n+1}$ if $m = n$ (I.A.S. 1969, 1981, U.P.T.U. 2001, 2002, 2004, 2008)

OR

$$\int_{-1}^{1} P_m(x) P_n(x) dx = \frac{2}{2n+1} \delta_{mn}$$

where δ_{mn} is the kronecker delta defined by

$$\delta_{mn} = \begin{cases} 0 & \text{if } m \neq n \\ 1 & \text{if } m = n \end{cases}$$

Proof (a) since we know that Legendre's equation is

$$(1 - x^2)\frac{d^2y}{dx^2} - 2x\frac{dy}{dx} + n(n+1)y = 0$$

$$\Rightarrow \frac{d}{dx}\left\{(1 - x^2)\frac{dy}{dx}\right\} + n(n+1)y = 0 \quad \text{(i)}$$

Now since $P_n(x)$ and $P_m(x)$ are solution are solution of (i), therefore, we have

$$\frac{d}{dx}\left\{(1 - x^2)\frac{dP_n}{dx}\right\} + n(n+1)P_n = 0 \quad \text{(ii)}$$

$$\text{and } \frac{d}{dx}\left\{(1 - x^2)\frac{dP_m}{dx}\right\} + m(m+1)P_m = 0 \quad \text{(iii)}$$

Multiplying (ii) by P_m and (iii) by P_n and then subtracting we get

$$P_m \frac{d}{dx}\left\{(1-x^2)\frac{dP_n}{dx}\right\} - P_n \frac{d}{dx}\left\{(1-x^2)\frac{dP_m}{dx}\right\} + P_n P_m\left[n(n+1)-m(m+1)\right] = 0$$

Integrating above within given limits, we get

$$\int_{-1}^{1} P_m \frac{d}{dx}\left\{(1-x^2)\frac{dP_n}{dx}\right\}dx - \int_{-1}^{1} P_n \frac{d}{dx}\left\{(1-x^2)\frac{dP_m}{dx}\right\}dx + (n-m)(n+m+1)\int_{-1}^{1} P_n P_m dx = 0$$

on integrating by parts, we get

$$\left[P_m(1-x^2)\frac{dP_n}{dx}\right]_{-1}^{1} - \int_{-1}^{1}\left[\frac{dP_m}{dx}\left\{(1-x^2)\frac{dP_n}{dx}\right\}\right]dx - \left[P_n(1-x^2)\frac{dP_m}{dx}\right]_{-1}^{1}$$

$$+\int_{-1}^{1}\left[\frac{dP_n}{dx}\left\{(1-x^2)\frac{dP_m}{dx}\right\}\right]dx + (n-m)(n+m+1)\int_{-1}^{1} P_n P_m dx = 0$$

$$\text{or } 0 - \int_{-1}^{1}\left\{\frac{dP_m}{dx}(1-x^2)\frac{dP_n}{dx}\right\}dx - 0 + \int_{-1}^{1}\left\{\frac{dP_n}{dx}(1-x^2)\frac{dP_m}{dx}\right\}dx + (n-m)(n+m+1)\int_{-1}^{1} P_n P_m dx = 0$$

$$\text{or } (n-m)(n+m+1)\int_{-1}^{1} P_n P_m dx = 0$$

$$\int_{-1}^{1} P_n P_m dx = 0 \text{ if } m \neq n$$

(b) we know that $(1-2xz+z^2)^{-1/2} = \sum_{n=0}^{\infty} z^n P_n(x)$

squaring both sides, we get

$$(1-2xz+z^2)^{-1} = \sum_{0}^{\infty} z^{2n}\left[P_n(x)\right]^2 + 2\sum_{0}^{\infty} z^{m+n} P_m(x) P_n(x)$$

Integrating both side w.r.t. "x" between the limits –1 to +1, we have

$$\int_{-1}^{1}\frac{dx}{(1-2xz+z^2)} = \sum_{0}^{\infty}\int_{-1}^{1} z^{2n}\left[P_n(x)\right]^2 dx + 2\sum_{0}^{\infty}\int_{-1}^{1} z^{m+n} P_m(x) P_n(x) dx$$

$$\text{or } \int_{-1}^{1}\frac{dx}{(1-2xz+z^2)} = \sum_{0}^{\infty}\int_{-1}^{1} z^{2n}\left[P_n(x)\right]^2 dx + 0 \text{ by part (a)}$$

$$\text{or } -\frac{1}{2z}\left[\log(1-2xz+z^2)\right]_{-1}^{1} = \sum_{0}^{\infty}\int_{-1}^{1} z^{2n}\left[P_n(x)\right]^2 dx$$

$$\text{or } -\frac{1}{2z}\left[\log(1-z^2) - \log(1+z)^2\right] = \sum_{0}^{\infty}\int_{-1}^{1} z^{2n}\left[P_n(x)\right]^2 dx$$

$$\text{or } \frac{1}{z}\left[\log(1+z) - \log(1-z)\right] = \sum_{0}^{\infty}\int_{-1}^{1} z^{2n}\left[P_n(x)\right]^2 dx$$

$$\text{or } \frac{2}{z}\left[z+\frac{z^3}{3}+\frac{z^5}{5}+\ldots\ldots\ldots\ldots\ldots+\frac{z^{2n+1}}{2n+1}+\ldots\ldots\ldots\ldots\right]=\sum_0^{\infty}\int_{-1}^{1} z^{2n}\left[P_n(x)\right]^2 dx$$

$$\text{or } 2\left[1+\frac{z^2}{3}+\frac{z^4}{5}+\ldots\ldots\ldots\ldots\ldots+\frac{z^{2n}}{2n+1}+\ldots\ldots\ldots\ldots\right]=\sum_0^{\infty}\int_{-1}^{1} z^{2n}\left[P_n(x)\right]^2 dx$$

Equating the coefficient of z^{2n} from both sides, we get

$$\frac{2}{2n+1}=\int_{-1}^{1}\left[P_n(x)\right]^2 dx$$

$$\text{or } \int_{-1}^{1}\left[P_n(x)\right]^2 dx=\frac{2}{2n+1}$$

Combining (a) and (b) we have $\int_{-1}^{1} P_n(x)P_m(x)dx=\frac{2}{2n+1}\delta_{mn}$

(Raj S.L.E.T 2001; M.P.S.L.E.T 2000)

RECURRENCE FORMULAE FOR $P_n(x)$

(i) $nP_n = (2n-1)x P_{n-1} - (n-1) P_{n-2}$ (I.A.S. 1973, 79, U.P.T.U. 2007)

Proof. we know that

$$\left(1-2xz+z^2\right)^{-1/2}=\sum_{n=0}^{\infty} z^n P_n(x) \tag{1}$$

Differentiating both sides of (1) w.r.t. "z", we get

$$-\frac{1}{2}\left(1-2xz+z^2\right)^{-3/2}(-2x+2z)=\sum_{n=0}^{\infty} nz^{n-1}P_n(x) \tag{2}$$

Multiplying both sides of (2) by $(1-2xz + z^2)$, we have

$$(x-z)\left(1-2xz+z^2\right)^{-3/2}\left(1-2xz+2z^2\right)=(1-2xz+z^2)\sum_{n=0}^{\infty} nz^{n-1}P_n(x)$$

$$\text{or } (x-z)\sum_{n=0}^{\infty} z^n P_n(x)=\left(1-2xz+z^2\right)\sum_{n=0}^{\infty} nz^{n-1}P_n(x) \text{ by (1)}$$

$$\text{or } x\sum_{n=0}^{\infty} z^n P_n-\sum_{n=0}^{\infty} z^{n+1}P_n(x)=\sum_{n=0}^{\infty} nz^{n-1}p_n-2x\sum_{n=0}^{\infty} nz^n p_n+\sum_{n=0}^{\infty} nz^{n-1}p_n$$

Equating coefficients of z^n from both sides, we get

$x P_n - P_{n-1} = (n+1) P_{n+1} - 2xn P_n + (n-1) P_{n-1}$

or $(n+1) P_{n+1} = (2n + 1) x P_n - nP_{n-1}$ (U.P.T.U. 2007) (3)

Replacing n by (n −1) in (3), we get

$nP_n = (2n-1) \times P_{n-1} - (n-1)P_{n-2}$ (4)

Again (3) can be rearrange to give another form

$$xP_n = \frac{n+1}{2n+1}P_{n+1} + \frac{n}{2n+1}P_{n-1}$$

(ii) $nP_n = xP'_n - P'_{n-1}$ (U.P.T.U. 2006, 2009)

Proof. we know that

$$(1-2xz+z^2)^{-1/2} = \sum_{n=0}^{\infty} z^n P_n(x) \quad (1)$$

Differentiating (1) w.r.t. z, we get

$$-\frac{1}{2}(1-2xz+z^2)^{-3/2}(-2x+2z) = \sum_{n=0}^{\infty} nz^{n-1}P_n(x)$$

$$\text{or } (x-z)(1-2xz+z^2)^{-3/2} = \sum_{n=0}^{\infty} nz^{n-1}P_n(x) \quad (2)$$

Now differentiating (1) w.r.t. x we get

$$z(1-2xz+z^2)^{-3/2} = \sum_{n=0}^{\infty} z^n P'_n(x) \quad (3)$$

Now on dividing (2) by (3) we get

$$\frac{x-z}{z} = \frac{\sum nz^{n-1}P_n(x)}{\sum z^n P'_n(x)}$$

$$\text{or } (x-z)\sum z^n p'_n(x) = z\sum nz^{n-1}p_n(x)$$

on equating the coefficients of z^n from both sides, we have

$$xP'_n(x) - P'_{n-1}(x) = nP_n(x)$$

(iii) $(2n+1)P_n = P'_{n+1} - P'_{n-1}$

Proof. From 1st recurrence formula we have

$(n+1)P_{n+1} = (2n+1)x\, P_n - n\, P_{n-1}$

on differentiating above w.r.t. "x" we have

$$(n+1)P'_{n+1} = (2n+1)P_n + (2n+1)x\,P'_n - nP'_{n-1} \quad (1)$$

Again from 2nd recurrence formula we have

$$nP_n = xP'_n - P'_{n-1}$$

or $xP'_n = nP_n + P'_{n-1}$ (2)

Now on putting the value of xP'_n from (2) in (1) we get

$$(n+1)P'_{n+1} = (2n + 1)P_n + (2n+1)(nP_n + P'_{n-1}) - nP'_{n-1}$$

or $(n+1)P'_{n+1} - (n + 1)P'_{n-1} = (2n+1)(n+1)P_n$

or $P'_{n+1} - P'_{n-1} = (2n+1)P_n$

Replacing (n–1) for n in the above relation, we have

$P'_n - P'_{n-2} = (2n-1)P_{n-1}$

(iv) $P'_{n+1} - xP'_n = (n+1)P_n$ (I.A.S. 1983)

Proof. From 1st recurrene formula, we have

$(n+1)P_{n+1} = (2n + 1)x\, P_n - n\, P_{n-1}$

Differentiating above w.r.t. x we get

$$\begin{aligned}(n+1)P'_{n+1} &= (2n+1)P_n + (2n+1)xP'_n - nP'_{n-1}\\ &= (2n+1)P_n + (n+1)xP'_n + n(xP'_n - P'_{n-1})\\ &= (2n+1)P_n + (n+1)xP'_n + n(nP_n) \because xP'_n - P'_{n-1} = nP_n\end{aligned}$$

$= (2n+1)P_n + (n+1)xP'_n + n^2P_n$

$= (n+1)xP'_n + (n^2 + 2n + 1)P_n$

$= (n+1)xP'_n + (n+1)^2 P_n$

or $P'_{n+1} = xP'_n + (n+1)P_n$

or $P'_{n+1} - xP'_n = (n+1)P_n$

Replacing (n–1) for n in the above we have

$P'_n - xP'_{n-1} = nP_{n-1}$ (A)

(v) $(1-x^2)P'_n = n(P_{n-1} - xP_n)$ (U.P.T.U. 2007)

Proof. From relation (A) of 4th recurrence formula, we have

$nP_{n+1} = P'_n - xP'_{n-1}$ (1)

Again from 2nd recurrence formula, we have

$nP'_n = xP'_n - P'_{n-1}$ (2)

Now on multiplying (2) by x and then subtracting from (1) we get

$x(P_{n-1} - xP_n) = (1 - x^2)\, P_n'$

or $\left(1-x^2\right)P_n' = n\left(P_{n-1} - xP_n\right)$

(vi) $\left(1-x^2\right)P_n' = (n+1)\left(xP_n - P_{n+1}\right)$

Proof. From first recurrence formula, we have

$(2n+1)xP_n = (n+1)\, P_{n+1} + n\, P_{n-1}$

$\Rightarrow \quad (n + 1 + n)x\, P_n = (n+1)P_{n+1} + nP_{n-1}$

$\Rightarrow \quad (n+1)\,(x\, P_n - P_{n+1}) = n\,(P_{n-1} - x\, P_n)$ (1)

From fifth recurrence formula we have

$(1 - x^2)\, P_n' = n\,(P_{n-1} - x\, P_n)$ (2)

From (1) and (2), we have

$(1 - x^2)\, P_n' = (n+1)\,(xP_n - P_{n+1})$

Legendre Polynomials

From Rodrigue's formula, we have

$$P_n(x) = \frac{1}{2^n \lfloor n} \frac{d^n}{dx^n}\left(x^2-1\right)^n$$

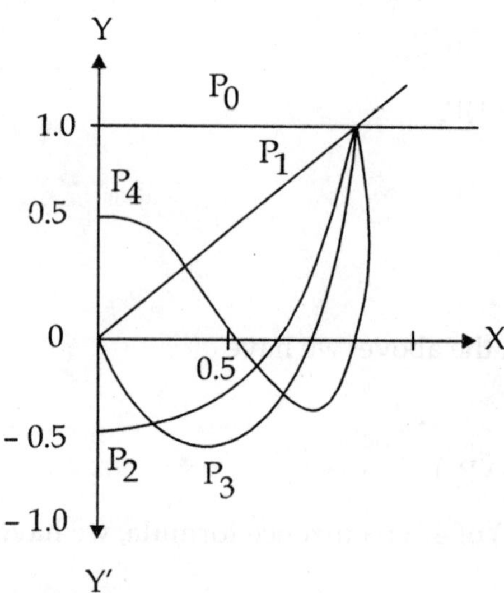

For n =0, $P_0 = \dfrac{1}{2^0 \lfloor 0} = 1 \Rightarrow P_0 = 1$

For $n=1, P_1 = \frac{1}{2^1\lfloor 1}\frac{d}{dx}(x^2-1) = \frac{1}{2}(2x) = x \Rightarrow P_1 = x$

For $n=2, P_2 = \frac{1}{2^2\lfloor 2}\frac{d^2}{dx^2}(x^2-1)^2 = \frac{1}{8}\frac{d}{dx}\left[2(x^2-1)(2x)\right]$

$= \frac{1}{2}\left[(x^2-1).1 + 2x.x\right]$

$\Rightarrow \quad P_2 = \frac{1}{2}(3x^2-1)$

Similarly

$P_3 = \frac{1}{2}(5x^3-3x)$, $P_4 = \frac{1}{8}(35x^4-30x^2+3)$, $P_5 = \frac{1}{8}(63x^5-70x^3+15x)$ and so on......

EXAMPLE 7 Express $P(x) = x^4 + 2x^3 + 2x^2 - x - 3$ in the terms of Legendre's polynomials.

Solution Let $x^4 + 2x^3 + 2x^2 - x - 3 = aP_4(x) + bP_3(x) + cP_2(x) + d\,P_1(x) + eP_0(x)$ (i)

$\Rightarrow x^4 + 2x^3 + 2x^2 - x - 3 = a.\ \frac{1}{8}(35x^4-30x^2+3) + \frac{b}{2}(5x^3-3x) + \frac{c}{2}(3x^2-1) + d.x + e$

Equating the coefficients of like power of x, we have

$1 = \frac{35}{8}a \quad \Rightarrow \quad a = \frac{8}{35}$

$2 = \frac{5b}{2} \quad \Rightarrow \quad b = \frac{4}{5}$

$2 = -\frac{30}{8}a + \frac{3c}{2} \Rightarrow -\frac{30}{8}x\frac{8}{35} + \frac{3c}{2} = 2 \Rightarrow c = \frac{40}{21}$

$-1 = -\frac{3b}{2} + d \Rightarrow \frac{-3}{2}x\frac{4}{5} + d = -1 \Rightarrow d = \frac{1}{5}$

and $-3 = \frac{3a}{8} - \frac{c}{2} + e \Rightarrow \frac{3}{8}x\frac{8}{35} - \frac{1}{2}x\frac{40}{21} + e = -3$

$\Rightarrow e = -\frac{224}{105}$

Putting these values of a, b, c, d and e in equation (i) we get

$x^4 + 2x^3 + 2x^2 - x - 3 = \frac{8}{35}P_4(x)\frac{4}{5}P_3(x) + \frac{40}{21}P_2(x) + \frac{1}{5}P_1(x) - \frac{224}{105}P_0(x)$

EXAMPLE 3 Express the polynomial $f(x)= 4x^3-2x^2 -3x + 8$ in terms of Legendre's polynomials (U.P.T.U. 2009)

Solution

Let $4x^3-2x^2 -3x + 8 = ap_3(x) + bp_2(x) + cp_1(x) + dp_0(x)$ (i)

$$\Rightarrow 4x^3-2x^2-3x+8 = a.\frac{1}{2}(5x^3-3x)+b\frac{1}{2}(3x^2-1)+cx+d.1$$

$$\Rightarrow 4x^3-2x^2-3x+8 = \frac{5}{2}x^3a+\frac{3}{2}x^2b+\left(c-\frac{3}{2}a\right)x+\left(d-\frac{b}{2}\right)$$

Equating the coefficients of like powers of x, we have

$$a=\frac{8}{5}, b=-\frac{4}{3}, c=-\frac{3}{5}, d=\frac{22}{3}$$

putting these values of a,b,c,d in equation (i) we get

$$4x^3-2x^2-3x+8 = \frac{8}{5}p_3(x)-\frac{4}{3}p_2(x)-\frac{3}{5}P_1(x)+\frac{22}{3}p_0(x)$$

EXAMPLE 4. Show that

$$\frac{2}{5}p_3(x)+\frac{3}{5}p_1(x)=x^3$$

Solution we have $p_1(x)=x,\ p_3(x)=\frac{1}{2}(5x^3-3x)$

Substituting in $\frac{2}{5}p_3(x)+\frac{3}{5}p_1(x)$, we get $\frac{2}{5}p_3(x)+\frac{3}{5}p_1(x)=\frac{2}{5}.\frac{1}{2}[5x^3-3x]+\frac{3}{5}(x)$

$$=\frac{1}{5}(5x^3-3x)+\frac{3x}{5}$$

$$=\frac{1}{5}[5x^3-3x+3x]$$

$$=x^3$$

EXAMPLE 5. Show that

$$\int_{-1}^{1}(1-x^2)p'_m p'_n dx=\begin{cases}0 & \text{when } m\neq n\\ \dfrac{2n(n+1)}{2n+1} & \text{when } m=n\end{cases}$$ (U.P.T.U. 2006)

where dashes denote differentiation w.r.t. x.

Solution we have

$$\int_{-1}^{1}(1-x^2)p_m' p_n' dx = \left[(1-x^2 p_m' p_n)\right]_{-1}^{1} - \int_{-1}^{1}\left[p_n \frac{d}{dx}\{(1-x^2)p_m'\}\right]dx$$

$$= 0 - \int_{-1}^{1} p_n \frac{d}{dx}\{(1-x^2)p_m'\}dx$$

$$\Rightarrow \int_{-1}^{1}(1-x^2)p_m' p_n' dx = -\int_{-1}^{1} p_n \frac{d}{dx}\{(1-x^2)p_m'\}dx \qquad \text{(i)}$$

From Legendre's equation, we have

$$\frac{d}{dx}\left\{(1-x^2)\frac{dy}{dx}\right\} + n(n+1)y = 0$$

$$\therefore \frac{d}{dx}\{(1-x^2)p_m'\} = -m(m+1)P_m \text{ As replace n by m so } P_n \to P_m \qquad \text{(ii)}$$

From (i) and (ii), we get

$$\int_{-1}^{1}(1-x^2)p_m' p_n' dx = -\int_{-1}^{1}[-p_n m(m+1)P_m]dx$$

$$\Rightarrow \int_{-1}^{1}(1-x^2)p_m' p_n' dx = m(m+1)\int_{-1}^{1} p_m p_n dx \qquad \text{(iii)}$$

$$\Rightarrow \int_{-1}^{1}(1-x^2)p_m' p_n' dx = 0,\ m \neq n \quad \text{As } \int_{-1}^{1} p_m p_n dx = 0,\ m \neq n$$

Again from (iii) for m = n

$$\int_{-1}^{1}(1-x^2)(p_n')^2 dx = n(n+1)\int_{-1}^{1} p_n^2 dx$$

$$= n(n+1)\frac{2}{2n+1} \text{As} \int_{-1}^{1} p_n^2 dx = \frac{2}{2n+1}$$

$$\Rightarrow \int_{-1}^{1}(1-x^2)(p_n')^2 dx = \frac{2n(n+1)}{2n+1}$$

Thus, we have

$$\int_{-1}^{1}(1-x^2)p_m' p_n' dx = \begin{cases} 0 & \text{if } m \neq n \\ \dfrac{2n(n+1)}{2n+1} & \text{if } m = n \end{cases}$$

EXAMPLE 6 Let $P_n(x)$ be the Legendre polynomial of degree n. Show that for any function f(x) for which nth derivative continuous.

$$\int_{-1}^{1} f(x)p_n dx = \frac{(-1)^n}{2^n \lfloor n} \int_{-1}^{1}(x^2-1)f^{(n)}(x)dx \qquad \text{(U.P.T.U. 2004)}$$

Solution

$$\int_{-1}^{1} f(x)p_n(x)dx = \frac{1}{2^n \lfloor n} \int_{-1}^{1} \frac{d^n}{dx^n}(x^2-1)dx \text{ By Rodrigue's formula}$$

$$= \frac{1}{2^n \lfloor n}\left[\left\{f(x)\frac{d^{n-1}}{dx^{n-1}}(x^2-1)^n\right\}_{-1}^{1} - \int_{-1}^{1} f'(x)\frac{d^{n-1}}{dx^{n-1}}(x^2-1)^n dx\right]$$

$$= \frac{1}{2^n \lfloor n}\left[0 - \int_{-1}^{1} f'(x)\frac{d^{n-1}}{dx^{n-1}}(x^2-1)^n dx\right]$$

$$= \frac{(-1)^1}{2^n \lfloor n}\int_{-1}^{1} f'(x)\frac{d^{n-1}}{dx^{n-1}}(x^2-1)^n dx$$

$$= \frac{(-1)^2}{2^n \lfloor n}\int_{-1}^{1} f''\frac{d^{n-2}}{dx^{n-2}}(x^2-1)^n dx \text{ As integrating by parts again}$$

$$\Rightarrow \int_{-1}^{1} f(x)p_n dx = \frac{(-1)^n}{2^n \lfloor n}\int_{-1}^{1} f^n(x)(x^2-1)^n dx \text{ As integrating by parts till n times.}$$

EXAMPLE 7 Show that

(i) $P_n(0) = 0$ for n is odd

or

$P_{2n+1}(0) = 0$ (U.P.T.U. 2005)

(ii) $p_n(0) = \dfrac{(-1)^{n/2}\lfloor n}{2^n\{\lfloor n/2\}^2}$, for n is even

or $p_{2n}(0) = (-1)^n \dfrac{1.3.5.............(2n-1)}{2.4.6................2n}$ (U.P.T.U. 2008)

Solution (i) we know that

$$\sum_{n=0}^{\infty} z^n p_n(x) = (1 - 2xz + z^2)^{-1/2}$$

putting x = 0, we get

$$\sum_{n=0}^{\infty} z^n p_n(0) = (1+z^2)^{-1/2} = 1 - \frac{1}{2}z^2 + \frac{1.3}{2.4}z^4 + ... + (-1)^r \frac{1.3............(2r-1)}{2.4............(2r)} z^{2r} + ... \quad (i)$$

Equating the coefficient of z^{2n+1} on both sides, we get

$P_{2n+1}(0) = 0$ or $P_n(0) = 0$, when n is odd

(ii) Now equating the coefficient of z^{2m} on both sides in equation (i), we get

$$p_{2m}(0)=\frac{1.3............(2m-1)}{2.4............(2m)}(-1)^m=(-1)^m\frac{\lfloor 2m}{2^{2m}(\lfloor m)^2}$$

i.e. when n = 2m, then

$$p_n(0)=\frac{(-1)^{n/2}\lfloor n}{2^n\{\lfloor n/2\}^2}$$

Fourier - Legendre expansion of (x).

if f(x) be a function defined from x = -1 to x = 1, we can write

$$f(x)=\sum_{n=0}^{\infty}a_np_n(x)=a_0p_0(x)+a_1p_1(x)+a_2p_2(x)+.................. \quad \text{(i)}$$

To determine the coefficient a_n, multiply both sides by $p_n(x)$ and integrate from -1 to 1, Then

$$\int_{-1}^{1}p_m(x)p_n(x)dx=0 \qquad (m\neq n)$$

and $\int_{-1}^{1}p_n^2(x)dx=\dfrac{2}{2n+1} \qquad (m=n)$ give

$$\int_{-1}^{1}f(x)p_n(x)dx=a_n\int_{-1}^{1}p_n^2(x)dx=\frac{2a_n}{2n+1}$$

or $a_n=\left(n+\dfrac{1}{2}\right)\int_{-1}^{1}f(x)p_n(x)dx \qquad$ (ii)

Equation (ii) is known as Fourier Legendre expansion of f(x)

EXAMPLE 8 If $f(x)=\begin{cases}0, & -1<x<0\\ x, & 0<x<1\end{cases}$

show that $f(x)=\dfrac{1}{4}p_0(x)+\dfrac{1}{2}p_1(x)+\dfrac{5}{16}p_2(x)-\dfrac{3}{32}p_4(x)+............$ (U.P.T.U. 2003)

Solution Let $f(x)=\sum_{n=0}^{\infty}a_np_n(x) \qquad$ (i)

Then a_n is given by $a_n=\left(n+\dfrac{1}{2}\right)\int_{-1}^{1}f(x)p_n(x)dx$

$$=\left(n+\frac{1}{2}\right)\left[\int_{-1}^{0}0.p_n(x)dx+\int_{0}^{1}xp_n(x)dx\right]$$

$$=\left(n+\frac{1}{2}\right)\int_0^1 xp_n(x)dx$$

putting n =0, 1,2, 3, ……………..in (ii) we have

$$\therefore a_0=\frac{1}{2}\int_0^1 xp_0(x)dx=\frac{1}{2}\int_{-1}^1 x\,dx=\frac{1}{4} \text{ As } p_0(x)=1$$

$$a_1=\frac{3}{2}\int_0^1 xp_1(x)dx=\frac{3}{2}\left(\frac{x^3}{3}\right)_0^1=\frac{1}{2} \text{ As } p_1(x)=x$$

$$a_2=\frac{5}{2}\int_0^1 xp_2(x)dx=\frac{5}{2}\int_0^1 x.\left(\frac{3x^2-1}{2}\right)dx$$

$$=\frac{5}{4}\left[\frac{3x^4}{4}-\frac{x^2}{2}\right]_0^1=\frac{5}{16}$$

$$a_3=\frac{7}{2}\int_0^1 xp_3(x)dx=\frac{7}{2}\int_0^1 x.\frac{(5x^3-3x)}{2}dx$$

$$=\frac{7}{4}\left[\frac{5x^5}{5}-\frac{3x^3}{3}\right]_0^1=0$$

$$a_4=\frac{9}{2}\int_0^1 xP_4(x)\,dx=\frac{9}{2}\int_0^1 x.\frac{(35x^4-30x^2+3)}{8}dx$$

$$=\frac{9}{16}\left[35\frac{x^6}{6}-35\frac{x^4}{4}+3\frac{x^2}{2}\right]_0^1=-\frac{3}{32} \text{ and so on.}$$

putting these values in equation (i) we get

$$f(x)=\frac{1}{4}p_0(x)+\frac{1}{2}p_1(x)+\frac{5}{16}p_2(x)-\frac{3}{32}p_4(x)+\dots\dots\dots$$

EXAMPLE 9 Express the function

$$f(x)=\begin{cases}0 & -1\ x\le 0\\ x & 0<x<1\end{cases}$$

in Fourier Legendre expansion.

Solution Let $f(x)=\sum_{n=0}^{\infty}a_n p_n(x)$

where $a_n=\left(n+\frac{1}{2}\right)\int_{-1}^1 f(x)p_n(x)dx$

$$=\left(n+\frac{1}{2}\right)\int_{-1}^{0} 0.p_n(x)dx+\left(n+\frac{1}{2}\right)\int_{0}^{1} x.p_n(x)dx$$

$$=\left(n+\frac{1}{2}\right)\int_{0}^{1} x.p_n(x)dx$$

$$\therefore a_0=\frac{1}{2}\int_0^1 x.p_0(x)dx=\frac{1}{2}\int_0^1 x.1dx$$

$$=\frac{1}{2}\int_0^1 x.dx=\frac{1}{2}\left[\frac{x^2}{2}\right]_0^1=\frac{1}{4}$$

$$a_1=\frac{3}{2}\int_0^1 x.p_1(x)dx=\frac{3}{2}\int_0^1 x.xdx$$

$$=\frac{3}{2}\int_0^1 x^2.dx=\frac{3}{2}\left[\frac{x^3}{3}\right]_0^1=\frac{1}{2}$$

$$a_2=\frac{5}{2}\int_0^1 x.p_2(x)dx=\frac{5}{2}\int_0^1 x.\frac{3x^2-1}{2}dx$$

$$=\frac{5}{4}\int_0^1\left(3x^3-x\right)dx=\frac{5}{4}\left[\frac{3x^4}{4}-\frac{x^2}{2}\right]_0^1=\frac{5}{16}$$

$$a_3=\frac{7}{2}\int_0^1 x.p_3(x)dx=\frac{7}{2}\int_0^1 x.\frac{5x^3-3x}{2}dx$$

$$=\frac{7}{4}\int_0^1\left(5x^4-3x^2\right)dx=\frac{7}{4}\left(x^5-x^3\right)_0^1=0$$

$$a_4=\frac{9}{2}\int_0^1 x.p_4(x)dx=\frac{9}{2}\int_0^1 x.\frac{35x^4-30x^2+3}{8}dx$$

$$=\frac{9}{16}\int_0^1\left(35x^5-30x^3+3x\right)dx=\frac{9}{16}\left[\frac{35x^6}{6}-\frac{15x^4}{2}+\frac{3x^2}{2}\right]_0^1=-\frac{3}{32}$$

$$a_5=\frac{11}{12}\int_0^1 x.p_5(x)dx=\frac{11}{2}\int_0^1 x.\frac{63x^5-70x^3+15x}{8}dx$$

$$=\frac{11}{16}\int_0^1\left(63x^6-70x^4+15x^2\right)dx$$

$$=\frac{11}{16}\left[9x^7-14x^5-5x^3\right]_0^1$$

$$=0$$

$$\Rightarrow f(x)=\frac{1}{4}p_0(x)+\frac{1}{2}p_1(x)+\frac{5}{16}p_2(x)-\frac{3}{32}p_4(x)+\ldots\ldots\ldots\ldots$$

EXERCISE

Solve the following differential equations by power series method.

1. $y'' + xy' + x^2y = 0$

Ans. $y = a_0\left(1 - \frac{x^4}{12} + \frac{x^6}{90} - ....\right) + a_1\left(x - \frac{x^3}{2} + \frac{x^5}{40} + ...\right)$ (U.P.T.U. 2005)

2. $9x(1-x)\frac{d^2y}{dx^2} - 12\frac{dy}{dx} + 4y = 0$ (U.P.T.U. 2008)

$y = a_0\left(1 + \frac{1}{3}x + \frac{2}{9}x^2 + \frac{14}{81}x^3 + ...\right) + a_0x^{7/3}\left(1 + \frac{4}{5}x + \frac{44}{65}x^2 + ...\right)$

3. $2x(1-x)\frac{d^2y}{dx^2} + (1-x)\frac{dy}{dx} + 3y = 0$ (U.P.T.U. 2004)

$y = c_1\left(1 - 3x + \frac{3x^2}{1.3} + \frac{3x^3}{3.5} + \frac{3x^4}{5.7} + ...\right) + c_2\sqrt{x}(1-x)$

4. Using method of Frobenius, obtain series solution in power of x for.

$x(1+x)\frac{d^2y}{dx^2} + (5+x)\frac{dy}{dx} - 4y = 0$

Ans. $y = a_0 x^{-4}(1+4x+5x^2) + a_4\left(1 + \frac{4}{5}x + \frac{1}{5}x^2\right)$

5. Solve in series the following differential equation

$x\frac{d^2y}{dx^2} + \frac{dy}{dx} - y = 0$ (U.P.T.U. (C.O.) 2003)

Ans. $y(a + b\log x)\left[1 + x + \frac{x^2}{(21)^2} + \frac{x^3}{(31)^2} + ...\right]\left[-2bx + \frac{1}{(21)^2}\left(1 + \frac{1}{2}\right)x^2 + \frac{1}{(31)^2}\left(1 + \frac{1}{2} + \frac{1}{3}\right)x^3 + ...\right]$

6. Prove that when n is positive integer

(i) $J_{-n}(x) = (-1)^n J_n(x)$ (U.P.T.U. (C.O.) 2005)

(ii) $J_n(-x) = (-1)^n J_n(x)$ for positive and negative integer

7. Prove that

(i) $J_0' = -J_1$

(ii) $J_2 = J_0'' - x^{-1}J_0'$

(iii) $J_2 - J_0 = 2J_0''$

(iv) $J_3 + 3J_0' + 4J_0''' = 0$ (U.P.T.U. 2001)

8. Prove that

$$\frac{d}{dx}(xJ_nJ_{n+1}) = x\left(J_n^2 - J_{n+1}^2\right)$$

9. Prove that

$$\frac{1}{2}xJ_n = (n+1)J_{n+1} - (n+3)J_{n+3} + (n+5)J_{n+5} - \ldots\ldots\ldots\ldots$$

10. Prove that

(i) $p_n'(1) = \frac{n}{2}(n+1)$ (U.P.T.U. 2003)

(ii) $p_n'(-1) = (-1)^{n-1}\frac{1}{2}n(n+1)$

11. Prove that

(i) $\int_{-1}^{1} p_n(x)dx = 0,\ n \neq 0$

(ii) $\int_{-1}^{1} p_0(x)dx = 2$

12. Express $f(x) = 4x^3 + 6x^2 + 7x + 2$ in terms of Legendre polynomials.

Ans. $4x^3 + 6x^2 + 7x + 2 = \frac{8}{5}p_3(x) + 4p_2(x) + \frac{47}{5}p_1(x) + 4p_0(x)$

13. prove that $\frac{1+z}{z\sqrt{1-2xz+z^2}} - \frac{1}{z} = \sum_{n=0}^{\infty}(p_n + p_{n+1})z^n$

14. Evaluate $\int_{-1}^{1} x^2 p_n^2(x)dx$

Ans. $\frac{2(n+1)^2}{(2n+1)^2(2n+3)} + \frac{2n^2}{(2n+1)^2(2n-1)}$

15. Using Rodrigue's formula, obtain expressions for $p_0(x)$, $p_1(x)$, $p_2(x)$, $p_3(x)$, and $p_4(x)$. Hence express x, x^2, x^3 and x^4 in terms of Legendre polynomials.

Ans. $p_0(x) = 1$, $p_1(x) = x$, $P_2(x) = \frac{1}{2}(3x^2 - 1), p_3(x) = \frac{1}{2}(5x^3 - 3x), p_4(x) = \frac{1}{8}(35x^4 - 30x^2 + 3)$

$\& \; x^4 = \frac{8}{35}p_4(x) + \frac{4}{7}p_2(x) + \frac{1}{5}p_0(x)$

16. Show tat

(i) $p_2(\cos\theta) = \frac{1}{4}(1 + 3\cos 2\theta)$

(ii) $p_3(\cos\theta) = \frac{1}{8}(3\cos\theta + 5\cos 3\theta)$

17. Obtain the Fourier-Legendre expansion of $f(x) = 3x^2 - 3x + 1$

Ans. $f(x) = 2\,p_0(x) - 3\,p_1(x) + 2P_2(x)$

OBJECTIVE TYPE OF QUESTIONS

1. The singular points of the differential equation

$(1-x^2)\frac{d^2y}{dx^2} - 2x\frac{dy}{dx} + n(n+1)y = 0$ is

(a) $x = 0$ (b) $x = \pm 1$

(c) $x = 1$ (d) $x = 2$

Ans. (b)

2. Frobenius was a

(a) Geraman mathematician (b) Indian mathematician

(c) British mathematician (d) American mathematician

Ans. (a)

3. Prower series about a point x = a is defined as

(a) $\sum_{n=0}^{\infty} a_n x$ (b) $\sum_{n=0}^{\infty} a_n x^n$

(c) $\sum_{n=0}^{\infty} a_n (x-a)^n$ (d) $\sum_{n=0}^{\infty} a_n (x+a)^n$

Ans. (c)

4. If J_0 and J_1 are Bessel functions, then $J'_1(x)$ is given by

(a) $J_1(x) - \frac{1}{x}J_1(x)$ (b) $J_0(x) + \frac{1}{x}J_1(x)$

(c) $J_0(x) - \frac{1}{x^2}J_1(x)$ (d) None of these

Ans. (a)

5. If $J_{n+1}(x) = \frac{2}{x}J_n(x) - J_0(x)$, then n is

(a) 0 (b) 2
(c) –1 (d) None of these

Ans. (d)

6 $J_{1/2}(x)$ is given by

(a) $\sqrt{\left(\frac{2\pi}{x}\right)}.\sin x$ (b) $\sqrt{\left(\frac{2\pi}{x}\right)}.\cos x$

(c) $\sqrt{\left(\frac{\pi}{2x}\right)}.\cos x$ (d) $\sqrt{\left(\frac{2}{\pi x}\right)}.\sin x$

Ans. (d)

7. $J_{-1/2}(x)$ is given by

(a) $\sqrt{\left(\frac{\pi}{2x}\right)}.\cos x$ (b) $\sqrt{\left(\frac{2}{\pi x}\right)}.\sin x$

(c) $\sqrt{\left(\frac{\pi}{2x}\right)}.\sin x$ (d) $\sqrt{\left(\frac{2}{\pi x}\right)}.\cos x$

Ans. (d)

8. $J_{n+3} + J_{n+5}$ is equal to

(a) $(n+4)J_{n+4}$ (b) $\frac{2}{x}(n+4)J_{n+4}$

(c) nJ_{n+4} (d) $\frac{n}{x}J_{n+4}$

Ans. (b)

9. $J_0'(x)$ is equal to

(a) $-J_1(x)$ (b) $J_1'(x)$

(c) $-J_1(x)$ (d) $J_2(x)$

Ans. (a)

10. $p_n(1)$ is equal to

(a) $\frac{n}{2}(n+1)$ (b) $\frac{n}{2}(n-1)$

(c) $n(n+1)$ (d) $n(n-1)$

Ans. (a)

11. The value of the integral $\int_{-1}^{1}[p_4(x)]^2\,dx$ is

(a) $\frac{2}{7}$ (b) $\frac{2}{3}$

(c) $\frac{2}{9}$ (d) $\frac{7}{9}$

Ans. (c)

12. The Rodrigues formula for Legendre polynomial $p_n(x)$ is given by

(a) $p_n(x)=\frac{1}{\lfloor n \cdot 2^n}\frac{d^n}{dx^n}(x^2-1)^n$

(b) $p_n(x)=\frac{\lfloor n}{2^n}\frac{d^n}{dx^n}(x^2-1)^{n-1}$

(c) $p_n(x)=\frac{\lfloor n}{2^{n-1}}\frac{d^n}{dx^n}(x^2-1)^{n-1}$

(d) $p_n(x)=\frac{1}{\lfloor n}2^n(x^2-1)^n$ (U.P.T.U. 2009)

Ans. (a)

13. $P_n(-x)$ is equal to

(a) $(-1)^n p_n'(x)$ (b) $(-1)^n p_n(x)$

(c) $(-1)^n p_{n-1}(x)$ (d) $p_n(x)$

Ans. (b)

14. $p_n(1)$ is equal to

(a) 0 (b) $(-1)^n$

(c) 1 (d) 2

Ans. (c)

15. if $\int_{-1}^{1}p_n(x)dx=2$ then n is equal to

(a) 1 (b) 0

(c) -1 (d) 2

Ans. (b)

16. $p_n(-1)$ is equal to

(a) 0 (b) 1

(c) $(-1)^n$ (d) n

Ans. (c)

17. The polynomial $2x^2 + x + 3$ in terms of Legendre polynomial is

(a) $\frac{1}{3}(4p_2 - 3p_1 + 11p_0)$ (b) $\frac{1}{3}(4p_2 + 3p_1 - 11p_0)$

(c) $\frac{1}{3}(4p_2 + 3p_1 + 11p_0)$ (d) $\frac{1}{3}(4p_2 - 3p_1 - 11p_0)$

Ans. (c)

18. The Legendre polynomial $p_n(x)$ has

(a) n real zeros between 0 and 1

(b) n zeros of which only one is between -1 and 1

(c) 2n–1 real zeros between –1 and 1

(d) None of these

Ans. (a)

19. $P_{2n}(0)$ is equal to

(a) $\frac{(\lfloor 2)^2}{\lfloor 2n}$ (b) $(-1)^n \frac{1.3.5.................(2n-1)}{2.4.6..............2n}$

(c) 0 (d) $(-1)^n \frac{2.4.6..............n}{1.3.5...............(2n-1)}$

Ans. (b)

Fill up the blanks or choose the correct answer in the following problems.

20. Generating function of $p_n(x)$ is...................

Ans. $(1-2xz+z^2)^{-1/2} = \sum_{n=0}^{\infty} z^n p_n(x)$

21. $\frac{d}{dx}[J_0(x)] = $ ....

Ans. $-J_1(x)$

22. Bessel's equation of order zero is ....

Ans. $x\frac{d^2y}{dx^2} + \frac{dy}{dx} + xy = 0$

23. The value of $P_2(x)$ is..................

Ans. $\frac{1}{2}(3x^2 - 1)$

24. For $m \neq n, \int_{-1}^{1} p_m(x) p_n(x) dx =$ ....

Ans. zero

25. $p_{2n+1}(0)$..................

Ans. zero

26. Bessel was a German mathematician

(True/False)

Ans. True

27. Bessel's function are not cylindrical functions.

(True/False)

Ans. False

28. Bessel equation of order 4 is $x^2 y'' + xy' + (x^2 - 4)y = 0$

(True/False)

Ans. False

29. Legendre's polynomial of first degree =x

(True/False)

Ans. True

Match the correct value of the following

30.

(i)	$J_{-n}(x)$	(a)	$(-1)^n J_n(x)$
(ii)	$J_n(-x)$	(b)	$-J_1$
(iii)	J_0'	(c)	$(-1)^n J_n(x)$
(iv)	J_0''	(d)	$-J_1'$

Ans. (i) → c, (ii) → a, (iii) → b, (iv) → d

31.

(i)	xJ_n'	(a)	$x(J_{n-1} + J_{n+1})$
(ii)	$2J_n'$	(b)	$x^n J_{n-1}$
(iii)	$2nJ_n$	(c)	$nJ_n - x J_{n+1}$
(iv)	$\frac{d}{dx} x^n J_n$	(d)	$J_{n-1} - J_{n+1}$

Ans. (i) → c, (ii) → d, (iii) → a, (iv) → b

32.

(i)	$P_n(1)$	(a)	x
(ii)	P_1	(b)	1
(iii)	p_0	(c)	$(-1)^n P_n(x)$
(iv)	$p_n(-x)$	(d)	1

Ans. (i) → b, (ii) → a, (iii) → d, (iv) → c

33.

(i)	$P_2(x)$	(a)	$\frac{n}{2}(n+1)$
(ii)	$P'_n(1)$	(b)	$(-1)^n$
(iii)	$P_n(-1)$	(c)	2
(iv)	$\int_{-1}^{1} P_0(x)dx$	(d)	$\frac{1}{2}(3x^2-1)$

Ans. (i) → d, (ii) → a, (iii) → b, (iv) → c

34.

List I		List II	
(i)	$p_2(x)$	(a)	$\frac{1}{2}(5x^3-3)$
(ii)	$P_3(x)$	(b)	$\frac{2}{5}P_3(x)+\frac{3}{5}P_2(x)$
(iii)	x^3	(c)	1
(iv)	$P_n(1)$	(d)	$\frac{1}{2}(3x^2-1)$

Codes

	(i)	(ii)	(iii)	(iv)
(A)	a	b	d	c
(B)	d	a	b	c
(C)	a	d	c	b
(D)	a	b	c	d

Ans. (B)

35.

List I		List II	
(i)	$P_n(-x)$	(a)	$\frac{1}{2}n(n+1)$
(ii)	$\int_{-1}^{1} p_3^2(x)dx$	(b)	$(-1)^n$
(iii)	$P_n(-1)$	(c)	$\frac{2}{7}$
(iv)	$p_n'(1)$	(d)	$(-1)^n P_n(x)$

Codes

	(i)	(ii)	(iii)	(iv)
(A)	a	b	c	d
(B)	d	c	b	a
(C)	a	c	d	b
(D)	a	d	b	c

Ans. (B)

UNIT - III
LAPLACE TRANSFORMS

Chapter 6

Laplace Transforms

INTRODUCTION

French mathematician Pierre Simon marquis De Laplace (1749-1827) used Laplace transform. Laplace transform is widely used in science and engineering. It is particularly effective in solving differential equations (ordinary as well as partial). It reduces an ordinary differential equations into an algebric equation.

To every scientist and engineer Laplace transform is a very powerful technique in finding solutions to initial value problems involving homogeneous and non-homogeneous equations alike. The systems of differential equations, partial differential equations, and integral equations when subjected to Laplace transform get converted into algebric equations which are relatively much easier to solve. The powerful practical Laplace transform techniques were developed over a century later by the English electrical engineer oliver Heaviside (1850-1925) and were often called "Heaviside Calculus".

DEFINITION

Let F(t) be a function of t defined for all positive values of t. Then the Laplace transforms of F(t), denoted by L{F(t)} is defined by

$$L\{F(t)\} = \int_0^\infty e^{-st} F(t)\,dt$$

we also write $L\{F(t)\} = f(s) = \int_0^\infty e^{-st}F(t)dt$ provided that the integral exists, s is a parameter which may be real or complex number.

L is known as Laplace transform operator. The given function F(t) known as determining function depends on t, while the new function to determined f(s), called as generating function depends only on s. f(s) is known as the Laplace transform of F(t).

NOTATION

We follow two types of notations:

(i) Functions are denoted by capital letters

F(t), G(t), H (t),...........

and their Laplace transforms are denoted by corresponding lower case letters f(s), g(s), h(s), ... or by f(p), g(p), h(p), ...

(ii) Functions are denoted by lower case letters

f(t), g(t), h(t),...........

and their Laplace transforms are denoted by

$\bar{f}(s)$, $\bar{g}(s)$, $\bar{h}(s)$ .... respectively or by $\bar{f}(p)$, $\bar{g}(p)$, $\bar{h}(p)$, ...

TRANSFORMS OF ELEMENTARY FUNCTIONS

The direct application of the definition gives the following formulae:

(i) $L\{1\} = \frac{1}{s}, s > o$

(ii) $L\{t^n\} = \frac{\overline{|(n+1)}}{s^{n+1}}$ if s > o and n > - 1

(iii) $L\{e^{at}\} = \frac{1}{s-a}$ if s > a

(iv) $L\{\sin at\} = \frac{a}{s^2+a^2}$ if s > o

(v) $L\{\cos at\} = \frac{s}{s^2+a^2}$ if s > o

(iv) $L\{\sin hat\} = \frac{a}{s^2-a^2}$ if $s > |a|$

(vii) $L\{\cos h\, at\} = \frac{s}{s^2-a^2}$ if $s > |a|$

Proofs:

(i) $L\{1\} = \int_o^\infty e^{-st}\, 1dt = \left[\frac{-e^{-st}}{s}\right]_{t=o}^{\infty} = \frac{1}{s}$ if s > o

(ii) $L\{t^n\} = \int_o^\infty e^{-st}\, t^n dt$, now putting st = x

$$= \int_o^\infty e^{-x}\left(\frac{x}{s}\right)^n \frac{dx}{s}$$

$$= \frac{1}{s^{n+1}} \int_o^\infty e^{-x}\, x^n\, dx$$

$$= \frac{1}{s^{n+1}} \int_0^{\infty} e^{-x} x^{n+1-1} dx$$

$$= \frac{\overline{|(n+1)}}{s^{n+1}}$$

if s > o and n > - 1

In particular, if n is positive integer, then $\overline{|(n+1)} = \underline{|n}$ so that

$$L\{t^n\} = \frac{\underline{|n}}{s^{n+1}}$$

(iii) $$L\{e^{at}\} = \int_0^{\infty} e^{-st} e^{at} dt$$

$$= \int_0^{\infty} e^{-(s-a)t} dt$$

$$= \left[\frac{-e^{-(s-a)t}}{s-a}\right]_{t=o}^{\infty}$$

$$= \frac{1}{s-a}, \text{ if } s > a$$

(iv) and (v)

$$L\{e^{iat}\} = \frac{1}{s-ia}$$

$$= \frac{s+ia}{(s-ia)(s+ia)}$$

$$= \frac{s+ia}{s^2+a^2}$$

$$= \frac{s}{s^2+a^2} + i\frac{a}{s^2+a^2}$$

$$\text{or } L\{\text{Cos at} + i \sin at\} = \frac{s}{s^2+a^2} + i\frac{a}{s^2+a^2}$$

$$\text{or } L\{\text{Cos at}\} + L\{\text{Sin at}\} = \frac{s}{s^2+a^2} + i\frac{a}{s^2+a^2}$$

Equating real and imaginary parts we have

$$L\{\text{Cos at}\} = \frac{s}{s^2 + a^2} \ \& \ L\{\text{Sin at}\} = \frac{a}{s^2 + a^2}$$

(vi) We known that $L\{e^{at}\} = \frac{1}{s - a}$

$$L\{\text{Sin hat}\} = L\{\frac{e^{at} - e^{-at}}{2}\}$$

$$= \frac{1}{2}\left[L\{e^{at}\} - L\{e^{-at}\}\right]$$

$$= \frac{1}{2}\left[\frac{1}{s-a} - \frac{1}{s+a}\right] = \frac{a}{s^2 - a^2}, \text{ if } s > |a|$$

(vii) $L\{\cosh at\} = L\left\{\frac{e^{at} + e^{-at}}{2}\right\}$

$$= \frac{1}{2}\left[\frac{1}{s-a} + \frac{1}{s+a}\right]$$

$$= \frac{s}{s^2 - a^2}, \text{ if } s > |a|$$

PROPERTIES OF LAPLACE TRANSFORMS

I. LINEARITY PROPERTY.

Theorem 1. Let F (t) and G(t) be any two functions, whose Laplace transforms exists, and let α and β be any two Constants, then

$L\{\alpha F(t) \pm \beta G(t)\} = \alpha L\{F(t) \pm \beta L\{G(t)\}$

Proof :- Using the definition of the Laplace transform, use have

$$L\{\alpha F(t) \pm \beta G(t)\} = \int_0^\infty e^{-st}\{\alpha F(t) \pm \beta G(t)\}\, dt$$

$$= \alpha \int_0^\infty e^{-st} F(t)\, dt \pm \beta \int_0^\infty e^{-st} G(t)\, dt$$

$$= \alpha L\{F(t)\} \pm \beta L\{G(t)\}$$

II. First shifting theorem (First Translation Theorem 2. If $L\{F(t)\} = f(s)$, then $L\{e^{at} F(t)\} = f(s - a)$

Proof :-

Let $L\{F(t)\} = f(s) = \int_{o}^{\infty} e^{-st} F(t)\, dt$

Then $L\{e^{at} F(t)\} = \int_{o}^{\infty} e^{-st} e^{at} F(t)\, dt$

$$= \int_{o}^{\infty} e^{-(s-a)t} F(t)\, dt$$

$$= \int_{o}^{\infty} e^{-ut} F(t)\, dt, \text{ where } u = s - a > o$$

$$= f(u) = f(s-a)$$

III Second shifting Theorem (Second translation)

Theorem 3. If $L\{F(t)\} = f(s)$

And $G(t) = \begin{cases} F(t-a), & t > a \\ O, & t < a \end{cases}$

than $L\{G(t)\} = e^{-as} f(s)$ (U.P.T.U 2006)

Proof.

$$L\{G(t)\} = \int_{o}^{\infty} e^{-st} G(t)\, dt$$

$$= \int_{o}^{a} e^{-st} G(t) dt + \int_{a}^{\infty} e^{-st} G(t)\, dt$$

$$= \int_{o}^{a} e^{-st}\, o\, dt + \int_{a}^{\infty} e^{-st} F(t-a)\, dt$$

$$= O + \int_{a}^{\infty} e^{-st} F(t-a)\, dt$$

Put $t - a = u$ so that $dt = du$

If $t = a$ then $u = t - a = a - a = o$

If $t = \infty$ them $u = \infty - a = \infty$

$$L\{G(t)\} = \int_{O}^{\infty} e^{-s(u+a)} F(u)\, du$$

$$= e^{-sa} \int_0^\infty e^{-su} F(u)\,du$$

$$= e^{-sa} f(s)$$

(iv) Change of scale property

Theorem 4. If L { F (t) } = f (s)]

Then $L\{F(at)\} = \frac{1}{a} f\left(\frac{s}{a}\right)$

Proof. By definition, we have

$$L\{F(at)\} = \int_0^\infty e^{-st} F(at)\,dt$$

$$= \frac{1}{a} \int_0^\infty e^{-s\,(x/a)} F(x)\,dx$$

Putting at = x so that $dt = \frac{1}{a} dx$

$$= \frac{1}{a} \int_0^\infty e^{-\left(\frac{s}{a}\right)t} F(t)\,dt$$

$$= \frac{1}{a} f\left(\frac{s}{a}\right)$$

Laplace Transform of Derivatives.

Theorem

If L {F (t) } = f(s), then prove that

L { F′ (t) } = s f (s) – F (o)

& L { F ″ (t) }= s² f (s) – s F (o) – F′ (o)

Proof. Let L { F (t) } = f (s), then $L\{F'(t)\} = \int_0^\infty e^{-st} F'(t)\,dt$

$$= \left[e^{-st} F(t)\right]_{t=o}^{\infty} - \int_0^\infty e^{-st} (-s) F(t)\,dt$$

$$= -F(o) + s f(s)$$

$\therefore L\{F'(t)\} = s f(s) - F(o)$ .... (i)

In View of this

$$L\{G'(t)\} = s\,g(s) - G(o)$$

Taking G (t) = F′ (t), We get

$$L\{F''(t)\} = s\,L\{F'(t)\} - F'(o)$$

$$= s\,[s\,f(s) - F(o)] - F'(o)$$

$$= L\{F''(t)\} = s^2 f(s) - s\,F(o) - F'(o)$$

Generalising the result (i) and (ii), we get

$$L\{F^n(t)\} = s^n f(s) - s^{n-1} F(o) - s^{n-2} F'(o) - \text{-----------------} - sF^{(n-2)}(o) - F^{(n-1)}(o)$$

Laplace Transform of Integral Theorem. If L { F (t) } = f (s), then

$$L\left\{\int_o^t F(u)\,du\right\} = \frac{1}{s} f(s)$$

Proof.

$$\text{Let } G(t) = \int_o^t F(u)\,du \text{ ---------- (i)}$$

Then G′ (t) = F (t)

And from (i) it is clear that $G(o) = \int_o^o F(u)\,du = o$

Since we know

$$L\{G'(t)\} = s\,L\{G(t)\} - G(o)$$

$$\text{i.e. } L\{F(t)\} = s\,L\{G(t)\} - O = s\,L\{G(t)\}$$

$$\text{or } \frac{1}{s} f(s) = L\{G(t)\} = L\left\{\int_o^t F(u)\,du\right\}$$

Multiplication by powers of t. Theorem. If L { F (t) } = f (s)

Then $L\{t^n F(t)\} = (-1)^n \dfrac{d^n}{ds^n} f(s)$ For n = 1,2,3 ---------

Proof.

Let $L\{F(t)\} = f(s)$ and $\dfrac{d^n f(s)}{ds^n} = f^{(n)}(s)$

Then

$$L\{F(t)\} = f(s) = \int_0^\infty e^{-st} F(t)\,dt$$

$$f'(s) = \frac{d}{ds}\int_0^\infty e^{-st} F(t)\,dt$$

$$= \int_0^\infty \frac{\partial}{\partial s}\{e^{-st} F(t)\}\,dt$$

[By Leibnitz's rule for differentiation under the sign of integral]

$$= -\int_0^\infty te^{-st} F(t)\,dt$$

$$= -\int_0^\infty e^{-st}\{tF(t)\}\,dt$$

$$= -L\{tF(t)\}$$

$$\Rightarrow L\{tF(t)\} = -f'(s)$$

This proves that the theorem is true for n =1

Now, let us assume that the theorem is true for a particular value of n say r, then we have

$$L\{t^r F(t)\} = (-1)^r \frac{d^r}{ds^r} f(s)$$

$$\int_0^\infty e^{-st} t^r F(t)\,dt = (-1)^r \frac{d^r}{ds^r} f(s)$$

Now differentiating both side w.r.t.s we have

$$\frac{d}{ds}\int_0^\infty e^{-st} t^n F(t)\,dt = (-1)^r \frac{d^{r+1}}{ds^{r+1}} f(s)$$

$$or\ \int_0^\infty \frac{\partial}{\partial s}\{e^{-st} t^r F(t)\,dt = (-1)^r \frac{d^{r+1}}{ds^{r+1}} f(s)\}$$

by Leibnitz's rule for differentiation under the sign of integration

$$or\ -\int_0^\infty e^{-st} t^{r+1} F(t)\,dt = (-1)^r \frac{d^{r+1}}{ds^{r+1}} f(s)$$

or $$\int_0^\infty e^{-st}\{t^{r+1}F(t)\}\,dt=(-1)^{r+1}\frac{d^{r+1}}{ds^{r+1}}f(s)$$

or $$L\{t^{r+1}F(t)\}=(-1)^{r+1}\frac{d^{r+1}}{ds^{r+1}}f(s)$$

Which shows that if the theorem is true for any particular value of n, it is true for the next value of n. But it is also true for n=1. Hence by the method of mathematical induction it is true for every positive integral value of n.

Division by t.

Theorem. If L {F (t)} = f (s), then

$$L\left\{\frac{F(t)}{t}\right\}=\int_s^\infty f(x)\,dx \qquad \text{(U.P.T.U 2007,2005)}$$

Provided the integral exists.

Proof.

Let L { F (t) } = f (s). Then

$$f(s)=\int_0^\infty e^{-st}F(t)\,dt$$

Integrating this w.r.t.s from s =s to s = ∞,

$$\int_s^\infty f(s)\,ds=\int_s^\infty ds\int_0^\infty e^{-st}F(t)\,dt$$

s and t are independent variables and hence order of integration in the repeated integral can be interchanged

$$\therefore\quad \int_s^\infty f(s)\,ds=\int_0^\infty dt\int_s^\infty e^{-st}F(t)\,ds$$

$$=\int_0^\infty F(t)\,dt\int_s^\infty e^{-st}\,ds$$

$$=\int_0^\infty F(t)\,dt\left(\frac{e^{-st}}{-t}\right)_{s=s}^{\infty}$$

$$=\int_0^\infty \frac{e^{-st}}{t}F(t)\,dt$$

$$= \int_0^\infty e^{-st}\left\{\frac{F(t)}{t}\right\}dt$$

$$= L\left\{\frac{F(t)}{t}\right\}$$

thus $L\left\{\frac{F(t)}{t}\right\} = \int_s^\infty f(x)\,dx$

Example 1. Find the Laplace transform of the function F (t), where F (t) = $7e^{2t} + 9e^{-2t} + 5\cos t + 7t^3 + 5\sin 3t + 2$

Solution .

$$L\{F(t)\} = L\{7e^{2t} + 9e^{-2t} + 5\cos t + 7t^3 + 5\sin 3t + 2\}$$

$$= 7L\{e^{2t}\} + 9L\{e^{-2t}\} + 5L\{\cos t\} + 7L\{t^3\} + 5L\{\sin 3t\} + 2L\{1\}$$

$$= 7\frac{1}{s-2} + 9\frac{1}{s+2} + 5\frac{s}{s^2+1} + 7\frac{\lfloor 3}{s^4} + 5\frac{3}{s^2+9} + 2\frac{1}{s}$$

$$= \frac{16s-4}{s^2-4} + \frac{5s}{s^2+1} + \frac{2s^3+42}{s^4} + \frac{15}{s^2+9}$$

Example 2. Find $L\{t^3 e^{3t}\}$

Solution. We have

$$L\{t^3\} = \frac{\lfloor 3}{s^4} = \frac{6}{s^4} = f(s) \text{ say}$$

Then by first shifting property

$$L\{t^3 e^{2t}\} = \frac{6}{(s-2)^4}$$

Example 3. Find L {sin kt sin h kt} & L {cos kt sin hkt}

Solution. We have

$$L\{\sinh kt\} = \frac{k}{s^2 - k^2} = f(s) \text{ say}$$

$$L\{e^{ikt}\sinh kt\} = \frac{K}{(s-ik)^2 - k^2} \text{ by first shifting properties}$$

$$= \frac{K}{s^2 - 2k^2 - 2isk}$$

$$= \frac{k(s-2k^2+2isk)}{(s^2-2k^2)^2+4s^2k^2}$$

Equating real and imaginary parts, we get

$$L\{\cos kt \sin hkt\} = \frac{(s^2 - 2k^2)k}{s^4 + 4k^4}$$

and $$L\{\sin kt \sin h\, kt\} = \frac{2sk^2}{s^4 + 4k^4}$$

Example 4. Find L { e^{-t} (3 sin h2t - 5 cos h 2t) }

Solution :- We know that, if

L { F (t) } = f (s), then L { e^{at} F (t) } = f (s – a)

$$\text{Also } L\{\text{Sin h } 2t\} = \frac{2}{s^2-2^2}, L\{\cosh 2t\} = \frac{s}{s^2-2^2}$$

Therefore L {e^{-t} (3 sin h2t – 5 cos h 2t)}

$$= 3\frac{2}{(s+1)^2-2^2} - 5\frac{(s+1)}{(s+1)^2-2^2}$$

$$= \frac{1-5s}{s^2+2s-3}$$

Example 5. Find the Laplace transform of F (t), where

$$F(t) = \begin{cases} \cos\left(t - \frac{2\pi}{3}\right) \text{ if } t > \frac{2\pi}{3} \\ O \qquad \text{if } t < \frac{2\pi}{3} \end{cases}$$

Solution:- $L\{F(t)\} = \int_0^\infty e^{-st} F(t)\, dt$

$$= \int_0^{2\pi/3} e^{-st} F(t)\, dt + \int_{2\pi/3}^\infty e^{-st} F(t)\, dt$$

$$= \int_0^{2\pi/3} e^{-st} . o\, dt + \int_{2\pi/3}^\infty e^{-st} \cos\left(t - \frac{2\pi}{3}\right) dt$$

$$= o + \int_0^\infty e^{-st} \cos\left(t - \frac{2\pi}{3}\right) dt$$

$$= \int_0^{\infty} e^{-s(x+\frac{2\pi}{3})} \cos x \, dx \text{ where } t - \frac{2\pi}{3} = x$$

$$= e^{-2\pi s/3} \int_0^{\infty} e^{-sx} \cos x \, dx$$

$$= e^{-2\pi s/3} \, L\{\cos x\}$$

$$= e^{-2\pi s/3} \frac{s}{s^2+1}$$

Example 6. If $L\{\cos^2 t\} = \dfrac{s^2+2}{s(s^2+4)}$, find $L\{\cos^2 at\}$ (U.P.T.U 2006)

Solution :- we have

$$L\{\cos^2 t\} = \frac{s^2+2}{s(s^2+4)} = f(s)$$

By change of scale property we have

$$L\{\cos^2 at\} = \frac{1}{a} f(s/a)$$

$$= \frac{1}{a} \frac{\left(\frac{s}{a}\right)^2 + 2}{\frac{s}{a}\left(\frac{s^2}{a^2}+4\right)}$$

$$= \frac{1}{a}\left[\frac{s^2+2a^2}{\frac{s}{a}\left(s^2+4a^2\right)}\right]$$

$$= \frac{s^2+2a^2}{s(s^2+4a^2)}$$

Example :- find the Laplace transform of

(i) $\sin\sqrt{t}$ (ii) $\dfrac{\cos\sqrt{t}}{\sqrt{t}}$

Solution :- (i) $L\{\sin\sqrt{t}\} = L\{\sqrt{t} - \dfrac{(\sqrt{t})^3}{\lfloor 3} + \dfrac{(\sqrt{t})^5}{\lfloor 5} - \ldots\ldots\ldots\}$

$$= L\{t^{1/2}\} - \frac{1}{\underline{|3}} L\{t^{3/2}\} + \frac{1}{\underline{|5}} L\{t^{5/2}\} - \ldots\ldots$$

$$= \frac{\Gamma\left(\frac{3}{2}\right)}{s^{3/2}} - \frac{1}{\underline{|3}} \frac{\Gamma\left(\frac{5}{2}\right)}{s^{5/2}} + \frac{1}{\underline{|5}} \frac{\Gamma\left(\frac{7}{2}\right)}{s^{7/2}} - \ldots\ldots$$

$$= \frac{\frac{1}{2}\sqrt{\pi}}{s^{3/2}} - \frac{1}{6} \frac{\frac{3}{2}\,\frac{1}{2}\,\sqrt{\pi}}{s^{5/2}} + \frac{1}{120} \frac{\frac{5}{2}\,\frac{3}{2}\,\frac{1}{2}\,\sqrt{\pi}}{s^{7/2}} - \ldots\ldots$$

$$= \frac{\sqrt{\pi}}{2\,s^{3/2}} \left[1 - \left(\frac{1}{4s}\right) + \frac{1}{\underline{|2}}\left(\frac{1}{4s}\right)^2 - \frac{1}{\underline{|3}}\left(\frac{1}{4s}\right)^3 + \ldots\ldots\right]$$

$$= \frac{\sqrt{\pi}}{2s^{3/2}} e^{-1/4s}$$

Thus $L\{\sin\sqrt{t}\} = \sqrt{\pi}\,\frac{e^{-1/4s}}{2s^{3/2}} = f(s)$ say

(ii) Let $L\{F(t)\} = L\{\text{Sin}\sqrt{t}\}$

$$F'(t) = \frac{d}{dt}(\sin\sqrt{t})$$

$$= \frac{1}{2\sqrt{t}} \cos\sqrt{t}$$

Since we know that

$$L\{F'(t)\} = s\,f(s) - F(o)$$

$$L\left\{\frac{\cos\sqrt{t}}{2\sqrt{t}}\right\} = s\,f(s) - o \quad \because F(o) = \sin o = o$$

$$= \frac{\sqrt{\pi}}{2s^{1/2}} e^{-1/4s}$$

$$\Rightarrow \frac{1}{2} L\left\{\frac{\cos\sqrt{t}}{\sqrt{t}}\right\} = \frac{1}{2}\left(\frac{\pi}{s}\right)^{1/2} e^{-1/4s}$$

$$\Rightarrow L\left\{\frac{\cos\sqrt{t}}{\sqrt{t}}\right\} = \left(\frac{\pi}{s}\right)^{1/2} e^{-1/4s}$$

Example 8. A function F (t) obeys the equation $F(t)+2\int_0^t F(t)\,dt = \text{Cosh } 2t$ find the Laplace transform of F (t). (U.P.T.U 2006)

Solution. We have

$$F(t) + 2\int_0^t F(t)\,dt = \cos h\, 2t$$

Taking Laplace transform of both sides, we get

$$L\{F(t)\} + 2L\int_0^t F(t)\,dt = L\{\cos h\, 2t\}$$

$$\Rightarrow f(s) + 2\,\frac{1}{s}\,f(s) = \frac{s}{s^2-4}$$

$$\Rightarrow f(s)\left(1+\frac{2}{s}\right) = \frac{s}{s^2-4}$$

$$\Rightarrow f(s)\left(\frac{s+2}{s}\right) = \frac{s}{s^2-4}$$

$$\Rightarrow f(s) = \left(\frac{s}{s^2-4}\right)\left(\frac{s}{s+2}\right)$$

$$\Rightarrow f(s) = \frac{s^2}{(s^2-4)(s+2)}$$

Example 9. Show that $\int_0^\infty t e^{-3t}\,\text{Sin } t\,dt = \frac{3}{50}$

Solution :- we have

$$L\{\sin t\} = \frac{1}{s^2+1} = f(s) \text{ say}$$

and so $L\{t \sin t\} = (-1)^1 \frac{d}{ds}\left(\frac{1}{s^2+1}\right)$

$$= \frac{2s}{(s^2+1)^2}$$

This $\Rightarrow \int_0^\infty e^{-st}\,(t \operatorname{Sin} t)\,dt = \dfrac{2s}{\left(s^2+1\right)^2}$

Putting s = 3, we have

$$\int_0^\infty e^{-3t}\, t \sin t\, dt = \frac{2\times 3}{\left(3^2+1\right)^2}$$

$$= \frac{3}{50}$$

Example 10. Prove that

$$\int_0^\infty t^3\, e^{-t} \sin t\, dt = O$$

Solution. We have

$$L\{\operatorname{Sin} t\} = \frac{1}{s^2+1} = f(s) \text{ say}$$

$$L\left\{t^3 \sin t\right\} = (-1)^3 \frac{d^3}{ds^3}\left(\frac{1}{s^2+1}\right)$$

$$= -\frac{d^2}{ds^2}\left[\frac{-2s}{\left(s^2+1\right)^2}\right]$$

$$= 2\,\frac{d}{ds}\left[\frac{1-3s^2}{\left(1+s^2\right)^3}\right]$$

$$= \frac{2.12s\left(1-s^2\right)}{\left(1+s^2\right)^4}$$

$$= \frac{24s\left(1-s^2\right)}{\left(1+s^2\right)^4}$$

$$\Rightarrow \int_0^\infty e^{-st}\, t^3 \sin t\, dt = \frac{24s\,(1-s^2)}{(1+s^2)^4}$$

putting s = 1, we have

$$\int_0^\infty e^{-t}\, t^3 \sin t\, dt = \frac{24(1-1)}{(1+1)^4} = 0$$

Example 11. obtain $L\left\{\frac{\text{Sin } t}{t}\right\}$

Solution. We have

$$L\{\sin t\} = \frac{1}{s^2+1} = f(s) \text{ say}$$

Since we know $L\left\{\frac{F(t)}{t}\right\} = \int_s^\infty f(x)\,dx$

Provided $\lim_{t\to\infty} \frac{F(t)}{t}$ exists

$$\therefore L\left\{\frac{\sin t}{t}\right\} = \int_s^\infty \frac{1}{x^2+1}\,dx$$

$$= \left[\tan^{-1} x\right]_s^\infty = \frac{\pi}{2} - \tan^{-1} s$$

$$= \cot^{-1} s$$

Example 12. Show that $\int_0^\infty \frac{\sin t}{t}\,dt = \pi/2$ **(U.P.T.U. 2009)**

Solution. We have

$$L\{\sin t\} = \frac{1}{s^2+1} = f(s) \text{ say}$$

$$L\left\{\frac{\sin t}{t}\right\} = \int_s^\infty \frac{1}{x^2+1}\,dx$$

$$= \left[\tan^{-1} x\right]_s^\infty$$

$$\Rightarrow \int_0^\infty e^{-st}\frac{\sin t}{t}\,dt = \frac{\pi}{2} - \tan^{-1} s$$

Taking limit as $s \to 0$, we have

$$\Rightarrow \int_0^\infty \frac{\sin t}{t}\, dt = \pi / 2$$

Example 13. Find the Laplace transform of $\frac{\sin at}{t}$. Does the transform of $\frac{\cos at}{t}$ exist ?

Solution. We have

$$L\{\sin at\} = \frac{a}{s^2 + a^2} \;\&\; L\{\cos at\} = \frac{s}{s^2 + a^2}$$

By theorem

$$L\left\{\frac{F(t)}{t}\right\} = \int_s^\infty f(x)\, dx$$

$$\Rightarrow L\left\{\frac{\sin at}{t}\right\} = \int_s^\infty \frac{a}{x^2 + a^2}\, dx$$

$$= \left[\tan^{-1}\frac{x}{a}\right]_{x=s}^\infty$$

$$= \tan^{-1}\infty - \tan^{-1}\frac{s}{a}$$

$$= \frac{\pi}{2} - \tan^{-1}\frac{s}{a} = \cot^{-1}\left(\frac{s}{a}\right)$$

$$\text{Now } L\left\{\frac{\cos at}{t}\right\} = \int_s^\infty \frac{x}{x^2 + a^2}\, dx$$

$$= \frac{1}{2}\left[\log(x^2 + a^2)\right]_{x=s}^\infty$$

$$= \frac{1}{2}\left[\lim_{x\to\infty} \log\left(x^2 + a^2\right) - \log\left(s^2 + a^2\right)\right]$$

Which does not exist, $\lim_{x\to\infty} \log\left(x^2 + a^2\right)$ is infinite

Hence $L\left\{\frac{\cos at}{t}\right\}$ does not exist.

Example 14. Evaluate $\int_0^\infty \frac{e^{-t} - e^{-3t}}{t} dt$

Solution. Here $F(t) = e^{-t} - e^{-3t}$

$$L\{F(t)\} = L\{e^{-t} - e^{-3t}\}$$

$$= L\{e^{-t}\} - L\{e^{-3t}\}$$

$$L\{e^{-t} - e^{-3t}\} = \frac{1}{s+1} - \frac{1}{s+3}$$

Since we know

$$L\left\{\frac{F(t)}{t}\right\} = \int_s^\infty f(x)\, dx$$

where $f(s) = L\{F(t)\}$

$$L\left\{\frac{e^{-t} - e^{-3t}}{t}\right\} = \int_s^\infty \left(\frac{1}{x+1} - \frac{1}{x+3}\right) dx$$

$$= \left[\log\left(\frac{x+1}{x+3}\right)\right]_{x=s}^\infty$$

$$= -\log\left(\frac{s+1}{s+3}\right)$$

$$\Rightarrow \int_0^\infty e^{-st} \frac{e^{-t} - e^{-3t} dt}{t} = -\log\left(\frac{s+1}{s+3}\right)$$

putting s = o, we get

$$\int_0^\infty \frac{e^{-t} - e^{-3t}}{t} dt = -(\log 1 - \log 3) = \log 3$$

Example 15. Using Laplace transform, evaluate

$$\int_0^\infty e^{-t} \frac{\sin^2 t}{t} dt$$ **(U.P.T.U. 2008)**

Solution. We have

$$L\{\sin^2 t\} = L\left\{\frac{1 - \cos 2t}{2}\right\}$$

$$= \frac{1}{2}\left(\frac{1}{s} - \frac{s}{s^2+4}\right)$$

$$\therefore L\left\{\frac{\sin^2 t}{t}\right\} = \frac{1}{2}\int_s^{\infty}\left(\frac{1}{x} - \frac{x}{x^2+4}\right)dx$$

$$= \frac{1}{2}\left[\log x - \log\sqrt{x^2+4}\right]_s^{\infty}$$

$$= \frac{1}{2}\left[\log\frac{s}{\sqrt{s^2+4}}\right]_s^{\infty}$$

$$= \frac{1}{2}\left[\log\frac{1}{\sqrt{1+\frac{4}{s^2}}}\right]_s^{\infty}$$

$$= \frac{1}{2}\left[o - \log\frac{1}{\sqrt{1+\frac{4}{s^2}}}\right]$$

$$= \frac{1}{2}\log\sqrt{\left(\frac{s^2+4}{s}\right)}$$

$$\Rightarrow \int_o^{\infty} e^{-st}\,\frac{\sin^2 t}{t}\,dt = \frac{1}{2}\log\left(\frac{\sqrt{s^2+4}}{s}\right)$$

putting s = 1, we get

$$\int_o^{\infty} e^{-t}\,\frac{\sin^2 t}{t}\,dt = \frac{1}{2}\log\frac{\sqrt{5}}{1}$$

$= \log_e 5$

Example 16. Find

(i) $L\left\{\frac{\sin ht}{t}\right\}$ (ii) $L\ \{t \cos ht\}$

(iii) $L\ \{t^2 e^t \operatorname{Sin} 4t\}$ **(U.P.T.U. 2002)**

Solution. (i) we have

$$L\{\sin h\, at\} = \frac{a}{s^2 - a^2}$$

& $$L\{\cos h\, at\} = \frac{s}{s^2 - a^2}$$

Since we know

$$L\left\{\frac{F(t)}{t}\right\} = \int_s^{\infty} f(x)\,dx$$

$$L\left\{\frac{\sin ht}{t}\right\} = \int_s^{\infty} \frac{1}{x^2 - 1}\,dx$$

$$= \frac{1}{2}\left\{\log\left(\frac{x-1}{x+1}\right)\right\}_{x=s}^{\infty}$$

$$= \frac{1}{2}\left\{\log\left(\frac{1-\frac{1}{x}}{1+\frac{1}{x}}\right)\right\}_s^{\infty}$$

$$= -\frac{1}{2}\log\left(\frac{s-1}{s+1}\right)$$

(ii) $$L\{t\cos ht\} = (-1)^1 \frac{d}{ds}\left(\frac{s}{s^2-1}\right)$$

$$= -\frac{d}{ds}\left(\frac{s}{s^2-1}\right)$$

$$= \frac{(s^2+1)}{(s^2-1)^2}$$

(iii) we have

$$L\{\sin 4t\} = \frac{4}{s^2 + 16}$$

$$L\{e^t \sin 4t\} = \frac{4}{(s-1)^2 + 16}$$ by using first shifting theorem

Now $$L\{te^t \sin 4t\} = -\frac{d}{ds}\left(\frac{4}{s^2 - 2s + 17}\right)$$

$$= \frac{4(2s-2)}{(s^2-2s+17)^2}$$

Again $L\left\{t^2e^t \text{ Sin } 4t\right\} = -\frac{d}{ds}\left\{\frac{4(2s-2)}{(s^2-2s+17)^2}\right\}$

$$= -4\left[\frac{(s^2-2s+17)^2.2-(2s-2).2(s^2-2s+17)(2s-2)}{(s^2-2s+17)^4}\right]$$

$$= -\frac{-4(2s^2-4s+34-8s^2+16s-8)}{(s^2-2s+17)^3}$$

$$= -4\frac{(-4s^2+12s+26)}{(s^2-2s+17)^3}$$

$$= \frac{8(3s^2-6s-13)}{(s^2-2s+17)^3}$$

Example 17. If F (t) = (e^{at} – cos bt)/t find the Laplace transform F (t)

(U.P.T.U. 2003)

Solution. Given that

$$F(t) = \frac{e^{at} - \cos bt}{t}$$

But the have

$$L\left\{e^{at} - \cos bt\right\} = \frac{1}{s-a} - \frac{s}{s^2+b^2}$$

Therefore

$$L\left\{\frac{e^{at} - \cos bt}{t}\right\} = \int_s^\infty \left(\frac{1}{s-a} - \frac{s}{s^2+b^2}\right) ds$$

$$= \left[\log(s-a) - \frac{1}{2}\log(s^2+b^2)\right]_s^\infty$$

$$= \frac{1}{2}\left[\log\frac{(s-a)^2}{(s^2+b^2)}\right]_s^\infty$$

$$= \frac{1}{2}\left[\log \frac{(1-a/s)^2}{\left(1+b^2/s^2\right)}\right]_s^{\infty}$$

$$= \frac{1}{2}\left[o - \log \frac{(1-a/s)^2}{\left(1+b^2/s^2\right)}\right]$$

$$= \frac{1}{2}\left[\log \frac{\left(s^2+b^2\right)}{(s-a)^2}\right]$$

PERIODIC FUNCTIONS

If f (t) be a periodic function with period T>o, then f (t + T) = f (t + 2 T) = … = f (t)

Theorem. Let f (t) be a periodic function with period T then

$$L\{f(t)\} = \frac{\int_0^T e^{-st} f(t)\,dt}{1-e^{-sT}}$$

Proof. Use have

$$L\{f(t)\} = \int_0^\infty e^{-st} f(t)\,dt$$

$$= \int_0^T e^{-st} f(t)\,dt + \int_T^\infty e^{-st} f(t)\,dt \text{ ------- (i)}$$

$$= \int_0^T e^{-st} f(t)\,dt + \int_0^\infty {}^{-s(z+T)} f(z+T)\,dz \text{ on putting } t = z + T \text{ in second integral}$$

$$= \int_0^T e^{-st} f(t)\,dt + e^{-sT} \int_0^\infty e^{-st} f(z+T)\,dt$$

$$= \int_0^T e^{-st} f(t)\,dt + e^{-sT} \int_0^\infty e^{-st} f(t)\,dt$$

$$\Rightarrow L\{f(t)\} = \int_0^T e^{-st} f(t)\,dt + e^{-sT} L\{f(t)\}$$

$$\Rightarrow \left(1-e^{-sT}\right) L\{f(t)\} = \int_0^T e^{-st} f(t)\,dt$$

$$\Rightarrow L\{f(t)\} = \frac{\int_o^T e^{-st} f(t)\, dt}{1 - e^{-st}}$$

Example 18. Find the Laplace transform of the function (half ware rectifies)

$$f(t) = \begin{cases} \sin \omega t & \text{for } o < t < \dfrac{\pi}{\omega} \\ O & \text{for } \dfrac{\pi}{\omega} < t < \dfrac{2\pi}{\omega} \end{cases}$$ **(U.P.T.U. 2002)**

Solution. We know that

$$L\{f(t)\} = \frac{1}{1 - e^{-sT}} \int_o^T e^{-st} f(t)\, dt \text{ where T is period}$$

$$= \frac{1}{1 - e^{\left(\frac{-2\pi}{\omega}\right)s}} \left[\int_o^{2\pi/\omega} e^{-st} f(t)\, dt \right]$$

$$= \frac{1}{1 - e^{(-2\pi/\omega)s}} \left[\int_o^{2\pi/\omega} e^{-st} \sin \omega t\, dt + \int_{\pi/\omega}^{2\pi/\omega} e^{-st}.O\, dt \right]$$

$$= \frac{1}{1 - e^{(-2\pi s/\omega)}} \int_o^{\pi/\omega} e^{-st} \sin \omega t\, dt$$

$$= \frac{1}{1 - e^{(-2\pi s/\omega)}} \left[\frac{e^{-st}(-s \sin \omega t - w \cos \omega t)}{s^2 + \omega^2} \right]_o^{\pi/\omega}$$

$$\because \int e^{ax} \sin bx\, dx = \frac{e^{ax}}{a^2 + b^2} (a \sin bx - b \cos bx)$$

$$= \frac{1}{1 - e^{(-2\pi s/\omega)}} \left[\frac{\omega e^{-(\pi/\omega)s} + \omega}{s^2 + \omega^2} \right]$$

$$= \frac{\omega \left[1 + e^{-\pi s/\omega} \right]}{(s^2 + \omega^2)\left(1 - e^{(-2\pi s/\omega)}\right)}$$

$$= \frac{\omega}{(s^2 + \omega^2)\left(1 - e^{-\pi s/\omega}\right)}$$

Example 19. Draw the graph of the periodic function

$$f(t) = \begin{cases} t, & o < t < \pi \\ \pi - t, & \pi < t < 2\pi \end{cases}$$

Also find its Laplace transform. **(U.P.T.U. 2003)**

Solution. We know that

$$L\{f(t)\} = \frac{1}{1-e^{-sT}} \int_0^T e^{-st} f(t)\,dt \text{ Where T is the period}$$

$$= \frac{1}{1-e^{-2\pi s}} \int_0^{2\pi} e^{-st} f(t)\,dt \quad \because T = 2\pi$$

$$= \frac{1}{1-e^{-2\pi s}} \left[\int_o^{\pi} e^{-st}\, t\,dt + \int_{\pi}^{2\pi} e^{-st} (\pi - t)\,dt \right]$$

$$= \frac{1}{1-e^{-2\pi s}} \left[\left\{ \frac{te^{-st}}{-s} - 1\,\frac{e^{-st}}{(-s)^2} \right\}_O^{\pi} + \left\{ (\pi - t)\,\frac{e^{-st}}{-s} - (-1)\,\frac{e^{-st}}{(-s)^2} \right\}_{\pi}^{2\pi} \right]$$

$$= \frac{1}{1-e^{-2\pi s}} \left[\left(\frac{\pi e^{-s\pi}}{-s} - \frac{e^{-\pi s}}{s^2} - o + \frac{1}{s^2} \right) + \left\{ (\pi - 2\pi)\frac{e^{-2\pi s}}{-s} + \frac{e^{-2\pi s}}{s^2} (\pi - \pi)\,\frac{e^{-\pi s}}{-s} + \frac{e^{-\pi s}}{s^2} \right\} \right]$$

$$= \frac{1}{1-e^{-2\pi s}} \left\{ -\frac{\pi e^{-\pi s}}{s} - \frac{e^{-\pi s}}{s^2} + \frac{1}{s^2} + \pi\,\frac{e^{-2\pi s}}{s} + \frac{e^{-2\pi s}}{s^2} - O - \frac{e^{-\pi s}}{s^2} \right.$$

$$= \frac{1}{1-e^{-2\pi s}} \left[\frac{\pi}{s} \left(e^{-2\pi s} - e^{-\pi s} \right) + \frac{1}{s^2} \left(1 + e^{-2\pi s} - 2\,e^{-\pi s} \right) \right]$$

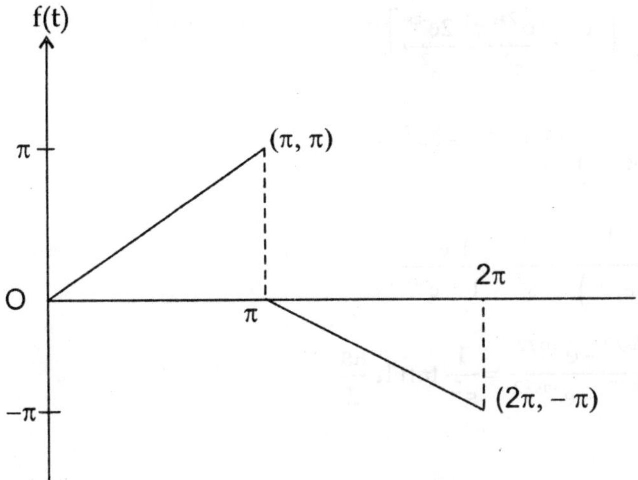

Figure. Graph of the function

Example 20. Draw the graph of he following periodic function. Also find its Laplace transform.

$$f(t) = \begin{cases} t & \text{for } o < t \le a \\ 2a - t & \text{for } a < t < 2a \end{cases}$$

(U.P.T.U 2002, 2007)

Solution . The given function is

$$f(t) \begin{cases} t \quad , & o < t \le a \\ 2a - t, & a < t < 2a \end{cases}$$

Which is periodic function with period 2a

Therefore,

$$L\{f(t)\} = \frac{1}{1-e^{-2as}} \int_o^{2a} f(t)\, e^{-st}\, dt$$

$$= \frac{1}{1-e^{-2as}} \left[\int_o^a te^{-st}\, dt + \int_a^{2a} (2a-t)e^{-st} dt \right]$$

$$= \frac{1}{1-e^{-2as}} \left[\left\{ \frac{te^{-st}}{-s} - 1 \frac{e^{-st}}{(-s)^2} \right\}_o^a + \left\{ \frac{(2a-t)\, e^{-st}}{-s} - \frac{(-1)\, e^{-st}}{(-s)^2} \right\}_a^{2a} \right]$$

$$= \frac{1}{1-e^{-2as}} \left[-\frac{ae^{-as}}{s} - \frac{e^{-as}}{s^2} + \frac{1}{s^2} + \frac{e^{-2as}}{s^2} + \frac{ae^{-as}}{s} - \frac{e^{-as}}{s^2} \right]$$

$$= \frac{1}{1-e^{-2as}}\left[\frac{1}{s^2}+\frac{e^{-2as}}{s^2}-\frac{2e^{-as}}{s^2}\right]$$

$$= \frac{1}{s^2\left(1-e^{-2as}\right)}\left(1+e^{-2as}-2e^{-as}\right)$$

$$= \frac{\left(1-e^{-as}\right)^2}{s^2\left(1-e^{-2as}\right)} = \frac{1}{s^2}\cdot\frac{1-e^{-as}}{1+e^{-as}}$$

$$= \frac{1}{s^2}\frac{e^{as/2}-e^{-as/2}}{e^{as/2}+e^{-as/2}} = \frac{1}{s^2}\tan h\frac{as}{2}$$

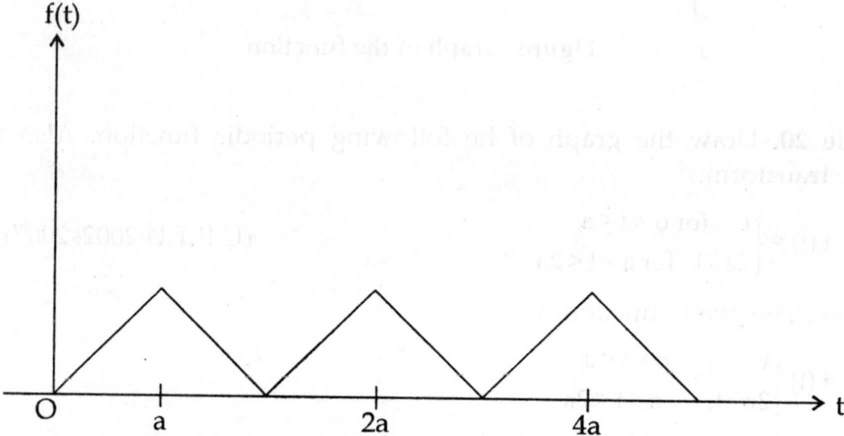

Unit step Function u (t -a)

By definition u (t -a) is o for t < a, has a jump of size1 at t = a (where we can leave it undefined) and is 1 for t >a

$$u(t-a) = \begin{cases} O & \text{if } t < a \\ 1 & \text{if } t \geq a \quad \text{where } a \geq o \end{cases}$$

Figure (a) shows the special case u (t) , which has jump at zero and figure (b) the general case u (t - a) for any arbitrary positive a. The unit step function is also called the Heaviside function.

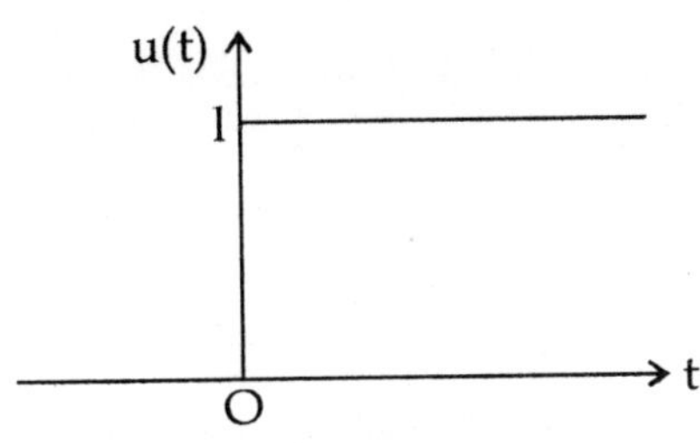

Unit step function u (t)
figure (a)

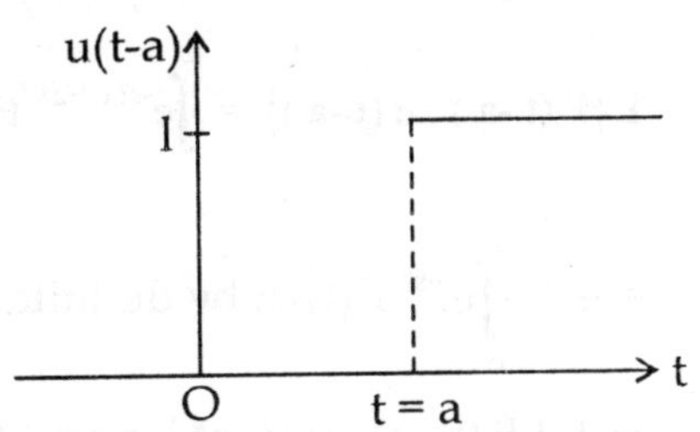

Unit step function u (t -a)
figure (b)

Laplace transform of unit step function

$$L\{u(t-a)\} = \int_0^\infty e^{-st}\, u(t-a)\, dt$$

$$= \int_0^a e^{-st}\, O\, dt + \int_a^\infty e^{-st}\, 1 dt$$

$$= O + \left[\frac{e^{-st}}{-s}\right]_a^\infty$$

$$= \frac{e^{-as}}{s}$$

Second shifting theorem in unit step's term **(U.P.T.U 2008)**

Theorem. If L { F (t) } = f (s), then

$$L\{F(t-a) \,.\, u(t-a)\} = e^{-as} f(s)$$

Proof. $L\{F(t-a).u(t-a)\} = \int_0^\infty e^{-st} F(t-a).u(t-a)\, dt$

$$= \int_0^a e^{-st} F(t-a).o dt + \int_a^\infty e^{-st}.F(t-a).1 dt$$

$$= \int_a^\infty e^{-st}.F(t-a)\, dt$$

putting $t - a = x \Rightarrow t = x + a$, we get

$$L\{F(t-a).u(t-a)\} = \int_0^\infty e^{-s(x+a)} F(x)\,dx$$

$$= e^{-sa}\int_0^\infty e^{-st} F(t)\,dt \text{ by definite integral}$$

$$\Rightarrow L\{F(t-a).u(t-a)\} = e^{-sa} f(s)$$

Example 21. Using unit step function. Find the Laplace transform of

(i) $(t-1)^2 . u(t-1)$

(ii) $\text{Sin}t . u(t-\pi)$ **(U.P.T.U. 2008)**

Solution (i) using second shifting theorem

$$L\left((t-1)^2 . u(t-1)\right) = e^{-s} L\{t^2\}$$

$$= e^{-s}\frac{\lfloor 2}{s^{2+1}}$$

$$= \frac{2e^{-s}}{s^3}$$

(ii) $\sin t. u(t-\pi) = \sin(t-\pi+\pi). u(t-\pi)$

$$= \{\sin(t-\pi)\cos\pi + \cos(t-\pi)\sin\pi\} . u(t-\pi)$$

$$= -\text{Sin}(t-\pi) . u(t-\pi) \text{ as } \text{Sin}\,\pi = O$$

$$\therefore L\{\sin t. u(t-\pi)\} = -L\{\sin(t-\pi). u(t-\pi)\}$$

$$= -e^{\pi s} L\{\text{Sin}\, t\}$$

$$= -\frac{e^{-\pi s}}{s^2+1}$$

Example 22. Express the following function in terms of unit step function and find its Laplace transform **(U.P.T.U. 2002)**

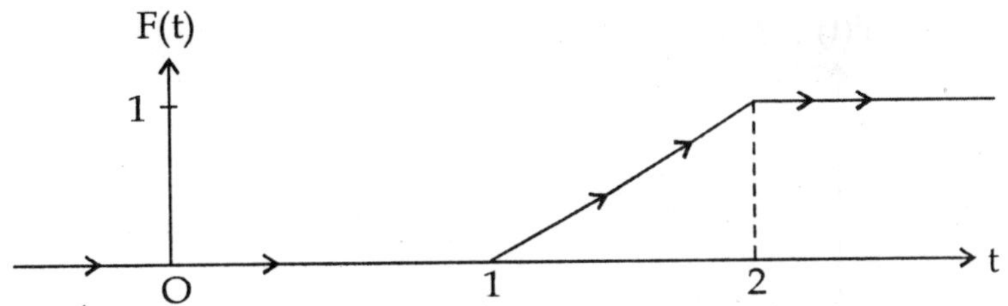

Solution. The above function shown in the figure is expressed in algebric form

$$F(t)=\begin{cases} O, & o<t<1 \\ t-1, & 1<t<2 \\ 1, & 2<t \end{cases}$$

Therefore, F (t) = (t - 1) [u (t - 1) – u (t - 2)] + u (t – 2)

$= (t - 1) u (t - 1) – u (t - 2) \{ t – 1-1\}$

$= (t - 1) u (t - 1) – (t - 2) u (t-2)$

$\therefore \quad L \{ F (t) \} = L \{ (t - 1) u (t - 1)\} – L \{ (t - 2) u (t - 2)\}$

$$= \frac{e^{-s}}{s^2} - \frac{e^{-2s}}{s^2}$$

$$= \frac{e^{-s} - e^{-2s}}{s^2}$$

Example 23. Find the Laplace transform of the function

$$F(t) = \begin{cases} 2 & \text{if } o < t < \pi \\ O & \text{if } \pi < t < 2\pi \\ \sin t & \text{if } t > 2\pi \end{cases}$$

Solution. We write F (t) in term of unit step function. For o < t < π, we take 2 u (t). For t > π we want O, so we must subtract the step function 2 u (t - π) with step at π . Then we have 2 u (t) – 2 u (t – π) = o when t > π . This is fine until we reach 2π where we want sin t to come in, so we add u (t – 2π) sin t. Together F (t) = 2 u (t) – 2u (t - π) + u (t – 2π) sin t The last term equals u (t – 2π) sin (t – 2π) because of the periodicity. So

$$L \{ F(t) \} = \frac{2}{s} - \frac{2e^{-\pi s}}{s} + \frac{e^{-2\pi s}}{s^2+1}$$

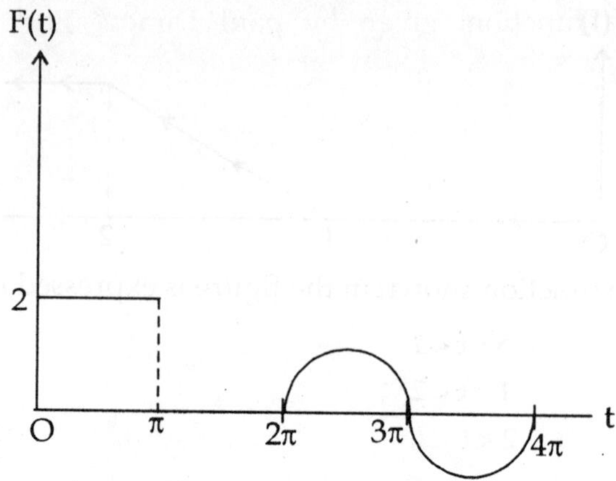

Example 24. Express the following function in terms of unit step function

$$F(t) = \begin{cases} t-1, & 1<t<2 \\ 3-t, & 2<t<3 \end{cases}$$ **(U.P.T.U 2009)**

and find its Laplace transform.

Solution. $F(t) = \begin{cases} t-1, & 1<t<2 \\ 3-t, & 2<t<3 \end{cases}$

$$F(t) = (t-1)\,[u(t-1) - u(t-2)] + (3-t)\,[u(t-2) - u(t-3)]$$
$$= (t-1)\,u(t-1) - (t-1)\,u(t-2) + (3-t)\,u(t-2) + (t-3)\,u(t-3)$$
$$= (t-1)\,u(t-1) - 2\,(t-2)\,u(t-2) + (t-3)\,u(t-3)$$

$$L\{F(t)\} = \frac{e^{-s}}{s^2} - 2\frac{e^{-2s}}{s^2} + \frac{e^{-3s}}{s^2}$$

DIRAC'S DELTA FUNCTION (or unit impulse function)

Phenomena of an impulsive nature, such as the action of very large forces (or voltages) over very short intervals of time, are of great practical interest, since they arise in various applications. This situation occurs, for instance, when a tennis ball is hit, a system is given a blow by a hammer, an airplane makes a "hard" landing, a ship is hit by a high signal wave, earthquake, and so on. Our present goal is to show how to solve problems involving short impulses by Laplace transformations.

In mechanics occasionally we come across problems where a very large force acts for a very short duration. Likewise in the study of bending of beams, a load acting at a point of a beam introduces a very large pressure on the beam acting over a very small area. The function which can handle this type of problem is called unit impulse function, given by paul Dirac (English Physicist, was awarded the Nobel Prize in 1933 for his work in quantum mechanics.)

The impulse of a force f (t) in the interval (a, a + ∈) = $\int_a^{a+\epsilon} f(t)\,dt$

Now define the function

$$f_\epsilon(t-a) = \begin{cases} O & \text{for } t < a \\ \dfrac{1}{\epsilon} & \text{for } a \leq t < a+\epsilon \\ O & \text{for } t > a \end{cases}$$

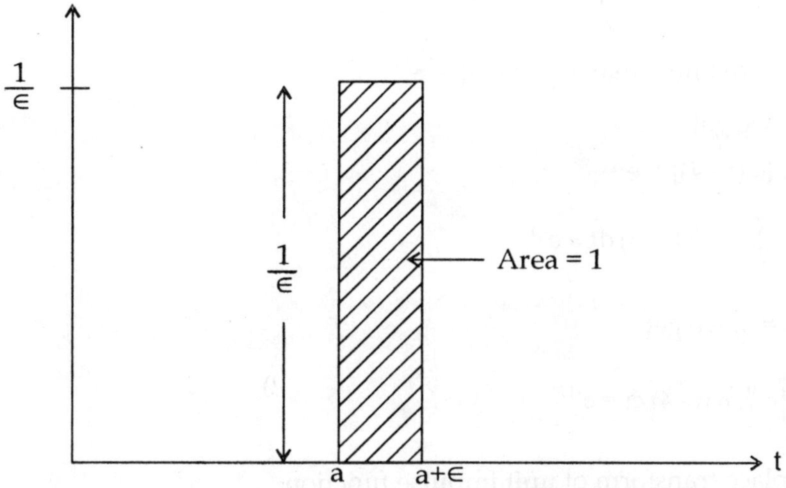

Or $f_\epsilon(t-a) = \frac{1}{\epsilon}\left[u(t-a) - u\{t-(a+\epsilon)\}\right]$

Taking Laplace transform, we have

$$L\{f_\epsilon(t-a)\} = \frac{1}{\epsilon} L\left[u(t-a) - u\{t-(a+\epsilon)\}\right]$$

$$= \frac{1}{\epsilon s}\left[e^{-as} - e^{-(a+\epsilon)s}\right]$$

$$L\{f_\epsilon(t-a)\} = e^{-as}\,\frac{1-e^{-\epsilon s}}{\epsilon s}$$

Dirac's delta function or unit impulse function denoted by δ (t - a) is defined as the limit of f_ϵ (t -a) as $\epsilon \to o$ i.e,

$$\delta(t-a) = \lim_{\epsilon \to 0} f_{\epsilon}(t-a)$$

$$= \lim_{\epsilon \to 0} \begin{cases} 0 & \text{for } t < a \\ \frac{1}{\epsilon} & \text{for } a \le t \le a+\epsilon \\ 0 & \text{for } t > a \end{cases} = \begin{cases} 0 & \text{for } t < a \\ \infty & \text{for } a = t = a \\ 0 & \text{for } t > a \end{cases}$$

$$\therefore \delta(t-a) = \begin{cases} \infty & \text{for } t = a \\ 0 & \text{for } t \ne a \end{cases}, \text{ subjected to } \int_0^\infty \delta(t-a)\,dt = 1$$

Example 25. Evaluate $\int_0^\infty e^{-3t}\,\delta(t-4)\,dt$

Solution :- we know that $L\{\delta(t-a)\} = e^{-as}$

For a = 4, we get

$$L\{\delta(t-4)\} = e^{-4s}$$

$$\Rightarrow \int_0^\infty e^{-st}\,\delta(t-4)\,dt = e^{-4s}$$

putting s = 3, we get

$$\int_0^\infty e^{-3t}.\,\delta(t-4)\,dt = e^{-12}$$

Note. Laplace transform of unit impulse function

$$L\{\delta(t-a)\} = \lim_{\epsilon \to 0} L\{f_{\epsilon}(t-a)\}$$

$$= \lim_{\epsilon \to 0} \left\{ e^{-as}\,\frac{1-e^{-\epsilon s}}{\epsilon s} \right\}$$

$$= e^{-as} \lim_{\epsilon \to 0} \left[\frac{1}{\epsilon s} \left\{ 1 - 1 + \frac{\epsilon s}{\underline{|1}} - \frac{(\epsilon s)^2}{\underline{|2}} + \text{-----} \right\} \right]$$

$$\Rightarrow L\{\delta(t-a)\} = e^{-as}$$

SOME SPECIAL FUNCTIONS

Example 26. Prove that $\int_0^\infty J_0(t)\,dt = 1$

Solution. We know that

$$J_n(x) = \frac{x^n}{2^n \, \Gamma(n+1)}\left[1 - \frac{x^2}{2(2n+2)} + \frac{x^4}{2 \cdot 4 (2n+2)(2n+4)} - \dots\dots\dots\right]$$

putting n = o , we get

$$J_o(x) = \left\{1 - \frac{x^2}{2 \times 2} + \frac{x^4}{2 \times 4 \times 2 \times 4} - \dots\dots\dots\right\}$$

$$\therefore L\{J_o(x)\} = \left\{\frac{1}{s} - \frac{\lfloor 2}{2^2 s^3} + \frac{\lfloor 4}{2^2 \cdot 4^2 \cdot s^5} - \dots\dots\dots\right\}$$

$$= \frac{1}{s}\left\{1 - \frac{1}{2 \cdot s^2} + \frac{(3/2)\,(1/2)}{\lfloor 2}\left(\frac{1}{s^2}\right)^2 - \dots\dots\dots\right\}$$

$$= \frac{1}{s}\left\{\left(1 + \frac{1}{s^2}\right)^{-1/2}\right\}$$

$$= \frac{1}{s}\left(\frac{s^2+1}{s^2}\right)^{-1/2} = \frac{1}{\sqrt{(s^2+1)}}$$

$$\Rightarrow \int_o^\infty e^{-st} J_o(t)\, dt = \frac{1}{(s^2+1)^{1/2}}$$

putting s = o , we get

$$\int_o^\infty J_o(t)\, dt = 1$$

Example 27. Find L {t J₀ (at)}

Solution. From last example, we have

$$L\{J_o(t)\} = \frac{1}{\sqrt{(s^2+1)}}$$

by change of scale property, we have

$$L\{J_o(at)\} = \frac{1}{a}\frac{1}{\sqrt{\{(s/a)^2 + 1\}}}$$

$$= \frac{1}{\sqrt{(s^2+a^2)}}$$

$$\therefore L\{t\,J_o(at)\} = (-1)\,\frac{d}{ds}\,L\{J_o(at)\}$$

$$= -\frac{d}{ds}\left\{\frac{1}{\sqrt{(s^2+a^2)}}\right\}$$

$$= \frac{s}{(s^2+a^2)^{3/2}}$$

Example 29. The error function is defined as

$$\text{erf }(t) = \frac{2}{\sqrt{\pi}}\int_0^t e^{-x^2}\,dx$$

then show that

$$L\{\text{erf}\sqrt{t}\} = \frac{1}{s\sqrt{(s+1)}}$$

Solution. We have

$$\text{erf }(\sqrt{t}) = \frac{2}{\sqrt{\pi}}\int_0^{\sqrt{t}} e^{-x^2}\,dx$$

$$= \frac{2}{\sqrt{\pi}}\int_0^{\sqrt{t}}\left(1 - x^2 + \frac{x^4}{\lfloor 2} - \frac{x^6}{\lfloor 3} + \text{--------}\right)dx$$

$$= \frac{2}{\sqrt{\pi}}\left[x - \frac{x^3}{3} + \frac{x^5}{5\lfloor 2} - \frac{x^7}{7\,.\lfloor 3} + \text{-----}\right]_0^{\sqrt{t}}$$

$$= \frac{2}{\sqrt{\pi}}\left[t^{1/2} - \frac{t^{3/2}}{3} + \frac{t^{5/2}}{5.\lfloor 2} - \frac{t^{7/2}}{7.\lfloor 3} + \text{-----}\right]$$

Taking Laplace transform of both the sides, we get

$$L\left\{\operatorname{erf}\left(\sqrt{t}\right)\right\} = \frac{2}{\sqrt{\pi}}\left[\frac{\lceil(3/2)}{s^{3/2}} - \frac{\lceil(5/2)}{3\,s^{5/2}} + \frac{\lceil\left(\frac{7}{2}\right)}{5.\lfloor 2\, s^{7/2}} - \frac{\lceil 9/2}{7.\lfloor 3\, s^{9/2}} + \text{-----}\right]$$

$$= \frac{1}{s^{3/2}}\left[1 - \frac{1}{2}\frac{1}{s} + \frac{1}{2}\times\frac{3}{4}\times\frac{1}{s^2} - \frac{1}{2}\times\frac{3}{4}\times\frac{5}{6}\frac{1}{s^3} + \dots\dots\dots\right]$$

$$= \frac{1}{s^{3/2}}\left(1+\frac{1}{s}\right)^{-1/2} = \frac{1}{s(s+1)^{1/2}}$$

Inverse Laplace Transform

Definition. If L{F(t)} = f(s), then

F(t) is called on inverse Laplace transform of f(s) and we write

$$F(t) = L^{-1}\{f(s)\}$$

Hence L^{-1} is called the inverse Laplace transformation operator.

Thus $L^{-1}\left\{\frac{1}{s^2}\right\} = t$, $L^{-1}\left\{\frac{a}{s^2+a^2}\right\} = \sin at$

Example 1. Find $L^{-1}\left\{\frac{1}{s-2} + \frac{2}{s+5} + \frac{6}{s^4}\right\}$

Solution : $L^{-1}\left\{\frac{1}{s-2} + \frac{2}{s+5} + \frac{6}{s^4}\right\}$

$$= L^{-1}\left\{\frac{1}{s-2}\right\} + 2L^{-1}\left\{\frac{1}{s+5}\right\} + L^{-1}\left\{\frac{6}{s^4}\right\}$$

$$= e^{2t}\, L^{1}\left\{\frac{1}{s}\right\} + 2e^{-5t}\, L^{-1}\left\{\frac{1}{s}\right\} + t^3$$

$$= e^{2t} + 2\,e^{-5t} + t^3$$

Example 2. Find $L^{-1}\left\{\frac{s-2}{(s-2)^2+5^2} + \frac{s+4}{(s+4)^2+9^2} + \frac{1}{(s+2)^2+3^2}\right\}$

Solution.

$$L^{-1}\left\{\frac{s-2}{(s-2)^2+5^2} + \frac{s+4}{(s+4)^2+9^2} + \frac{1}{(s+2)^2+3^2}\right\}$$

$$= L^{-1}\left\{\frac{s-2}{(s-2)^2+5^2}\right\} + L^{-1}\left\{\frac{s+4}{(s+4)^2+9^2}\right\} + L^{-1}\left\{\frac{1}{(s+2)^2+3^2}\right\}$$

$$= e^{2t} L^{-1}\left\{\frac{s}{s^2+5^2}\right\} + e^{-4t} L^{-1}\left\{\frac{s}{s^2+9^2}\right\} + e^{-2t} L^{-1}\left\{\frac{1}{s^2+3^2}\right\}$$

$$= e^{2t}\cos 5t + e^{-4t}\cos 9t + \frac{e^{-2t}}{3}\sin 3t$$

Example 3. Evaluate

$$L^{-1}\left\{\frac{1}{(s-4)^5} + \frac{5}{(s-2)^2+5^2} + \frac{s+3}{(s+3)^2+6^2}\right\}$$

Solution.

$$L^{-1}\left\{\frac{1}{(s-4)^5} + \frac{5}{(s-2)^2+5^2} + \frac{s+3}{(s+3)^2+6^2}\right\}$$

$$= e^{4t} L^{-1}\left\{\frac{1}{s^5}\right\} + e^{2t} L^{-1}\left\{\frac{5}{s^2+5^2}\right\} + e^{-3t} L^{-1}\left\{\frac{s}{s^2+6^2}\right\}$$

$$= e^{4t}\frac{t^4}{\lfloor 4} + e^{2t}\sin 5t + e^{-3t}\cos 6t$$

Example 4. Find $L^{-1}\left\{\frac{1}{s^{7/2}}\right\}$

Solution. $L^{-1}\left\{\frac{1}{s^{7/2}}\right\} = \frac{t^{\frac{7}{2}-1}}{\left|\left(\frac{7}{2}\right)\right.}$

$$= \frac{t^{5/2}}{\frac{5}{2}\cdot\frac{3}{2}\cdot\frac{1}{2}\left|\frac{1}{2}\right.}$$

$$= \frac{8}{15} t^2 \sqrt{\left(\frac{t}{\pi}\right)}$$

Example 5. Find

$$L^{-1}\left\{\frac{6}{2s-3} - \frac{3+4s}{9s^2-16} + \frac{8-6s}{16s^2+9}\right\}$$ **(U.P.T.U. 2001)**

Solution.

$$L^{-1}\left\{\frac{6}{2s-3}-\frac{3+4s}{9s^2-16}+\frac{8-6s}{16s^2+9}\right\}$$

$$=3\,L^{-1}\left\{\frac{1}{s-\frac{3}{2}}\right\}-3\,L^{-1}\left\{\frac{1}{9s^2-16}\right\}-4\,L^{-1}\left\{\frac{s}{9s^2-16}\right\}+8\,L^{-1}\left\{\frac{1}{16\,s^2+9}\right\}-6\,L^{-1}\left\{\frac{s}{16\,s^2+9}\right\}$$

$$=3\,L^{-1}\left\{\frac{1}{s-\frac{3}{2}}\right\}-\frac{1}{3}\,L^{-1}\left\{\frac{1}{s^2-(4/3)^2}\right\}$$

$$-\ \frac{4}{9}L^{-1}\left\{\frac{s}{s^2-(4/3)^2}\right\}+\frac{1}{2}L^{-1}\left\{\frac{1}{s^2+(3/4)^2}\right\}-\frac{3}{8}L^{-1}\left\{\frac{s}{s^2+(3/4)^2}\right\}$$

$$=3e^{3t/2}-\frac{1}{3}\,\frac{1}{4/3}\sinh\frac{4}{3}t-\frac{4}{9}\cosh\frac{4t}{3}+\frac{1}{2}\,\frac{1}{3/4}\sin\frac{3t}{4}-\frac{3}{8}\cos\frac{3t}{4}$$

$$=3e^{3t/2}-\frac{1}{4}\sinh\frac{4t}{3}-\frac{4}{9}\cosh\frac{4t}{3}+\frac{2}{3}\sin\frac{3t}{4}-\frac{3}{8}\cos\frac{3t}{4}$$

Example 6. Evaluate

$$L^{-1}\left\{\frac{s+1}{s^2+6s+25}\right\}$$

Solution. $L^{-1}\left\{\frac{s+1}{s^2+6s+25}\right\}$

$$=L^{-1}\left\{\frac{(s+3)-2}{(s+3)^2+16}\right\}$$

$$=e^{-3t}\,L^{-1}\left\{\frac{s-2}{s^2+16}\right\}$$

$$=e^{-3t}\left[L^{-1}\left\{\frac{s}{s^2+4^2}\right\}-2\,L^{-1}\left\{\frac{1}{s^2+4^2}\right\}\right]$$

$$=e^{-3t}\left(\cos 4t-\frac{1}{2}\sin 4t\right)$$

Example 7. Find the inverse Laplace transform of $\dfrac{s+2}{s^2-2s+5}$

Solution.

$$L^{-1}\left\{\frac{s+2}{s^2-2s+5}\right\}=L^{-1}\left\{\frac{(s-1)+3}{(s-1)^2+2^2}\right\}$$

$$=L^{-1}\left\{\frac{s-1}{(s-1)^2+2^2}\right\}+L^{-1}\left\{\frac{3}{(s-1)^2+2^2}\right\}$$

$$=e^t L^{-1}\left\{\frac{s}{s^2+2^2}\right\}+\frac{3}{2}e^t L^{-1}\left\{\frac{2}{s^2+2^2}\right\}$$

$$=e^t\cos 2t+\frac{3}{2}e^t\sin 2t$$

Example 8. Find $L^{-1}\left\{\frac{3s+7}{s^2-2s-3}\right\}$

Solution. $L^{-1}\left\{\frac{3s+7}{s^2-2s-3}\right\}=L^{-1}\left\{\frac{3(s-1)+10}{(s-1)^2-4}\right\}$

$$=3L^{-1}\left\{\frac{s-1}{(s-1)^2-2^2}\right\}+10L^{-1}\left\{\frac{1}{(s-1)^2-4}\right\}$$

$$=3e^t L^{-1}\left\{\frac{s}{s^2-2^2}\right\}+10e^t L^{-1}\left\{\frac{1}{s^2-2^2}\right\}$$

$$=3e^t\cosh 2t+5e^t\sin h\,2t$$

$$=4e^{3t}-e^{-t}$$

Another method.

$$\frac{3s+7}{s^2-2s-3}=\frac{3s+7}{(s-3)(s+1)}$$

$$=\frac{A}{s-3}+\frac{B}{s+1}$$

$$\Rightarrow 3s+7=A(s+1)+B(s-3)$$

$$\Rightarrow A=4,\ B=-1$$

$$\therefore \frac{3s+7}{s^2-2s-3}=\frac{4}{s-3}-\frac{1}{s+1}$$

$$\Rightarrow L^{-1}\left\{\frac{3s+7}{s^2-2s-3}\right\} = 4L^{-1}\left\{\frac{1}{s-3}\right\} - L^{-1}\left\{\frac{1}{s+1}\right\}$$

$= 4e^{3t} - e^{-t}$

Inverse Laplace transform of derivatives

If $L\{F(t)\} = f(s)$

$\Rightarrow L^{-1}\{f(s)\} = F(t)$

& $L\{t^n F(t)\} = (-1)^n f^n(s)$

$\Rightarrow L^{-1}\{f^n(s)\} = (-1)^n t^n F(t)$

$\Rightarrow L^{-1}\{f^n(s)\} = (-1)^n t^n L^{-1}\{f(s)\}$

Example 9. Find $L^{-1}\left\{\log\left(\frac{s+3}{s+2}\right)\right\}$

Solution.

Let $f(s) = \log\left(\frac{s+3}{s+2}\right) = \log(s+3) - \log(s+2)$

Then $f'(s) = \frac{1}{s+3} - \frac{1}{s+2}$

$$L^{-1}\{f'(s)\} = L^{-1}\left\{\frac{1}{s+3}\right\} - L^{-1}\left\{\frac{1}{s+2}\right\}$$

$= e^{-3t} - e^{-2t}$

But $L^{-1}\{f'(s)\} = (-1)^1 t^1 L^{-1}\{f(s)\} = -t L^{-1}\{f(s)\}$

$\Rightarrow L^{-1}\{f(s)\} = -\frac{1}{t} L^{-1}\{f'(s)\}$

$= -\frac{1}{t}\left(e^{-3t} - e^{-2t}\right)$

$= \frac{1}{t}\left(e^{-2t} - e^{-3t}\right)$

Example 10. Find $L^{-1}\left\{\log\left(1+\frac{1}{s^2}\right)\right\}$

Solution. Let $f(s) = \log\left(1+\frac{1}{s^2}\right)$

$$= \log\left(\frac{s^2+1}{s^2}\right)$$

$$= \log\left(s^2+1\right) - \log s^2$$

$$= \log(s^2+1) - 2\log s$$

$$\therefore f'(s) = \frac{2s}{s^2+1} - \frac{2s}{s^2}$$

$$\Rightarrow f'(s) = 2\left(\frac{s}{s^2+1} - \frac{1}{s}\right)$$

$$\Rightarrow L^{-1}\{f'(s)\} = 2\left[L^{-1}\left\{\frac{s}{s^2+1}\right\} - L^{-1}\left\{\frac{1}{s}\right\}\right]$$

$$= 2(\cos t - 1)$$

$$\Rightarrow -t\,L^{-1}\{f(s)\} = -2(1-\cos t)$$

$$\Rightarrow L^{-1}\left\{\log\left(1+\frac{1}{s^2}\right)\right\} = \frac{2}{t}(1-\cos t)$$

Example 11. Find $L^{-1}\left\{\frac{s}{\left(s^2+a^2\right)^2}\right\}$

Solution. We have $L\{\sin at\} = \frac{a}{s^2+a^2}$

$$\therefore L\left\{\frac{1}{s^2+a^2}\right\} = \frac{\sin at}{a}$$

$$\Rightarrow L^{-1}\left\{\frac{d}{ds}\left(\frac{1}{s^2+a^2}\right)\right\} = (-1)^1 t\,\frac{\sin at}{a}$$

$$\Rightarrow L^{-1}\left\{-\frac{2s}{\left(s^2+a^2\right)^2}\right\} = -\frac{t}{a}\sin at$$

$$\Rightarrow L^{-1}\left\{\frac{s}{\left(s^2+a^2\right)^2}\right\} = \frac{1}{2a}\,t\sin at$$

Division by s.

Theorem. If $L^{-1}\{f(s)\} = F(t)$, then

$$L^{-1}\left\{\frac{f(s)}{s}\right\} = \int_0^t F(u)\,du$$

Proof. From the property of Laplace transform, we know that

$$\frac{f(s)}{s} = L\left[\int_0^t F(u)\,du\right]$$

$$L^{-1}\left\{\frac{f(s)}{s}\right\} = \int_0^t F(u)\,du$$

Example 12. Find $L^{-1}\left\{\frac{1}{s^2(s+1)^2}\right\}$

Solution. We know that

$$L\{e^{at}\} = \frac{1}{s-a}$$

$$L\{e^{-at}\} = \frac{1}{s+a}$$

or $$L^{-1}\left\{\frac{1}{s+a}\right\} = e^{-at}$$

or $$L^{-1}\left\{\frac{1}{s+1}\right\} = e^{-t}$$

or $$L^{-1}\left\{\frac{d}{ds}\left(\frac{1}{s+1}\right)\right\} = (-1)^1 t^1 e^{-t} = -te^{-t}$$

or $$L^{-1}\left\{-\frac{1}{(s+1)^2}\right\} = -t e^{-t}$$

or $$L^{-1}\left\{\frac{1}{(s+1)^2}\right\} = te^{-t}$$

Since we know

$$L^{-1}\left\{\frac{f(s)}{s}\right\} = \int_0^t F(x)\,dx$$

Then we have

$$L^{-1}\left\{\frac{1}{s(s+1)^2}\right\} = \int_o^t x\, e^{-x}\, dx$$

$$= \left[-x\, e^{-x} + \int e^{-x}\, dx\right]_o^t$$

$$= \left[-e^{-x}\,(x+1)\right]_o^t$$

$$= 1 - e^{-t}.(t+1)$$

$$\therefore\; L^{-1}\left\{\frac{1}{s^2(s+1)^2}\right\} = \int_o^t \left[1-e^{-x}\,(x+1)\right]dx$$

$$= \left[x + e^{-x}\,(x+1) + e^{-x}\right]_o^t$$

$$= t + e^{-t} + e^{-t}(t+1) - (1+1)$$

$$= te^{-t} + 2e^{-t} + t-2$$

Example 13. Find the inverse Laplace transform of **(U.P.T.U. 2005)**

$$\frac{5s+3}{(s-1)\,(s^2+2s+5)}$$

Solution. $L^{-1}\left\{\dfrac{5s+3}{(s-1)\,(s^2+2s+5)}\right\}$

$$\frac{5s+3}{(s-1)\,(s^2+2s+5)} = \frac{A}{s-1} + \frac{Bs+c}{s^2+2s+5}$$

$$\Rightarrow 5s+3 = A\,(s^2+2s+5) + (B\,s+C)\,(s-1)$$

$$\Rightarrow 5s+3 = s^2\,(A+B) + s\,(2A-B+C) + (5A-C)$$

Comparing the coefficients

$A + B = o$ --------- (i)

$2\,A - B + C = 5$ ---------(ii)

$5\,A - C = 3$ --------(iii)

on adding equations (i) and (ii), we have,

$3\,A + C = 5$ -------(iv)

Adding equations (iii) and (iv)

$$8A = 8 \Rightarrow A = 1$$

putting A = 1 in (iii) we get

$$C = 2$$

Putting A = 1, C = 2 in (ii), we get

$$B = -1$$

Then $\dfrac{5s+3}{(s-1)(s^2+2s+5)} = \dfrac{1}{s-1} + \dfrac{-s+2}{s^2+2s+5}$

$$= \frac{1}{s-1} - \frac{s-2}{(s+1)^2+2^2}$$

$$= \frac{1}{s-1} - \frac{s+1}{(s+1)^2+(2)^2} + \frac{3}{(s+1)^2+2^2}$$

$$\Rightarrow L^{-1}\left\{\frac{5s+3}{(s-1)(s^2+2s+5)}\right\} = L^{-1}\left\{\frac{1}{s-1}\right\} + L^{-1}\left\{\frac{3}{(s+1)^2+2^2}\right\} - L^{-1}\left\{\frac{s+1}{(s+1)^2+2^2}\right\}$$

$$= e^t + 3e^{-t} L^{-1}\left\{\frac{1}{s^2+2^2}\right\} - e^{-t} L^{-1}\left\{\frac{s}{s^2+2^2}\right\}$$

$$= e^t + 3e^{-t}\frac{1}{2}\sin 2t - e^{-t}\cos 2t$$

Example 14. obtain the inverse Laplace transform of $\cot^{-1}\left(\dfrac{s+3}{2}\right)$

(U.P.T.U 2002)

Solution. We know that

$$L^{-1}\{f(s)\} = -\frac{1}{t} L^{-1}\left\{\frac{d}{ds} f(s)\right\}$$

$$\therefore L^{-1}\left\{\cot^{-1}\left(\frac{s+3}{2}\right)\right\} = -\frac{1}{t} L^{-1}\left\{\frac{d}{ds}\cot^{-1}\left(\frac{s+3}{2}\right)\right\}$$

$$= -\frac{1}{t} L^{-1}\left\{\frac{-\frac{1}{2}}{1+\left(\frac{s+3}{2}\right)^2}\right\}$$

$$= -\frac{1}{2t}L^{-1}\left\{\frac{4}{2^2 + (s+3)^2}\right\}$$

$$= \frac{1}{t}L^{-1}\left(\frac{-2}{2^2 + (s+3)^2}\right)$$

$$= \frac{1}{t}e^{-3t}L^{-1}\left(\frac{2}{2^2 + s^2}\right)$$

$$= \frac{e^{-3t}}{t}\sin 2t$$

Convolution Theorem.

If F (t) and G (t) be two functions of class A and $L^{-1}\{f(s)\} = F(t)$ and $L^{-1}\{g(s)\} = G(t)$,. Then

$$L^{-1}\{f(s)\, g(s)\} = \int_0^t F(x)\, G(t-x)\, dx$$ (U.P.T.U 2002)

$$= F * G$$

Proof. Let $\int_0^t F(x)\, G(t-x)\, dx = H(t)$

Then $L\{H(t)\} = \int_0^\infty e^{-st} H(t)\, dt$

$$= \int_0^\infty e^{-st}\left[\int_0^t F(t)\, G(t-x)\, dx\right] dt$$

$$= \int_0^\infty \left[\int_0^t e^{-st} F(x)\, G(t-x)\, dx\right] dt \quad \text{------------------------ (i)}$$

the integration being first with respect to x and then t. If the order of integration is changed, the strip will be taken parallel to OT so that the limits of t are from x to ∞ and of x from o to ∞

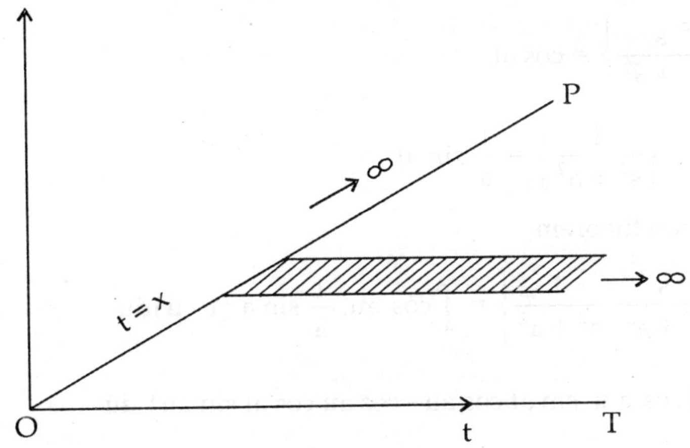

$$L\{H(t)\} = \int_0^\infty dx \int_x^\infty e^{-st}\, F(x)\, G(t-x)\, dt$$

$$= \int_0^\infty e^{-sx}\, F(x)\, dx \int_x^\infty e^{-s(t-x)}\, G(t-x)\, dt$$

putting $t - x = \theta$, we have $dt = d\theta$

$$L\{H(t)\} = \int_0^\infty e^{-sx}\, F(x) \int_0^\infty e^{-s\theta}\, G(\theta)\, d\theta\, dx$$

$$= \int_0^\infty e^{-sx}\, F(x)\, g(s)\, dx$$

$$= f(s)\, g(s)$$

putting the value of H (t), we get

$$L\left\{\int_0^t F(x)\, G(t-x)\, dx\right\} = f(s)\, g(s) \qquad \text{------------- (ii)}$$

$$\text{or} \quad \int_0^t F(x)\, G(t-x)\, dx = L^{-1}\{f(s)\, g(s)\}$$

$$= F * G$$

Example 15. Use convolution theorem to find

$$L^{-1}\left\{\frac{s}{\left(s^2 + a^2\right)^2}\right\}$$

Solution.

We have

$$L^{-1}\left\{\frac{s}{s^2+a^2}\right\} = \cos at$$

and $$L^{-1}\left\{\frac{1}{s^2+a^2}\right\} = \frac{1}{a}\sin at$$

∴ By Convolution theorem

$$L^{-1}\left\{\frac{s}{s^2+a^2}\ \frac{1}{s^2+a^2}\right\} = \int_0^t \cos au.\ \frac{1}{a}\sin a\,(t-u)du$$

$$= \frac{1}{a}\int_0^t (\cos a\,u.\ \sin at\cos au - \cos au\cos at\sin au)\ du$$

$$= \frac{1}{a}\int_0^t \left[\sin at\left(\frac{1+\cos 2\,a\,u}{2}\right) - \frac{1}{2}\cos at\sin 2au\right] du$$

$$= \frac{1}{a}\left[\frac{1}{2}\sin at\left(u+\frac{1}{2a}\sin 2au\right)_0^t + \frac{1}{4a}\cos at\,(\cos 2au)_0^t\right]$$

$$= \frac{1}{a}\left[\frac{t}{2}\sin at + \frac{1}{4a}(\sin at\sin 2\,at + \cos at\cos 2at) - \frac{1}{4a}\cos at\right]$$

$$= \frac{1}{a}\left[\frac{t}{2}\sin at + \frac{1}{4a}\cos\,(2at - at) - \frac{1}{4a}\cos at\right]$$

$$= \frac{t}{2a}\sin at$$

Example 16. Use the convolution theorem to find

$$L^{-1}\left\{\frac{1}{(s+1)\ (s-2)}\right\}$$

Solution. Since $L^{-1}\left\{\frac{1}{s+1}\right\} = e^{-t}$

and $L^{-1}\left\{\frac{1}{s-2}\right\} = e^{2t}$

By Convolution theorem, we get

$$L^{-1}\left\{\frac{1}{s+1}\cdot\frac{1}{(s-2)}\right\} = \int_0^t e^{-x}\ e^{2(t-x)}\ dx$$

$$= \int_0^t e^{2t} e^{-3x} dx$$

$$= e^{2t} \left\{ -\frac{e^{-3x}}{3} \right\}_0^t$$

$$= \frac{1}{3}\left(e^{2t} - e^{-t}\right)$$

HEAVISIDE'S EXPANSION THEOREM

Let f (s) and g (s) be two polynomials in s, where f(s) has degree less than that of g (s), if g (s) has n distinct zeros, $\alpha_r = 1, 2,$ --------------n i.e. g (s) = $(s - \alpha_1)(s - \alpha_2)$--- --------------- $(s - \alpha_n)$

$$\text{Then } L^{-1}\left\{\frac{f(s)}{g(s)}\right\} = \sum_{r=1}^{n} \frac{f(\alpha_r)}{g'(\alpha_r)} e^{\alpha_r t}$$

Example 17. Find $L^{-1}\left\{\frac{2s^2 - 6s + 5}{s^3 - 6s^2 + 11s - 6}\right\}$ **(U.P.T.U 2004)**

Solution. Here, we have

$f(s) = 2s^2 - 6s + 5$

$g(s) = s^3 - 6s^2 + 11s - 6$

$= (s - 1)(s - 2)(s - 3)$

$g'(s) = 3s^2 - 12s + 11$

g (s) has three distinct zeros i.e.,

$\alpha_1 = 1, \alpha_2 = 2, \alpha_3 = 3$

Therefore, by Heaviside's expansion formula, we have

$$L^{-1}\left\{\frac{2s^2 - 6s + 5}{s^3 - 6s^2 + 11s - 6}\right\} = \frac{f(1)}{g'(1)} e^{t} + \frac{f(2)}{g'(2)} e^{2t} + \frac{f(3)}{g'(3)} e^{3t}$$

$$= \frac{1}{2} e^{t} - e^{2t} + \frac{5}{2} e^{3t}$$

APPLICATION TO DIFFERENTIAL EQUATIONS

The Laplace transform method of solving differential equations yields particular solutions without the necessity of first finding the general solution and then evaluating the arbitrary constants. This method is, in general, shorter than our earlier methods and is especially useful for solving linear differential equations with constant coefficients. Thus Laplace transform is a very powerful technique to solve linear differential equations both ordinary and partial and system of simultaneous differential equations.

Example 1. Using Laplace transform, find the solution of the initial value problem

$$\frac{d^2y}{dt^2} + 9y = 6\cos 3t \qquad \text{(U.P.T.U. 2006)}$$

where y (o) = 2, y′ (o) = O

Solution. Given differential equations is

$$y''(t) + 9y(t) = 6\cos 3t$$

Taking Laplace transform on both sides, weget

$$L\{y''(t)\} + 9L\{y(t)\} = 6L\{\cos 3t\}$$

$$\Rightarrow s^2y(s) - sy(o) - y'(o) + 9y(s) = \frac{6s}{s^2+9}$$

$$\Rightarrow s^2y(s) - 2s - o + 9y(s) = \frac{6s}{s^2+9}$$

$$\Rightarrow (s^2+9)\,y(s) - 2s = \frac{6s}{s^2+9}$$

$$y(s) = \frac{6s}{(s^2+9)^2} + \frac{2s}{s^2+9}$$

Taking the inverse Laplace transform of both sides, we get

$$L^{-1}\{y(s)\} = L^{-1}\left\{\frac{6s}{(s^2+9)^2}\right\} + L^{-1}\left\{\frac{2s}{s^2+9}\right\}$$

$$= 6\,.\,\frac{t}{2\,.\,3}\sin 3t + 2\cos 3t \quad \because L^{-1}\left\{\frac{s}{(s^2+a^2)^2}\right\} = \frac{t}{2a}\sin at$$

$$= t\sin 3t + 2\cos 3t$$

Example 2. Solving by using the Laplace transform method.

$$y''(t) + 4y'(t) + 4y(t) = 6e^{-t} \qquad \text{(U.P.T.U. 2007)}$$

Where y (o) = - 2, y′ (o) = 8

Solution. Taking the Laplace transform of the given differential equation, we get

$L\{y''(t)\} + 4L'(t) + 4y(t)\} = 6L\{e^{-t}\}$

$$\Rightarrow L\{y''(t)\} + 4L\{y'(t)\} + 4L\{y(t)\} = \frac{6}{s+1}$$

$$\Rightarrow \left[s^2 y(s) - sy(o) - y'(o)\right] + 4\left[sy(s) - y(o)\right] + 4y(s) = \frac{6}{s+1}$$

$$\Rightarrow \left[s^2 y(s) - s(-2) - 8\right] + 4\left[sy(s) - (-2)\right] + 4y(s) = \frac{6}{s+1}$$

$$\Rightarrow \left(s^2 + 4s + 4\right) y(s) + 2s - 8 + 8 = \frac{6}{s+1}$$

$$\Rightarrow (s+2)^2 y(s) = \frac{6}{s+1} - 2s$$

$$\Rightarrow (s+2)^2\ y(s) = \frac{6 - 2s^2 - 2s}{s+1}$$

$$\Rightarrow y(s) = \frac{6 - 2s^2 - 2s}{(s+1)\ (s+2)^2}$$

$$= \frac{6}{s+1} - \frac{8}{s+2} - \frac{2}{(s+2)^2} \text{ by partial fractions}$$

. Taking the inverse Laplace transform of both sides, we get

$$L^{-1}\{y(s)\} = L^{-1}\left\{\frac{6}{s+1} - \frac{8}{s+2} - \frac{2}{(s+2)^2}\right\}$$

$$= 6L^{-1}\left\{\frac{1}{s+1}\right\} - 8L^{-1}\left\{\frac{1}{s+2}\right\} - 2L^{-1}\left\{\frac{1}{(s+2)^2}\right\}$$

$$= 6e^{-t} - 8e^{-zt} - 2e^{-2t}\ L^{-1}\left\{\frac{1}{s^2}\right\}$$

$$\Rightarrow y(t) = 6e^{-t} - 8e^{-2t} - 2t\,e^{-2t}$$

Example 3. Solve $(D^2 + 9)\,y = \cos 2t$ (U.P.T.U. 2002)

If $y(o) = 1$, $y\left(\frac{\pi}{2}\right) = -1$

Solution. The given differential equation is

$$y''(t) + 9y(t) = \cos 2t$$

Taking the Laplace transform of both sides of the given eq^n, we have

$$L\{y''(t)\} + 9L\{y(t)\} = L\{\cos 2t\}$$

Or $\left\{s^2y(s) - sy(o) - y'(o)\right\} + 9y(s) = \dfrac{s}{s^2+4}$

Or $\left(s^2+9\right)y(s) - s - A = \dfrac{s}{s^2+4}$ Where $y'(o) = A$

or $y(s) = \dfrac{s+A}{s^2+9} + \dfrac{s}{\left(s^2+9\right)\left(s^2+4\right)}$

$$= \frac{s}{s^2+9}\ \frac{A}{s^2+9} + \frac{s}{5\left(s^2+4\right)} - \frac{s}{5\left(s^2+9\right)}$$

Taking inverse Laplace transform on both sides, we have

$$L^{-1}\{y(s)\} = L^{-1}\left\{\frac{s}{s^2+3^2}\right\} + L^{-1}\left\{\frac{A}{s^2+3^2}\right\} + \frac{1}{5}L^{-1}\left\{\frac{s}{s^2+2^2}\right\} - \frac{1}{5}L^{-1}\left\{\frac{s}{s^2+3^2}\right\}$$

$$\Rightarrow y(t) = \cos 3t + \frac{A}{3}\sin 3t + \frac{1}{5}\cos 2t - \frac{1}{5}\cos 3t$$

$$\Rightarrow y(t) = \frac{4}{5}\cos 3t + \frac{A}{3}\sin 3t + \frac{1}{5}\cos 2t$$

Now, it is given that $y(\pi/2) = -1$

$$\therefore -1 = \frac{4}{5}\cos\frac{3\pi}{2} + \frac{A}{3}\sin\frac{3\pi}{2} + \frac{1}{5}\cos\pi$$

or $-1 = -\dfrac{A}{3} - \dfrac{1}{5}$

or $A = \dfrac{12}{5}$

Hence the required solution will be $y(t) = \dfrac{4}{5}\cos 3t + \dfrac{4}{5}\sin 3t + \dfrac{1}{5}\cos 2t$

Example 4. Use Laplace transform to solve the simultaneous differential equation

$$\frac{dx}{dt} + y = \sin t, \frac{dy}{dt} + x = \cos t$$

given that $x = 2$, $y = o$ at $t = o$ (U.P.T.U. 2004)

Solution. Taking Laplace transform on both sides of the given differential equations, be we get

$$L\{x'(t)\} + L\{y(t)\} = L\{\sin t\}$$

$$\& \; L\{y'(t)\} + L\{x(t)\} = L\{\cos t\}$$

$$\Rightarrow s\,x(s) - x(o) + y(s) = \frac{1}{s^2+1}$$

and $sy(s) - y(o) + x(s) = \dfrac{s}{s^2+1}$

Here $x(o) = 2$, $y(o) = o$

$$\Rightarrow sx(s) + y(s) = \frac{1}{s^2+1} + 2$$

$$\& \; x(s) + sy(s) = \frac{s}{s^2+1}$$

Eliminating x (s) in the above equations, we get

$$\left(s^2-1\right)y(s) = \frac{s^2}{s^2+1} - \frac{1}{s^2+1} - 2$$

$$= -\frac{s^2-3}{s^2+1}$$

or $y(s) = \dfrac{-s^2-3}{\left(s^2+1\right)\left(s^2-1\right)} = -\dfrac{2}{s^2-1} + \dfrac{1}{s^2+1}$

Taking inverse Laplace transform , we get

$$y(t) = -2\sin ht + \sin t$$

Eliminating y (s) we get

$$x(s) = \frac{s}{s^2+1} + \frac{2s}{s^2-1} - \frac{s}{s^2+1}$$

$$\text{or } x(s) = \frac{2s}{s^2-1}$$

Taking inverse Laplace transform, we get

$$x(t) = 2\cos ht$$

Thus, the required solution is

$$x = 2\cos ht,\; y = 2\sin ht + \sin t$$

Solution of ordinary differential equations with Variable coefficients :- The Laplace transform technique is very useful is solving the equations with variable coefficients. The method is found useful in case of the equations having the terms of the from $t^m y^n(t)$ whose Laplace transform is

$$(-1)^m \frac{d^m}{ds^m} L\left\{y^n(t)\right\}$$

Example 5. Solve $ty'' + y' + 4ty = o$ if $y(o) = 3, y'(o) = o$

Solution. Taking the Laplace transform of both sides of the given equation, we get

$$L\{t y''\} + L\{y'\} + 4L\{ty\} = o$$

$$\text{Or } \frac{-d}{ds} L\{y''\} + L\{y'\} + 4(-1)\frac{d}{ds} L\{y\} = o$$

$$\text{Or } \frac{-d}{ds}\left[s^2 L\{y\} - sy(o) - y'(o)\right] + \left[s L\{y\} - y(o)\right] - 4\frac{d}{ds}L\{y\} = o$$

$$\text{or } -\frac{d}{ds}\left(s^2 z - 3s\right) + (sz - 3) - 4\frac{dz}{ds} = o$$

where $z = L\{y\}$

$$\text{or } -\left(s^2+4\right)\frac{dz}{ds} - sz = o$$

$$\text{or } \frac{dz}{z} + \frac{s}{s^2+4}\,ds = o$$

Integrating , we get

$$\log z + \frac{1}{2}\log\left(s^2+4\right) = \log C_1$$

$$\text{or } z = \frac{C_1}{\sqrt{s^2+4}}$$

$$\text{or } L\{y\} = \frac{C_1}{\sqrt{s^2+4}}$$

$$y = C_1 L^{-1}\left\{\frac{1}{\sqrt{\left(s^2+4\right)}}\right\} \quad \because L\{J_o(t)\} = \frac{1}{\sqrt{1+s^2}}$$

$$\&L\ \{J_o\ (at)\} = \frac{1}{\sqrt{s^2 + a^2}}$$

or $y = C_1\ J_o\ (2t)$

Since $y\ (o) = 3 \therefore 3 = C_1 J_o\ (o) = C_1 \because J_o\ (o) = 1$

$\therefore y = 3\ J_o\ (2t)$

This is the required solution.

Example 6. Solve $[t\ D^2 + (1 - 2\ t)\ D - 2]\ y = o$ (U.P.T.U. 2002)

If $y\ (o) = 1, y'\ (o) = 2$

Solution. The given equation can be written as

$$ty'' + y' - 2ty' - 2\ y = o$$

Taking Laplace transform of both sides, we get

$L\{ty''\} + L\{y'\} - 2\ L\{t\ y'\} - 2\ L\{y\} = o$

$$\text{Or } -\frac{d}{ds}\ L\ \{y''\} + L\ \{y'\} + 2\ \frac{d}{ds}\ L\ \{y'\} - 2L\ \{y\} = o$$

$$\text{Or } -\frac{d}{ds}\left[s^2L\ \{y\} - sy\ (o) - y'\ (o)\right] + \left[s\ L\{y\} - y(o)\right] + 2\ \frac{d}{ds}\left[sL\{y\} - y(o)\right] - 2\ L\{y\} = o$$

$$\text{Or } -\frac{d}{ds}\ \left(s^2z - s - 2\right) + (sz - 1) + 2\frac{d}{ds}\ (sz - 1) - 2z = o \quad \text{Where } L\{y\} = z$$

$$\text{Or } -\left(s^2 - 2s\right)\frac{dz}{ds} - sz = o$$

$$\text{Or } \frac{dz}{z} + \frac{1}{s - 2}\ ds = o$$

Integrating, we get

$$\log z + \log (s - 2) = \log C_1$$

$$\text{Or } z = \frac{C_1}{s - 2}$$

$$\text{Or } L\ \{y\} = \frac{C_1}{s - 2}$$

$$\therefore y = C_1L^{-1}\left\{\frac{1}{s - 2}\right\} = C_1e^{2t}$$

But $y\ (o) = 1, \quad \therefore 1 = C_1$

$y = e^{2t}$

This is the required solution.

Example 7. A resistance R in series with inductance M is connected with e.m.f. E (t). The current i is given by $M\frac{di}{dt} + Ri = E(t)$ if the switch is connected at t=o and disconnected at t=a, find the current i in terms of t (U.P.T.U. 2001)

Solution.

Condition under which current i flows are i = o at t = o and $E(t) = \begin{cases} E & o < t < a \\ o & t > a \end{cases}$

$$\text{We have } M\frac{di}{dt} + Ri = E(t) \quad \text{--------(i)}$$

Taking Laplace transform of (i) we have

$$M\{L(di/dt)\} + RL\{i\} = L\{E(t)\}$$

$$M[sL\{i\} - i\{o\}] + RL\{i\} = \int_o^\infty e^{-st} E(t)\,dt$$

$$MsL\{i\} + RL\{i\} = \int_o^\infty e^{-st} E(t)\,dt \quad \because i(o) = O$$

$$(Ms + R)L\{i\} = \int_o^\infty e^{-st} E\,dt = \int_o^a e^{-st} E\,dt + \int_a^\infty e^{-st} E\,dt$$

$$\text{Now } E\int_o^a e^{-st}\,dt = E\left[\frac{e^{-st}}{-s}\right]_o^a$$

$$= \frac{E}{s}\left(1 - e^{-as}\right)$$

$$= \frac{E}{s} - \frac{E}{s}e^{-as}$$

$$\Rightarrow L\{i\} = \frac{E}{s(Ls + R)} - \frac{E\,e^{-as}}{s(Ls + R)}$$

on inversion, we obtain

$$i = L^{-1}\left\{\frac{E}{s(Ls + R)}\right\} - L^{-1}\left\{\frac{Ee^{-as}}{s(Ls + R)}\right\} \quad \text{-----------(ii)}$$

$$\text{But } L^{-1}\left\{\frac{E}{s(Ls+R)}\right\} = \frac{E}{L} L^{-1}\left\{\frac{1}{s\left(s+\frac{R}{L}\right)}\right\}$$

Resolving into partial fractions

$$= \frac{E}{L}\,\frac{L}{R}\, L^{-1}\left\{\frac{1}{s} + \frac{1}{s+\frac{R}{L}}\right\}$$

$$= \frac{E}{R}\left\{1 - e^{-(R/L)t}\right\}$$

$$\text{and } L^{-1}\left\{\frac{E\, e^{-as}}{s(Ls+R)}\right\} = \frac{E}{R}\left\{1 - e^{-\frac{R}{L}/(t-a)}\right\} u(t-a) \text{ By second shifting theorem}$$

Putting the values of the inverse transforms in (ii) we get

$$i = \frac{E}{R}\left[1 - e^{-(R/L)t}\right] - \frac{E}{R}\left[1 - e^{-\frac{R}{L}/(t-a)}\right] u\,(t-a)$$

$$\Rightarrow i = \frac{E}{R}\left[1 - e^{-(R/L)t}\right] \text{ for } o < t < a,\ u\,(t-a) = O$$

$$i = \frac{E}{R}\left[1 - e^{-(R/L)t}\right] - \frac{E}{R}\left\{1 - e^{-\frac{R}{L}/(t-a)}\right\} \text{ for } t > a,\ u\,(t-a) = 1$$

$$= \frac{E}{R}\left[1 - e^{-\frac{R}{L}/(t-a)} - e^{-(R/L)t}\right]$$

EXERCISE

Find the Laplace transform of the following:-

1. (a) $3t^4 - 2t^3 + 4e^{-3t} - 2\sin 5t + 3\cos 2t$

Ans. $\frac{72}{s^6} - \frac{12}{s^4} + \frac{4}{s+3} - \frac{10}{s^2+25} + \frac{3s}{s^2+4}$

(b) Find L {f(t)} where

$$f(t) = \begin{cases} \sin\left(t - \frac{\pi}{3}\right), & t > \frac{\pi}{3} \\ o & , t < \frac{\pi}{3} \end{cases}$$

Ans. $e^{-s\pi/3} \cdot \frac{1}{s^2+1}$, s>o

(c) Find L {1 + $te^{-t}\}^3$

Ans. $\frac{1}{s} + \frac{3}{(s+1)^2} + \frac{6}{(s+2)^3} + \frac{6}{(s+3)^4}$

(d) $t^3 e^{7t}$

Ans. $\frac{6}{(s-7)^4}$

(e) e^{-2t} (3 cos 6t – 5 sin 6t)

Ans. $\frac{3s - 24}{s^2 + 4s + 4o}$

2. (a) Express the following function in terms of unit step functions and find its Laplace transform.

$$f(t) = \begin{cases} 8, & t < 2 \\ 6, & t \geq 2 \end{cases}$$

Ans. H (t-2), $\frac{8}{s} - \frac{2e^{-2s}}{s}$

(b) Express the following functions in term of a unit step function and find its Laplace transform.

$$f(t) = \begin{cases} E, & a < t < b \\ o, & t \geq b \end{cases}$$

Ans. $H(t-a) - H(t-b), E\left(\frac{e^{-as}}{s} - \frac{e^{-bs}}{s}\right)$

(c) Express the function shown in the diagram in terms of unit step function and obtain its Laplace transform.

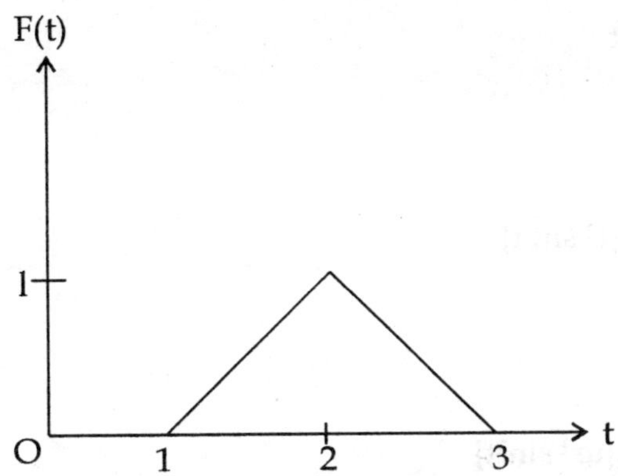

Ans. $(t-3)\ u\ (t-3)\ -2\ (t-2)\ u\ (t-2)+(t-1)\ u\ (t-1)$ & $\frac{e^{-s}\left(1-e^{-s}\right)^2}{s^2}$

(d) Find the Laplace transform of the function.

$$f(t) = \begin{cases} t-1 & ,1<t<2 \\ 3-t & 2<t<3 \end{cases}$$ (U.P.T.U. 2009)

Hint : $f(t) = (t-1)\left[u\ (t-1)-u\ (t-2)\right] + (3-t)\left[u(t-2)\ -\ u\ (t-3)\right]$

$\Rightarrow f(t) = (t\text{-}1)\ u\ (t\text{-}1) - 2\ (t\text{-}2)\ u\ (t\text{-}2) + (t\text{-}3)\ u\ (t\text{-}3)$

$\Rightarrow L\{f(t)\} = \frac{e^{-s}}{s^2} - \frac{2e^{-2s}}{s^2} + \frac{e^{-3s}}{s^2}$

$\Rightarrow L\{f(t)\} = \frac{1}{s^2}\left(e^{-s} - 2e^{-2s} + e^{-3s}\right)$

3. (a) If (t) = t^2, o<t<2, and f (t+2) = f(t), find L{f(t)}

Ans. $\frac{1}{1-e^{-2s}}\left[\frac{2}{s^3} - \frac{4e^{-2s}}{s} - \frac{4e^{-2s}}{s^2} - \frac{2e^{-2s}}{s^3}\right]$

(b) Find L{F(t)}, where F(t) is defined by

$$F(t) = \begin{cases} \sin t & o<t<\pi \\ o & \pi<t<2\pi \end{cases}$$

and $F(t + 2\pi) = F(t)$

Ans. $\frac{1}{\left(1-e^{-\pi s}\right)\left(1+s^2\right)}$

4. (a) Find the Laplace transform of the function.

$F(t) = te^{-t} \sin 2t$ (U.P.T.U. 2002)

Ans. $\frac{4(s+1)}{(s^2+2s+5)^2}$

(b) Evaluate $L\{t^2 \sin t\}$

Ans. $\frac{2(3s^2-1)}{(s^2+1)^3}$.

(c) Evaluate $L\{te^{-t} \sin^2 t\}$

Ans. $\frac{1}{2}\frac{1}{(s+1)^2}+\frac{1}{2}\left[\frac{4-3(s+1)^4}{\{(s+1)^4+4\}^2}\right]$

5. (a) Find the Laplace transform of

$f(t) = \frac{\sin^2 t}{t}$

Ans. $\frac{1}{4}\log\frac{s^2+4}{s^2}$

(b) If $F(t) = (\cos at - \cos bt)/t$, find the Laplace transform of F(t).

(U.P.T.U. 2004)

Ans. $\frac{1}{2}\left[\log\frac{s^2+b^2}{s^2+a^2}\right]$

(c) Find the Laplace transform of

$F(t) = \frac{1-\cos t}{t^2}$

Ans. $\cot^{-1} s + \frac{s}{2}\log\frac{s^2}{s^2+1}$

(d) Find the Laplace transform of

$F(t) = \frac{e^{-at}-e^{-bt}}{t}$

Ans. $\log\frac{s+b}{s+a}$

6. Find $L\left\{ \text{erf} \sqrt{t} \right\}$ and hence prove that

$$L\left\{t \text{ erf} \left(2\sqrt{t}\right)\right\} = \frac{3s + 8}{s^2 (s+4)^{3/2}}$$ (U.P.T.U. 2001)

Ans. $\frac{1}{s(s+1)^{1/2}}$

7. (a) Find $L^{-1}\left\{\frac{1}{s(s^2 + 4)}\right\}$

Ans. $\frac{1}{2}\ (1 - \cos\ 2t)$

(b) Find $L^{-1}\left\{\frac{1}{s^3\ (s^2 + 1)}\right\}$

Ans. $\frac{t^2}{2}$ + cost -1

(c) Find $L^{-1}\left\{\frac{4}{s^2 + 4s + 2o}\right\}$

Ans. $e^{-2t} \sin 4t$

(d) Find $L^{-1}\left\{\frac{(3s-1)}{s^2 - 6s + 2}\right\}$

Ans. $3e^{3t} \cos\left(\sqrt{7}t\right) + \frac{8}{\sqrt{7}}\ e^{3t}\ \sin h\ \sqrt{7}t$

(e) Find $L^{-1}\left\{\frac{s+8}{s^2 + 4s + 5}\right\}$

Ans. e^{-2t} (cos t + 6 sin t)

(f) Find $L^{-1}\left\{s / \left(s^2 + a^2\right)^2\right\}$

Ans $\frac{t}{2a}$ sin at

(g) Evaluate $L^{-1}\left\{\left(e^{-s} - 3e^{-3s}\right) / s^2\right\}$ (U.P.T.U. 2002)

Ans. (t-1) u (t-1) -3 (t-3) u (t-3)

7. Find the inverse Laplace transform of the following.

(a) $\frac{2s^2 - 1}{(s^2+1)(s^2+4)}$ (U.P.T.U. (c.o) 2004)

Ans. $-\sin t + \frac{3}{2}\sin 2t$

(b) $\frac{2s^2 - 6s + 5}{s^3 - 6s^2 + 11s - 6}$ (U.P.T.U. 2004)

Ans. $\frac{1}{2}e^t - e^{2t} + \frac{5}{2}e^{3t}$

(c) $\frac{e^{-cs}}{s^2(s+a)}$ (U.P.T.U. 2002)

Ans. $-\frac{1}{a^2}u(t-c)+\frac{1}{a}(t-c)u(t-c)+\frac{1}{a^2}e^{-a(t-c)}u(t-c)$

(d) $\frac{s}{s^4+s^2+1}$

Ans. $\frac{2}{\sqrt{3}}\sinh\frac{t}{2}\sin\frac{1}{2}\sqrt{3}\,t$

(e) $\log\left(\frac{s+a}{s+b}\right)$ (U.P.T.U. 2003)

Ans. $\frac{e^{-bt} - e^{-at}}{t}$

(f) $\log\left(1+\frac{1}{s}\right)$ (U.P.T.U. 2007)

Ans. $\frac{1-e^{-t}}{t}$

(g) $\tan^{-1}(s+1)$

Ans. $-\frac{1}{t}e^{-t}\sin t$

8. Using Convolution theorem, evaluate the following:

(a) $L^{-1}\left\{\frac{s}{(s^2+1)(s^2+4)}\right\}$ (U.P.T.U. 2002)

Ans. $\frac{1}{3}(\cos t - \cos 2t)$

(b) $L^{-1}\left\{\frac{1}{s^3(s^2+1)}\right\}$ (U.P.TU. 2005)

Ans. $\frac{t^2}{2} + \cos t - 1$

(c) $L^{-1}\left\{\frac{s^2}{(s^2+a^2)(s^2+b^2)}\right\}$ $a \neq b$ (U.P.T.U. 2004)

Ans. $\frac{1}{a^2-b^2}(a \sin at - b \sin bt)$

(d) $L^{-1}\left\{\frac{1}{(s-2)(s^2+1)}\right\}$

Ans. $\frac{1}{5}(e^{2t} - 2\sin t - \cos t)$

(e) $L^{-1}\left\{\frac{1}{s(s^2+4)^2}\right\}$

Ans. $\frac{1}{16}(1 - t\sin 2t - \cos 2t)$

(f) $L^{-1}\left\{\frac{1}{(s-1)^5(s+2)}\right\}$

Ans. $\frac{e^t}{72}\left(t^4 - \frac{4}{3}t^3 + \frac{4}{3}t^2 - \frac{8}{9}t + \frac{8}{27}\right) - \frac{e^{-2t}}{243}$.

(g) $L^{-1}\left\{\frac{1}{\left(s^2+a^2\right)^2}\right\}$ (U.P.T.U. 2009)

Hint – By convolution theorem

$$L^{-1}\{f(s)\,g(s)\} = \int_0^t F(u)\,G(t-u)\,du$$

$$f(s) = g(s) = \frac{1}{s^2 + a^2}$$

$$\Rightarrow f(t) = G(t) = \frac{1}{a}\sin\ at$$

$$\Rightarrow L^{-1}\left\{\frac{1}{\left(s^2 + a^2\right)^2}\right\} = \int_0^t \frac{1}{a}\ \sin au\ \frac{1}{a}\sin\ a\ (t-u)du$$

$$\Rightarrow L^{-1}\left\{\frac{1}{\left(s^2 + a^2\right)^2}\right\} = \frac{1}{2a^3}\ (\sin at - at \cos at)$$

9. Using Heaviside's expansion formula find

(a) $L^{-1}\left\{\frac{3s+1}{(s-1)\ (s^2+1)}\right\}$

Ans. $2e^t + \sin t - 2 \cos t$

(b) $L^{-1}\left\{\frac{s^2 - 6}{s^3 + 4s^2 + 3s}\right\}$

Ans. $-2 + \frac{5}{2}e^{-t} + \frac{1}{2}e^{-3t}$

10. Using Laplace transformation, solve the following differential equations.

(a) $\frac{d^2y}{dt^2} + y = t \cos 2t$, t>o (U.P.T.U. 2005)

given that $y = \frac{dy}{dt}$ =o for t = o

Ans. $y(t) = \frac{-5}{9}\ \sin t + \frac{4}{9}\sin 2t - \frac{t}{3}\cos 2t$

(b) $\frac{d^2y}{dx^2} + 2\frac{dy}{dx} + 5y = e^{-x} \sin\ x$

Where y (o) = o, y′(o) = 1 (U.P. T.U. 2004, 08)

Ans. $y = \frac{1}{3}e^{-x}(\sin\ x + \sin 2x)$

(c) $(D^2-2D+2)\ y = o,\ y = Dy = 1$ when $t = o$

Ans. $y = e^t \cos t$

11. (a) Solve the simultaneous equation

$$\frac{dx}{dt} - y = e^t, \frac{dy}{dt} + x = \sin t$$

given $x(0) = 1,\ y(0) = 0$ (U.P.T.U. 2006)

Ans. $x(t) = \frac{1}{2}(e^t + 2\sin t \cos t - t\cos t)$

$$y(t) = \frac{1}{2}(t\sin t - e^t + \cos t - \sin t)$$

(b) Using Laplace transformation, solve

$(D - 2)\ x - (D + 1)\ y = 6e^{3t}$

$(2D - 3)x + (D - 3)y = 6e^{3t}$

Given $x = 3,\ y = 0$ when $t = 0$ (U.P.T.U. 2001)

Ans. $x\ (t) = e^t + 2t\ e^t + 2e^{3t}$

$y(t) = \sin ht + \cos ht - e^{-3t} - te^t$

(c) $\frac{dx}{dt} + 4\frac{dy}{dt} - y = 0$

$\frac{dx}{dt} + 2y = e^{-t}$ with condition $x(0) = y(0) = 0$ (U.P.T.U. 2008)

Ans. $x(t) = \frac{1}{3} - \frac{5}{7}e^{-t} + \frac{8}{21}e^{3t/4},\ y(t) = \frac{1}{7}e^{-t} - \frac{1}{7}e^{\frac{3}{4}t}$

12. The coordinates (x, y) of a particle moving along a plane curve at any time t are given by

$$\frac{dy}{dt} + 2x = \sin 2t, \frac{dx}{dt} - 2y = \cos 2t, (t > 0)$$

It is given that at $t = 0$, $x = 1$ and $y = 0$, show using transforms that the particle moves along the curve $4x^2 + 4xy + 5y^2 = 4$ (U.P.T.U. 2003)

Ans. $x(t) = \frac{1}{2}\sin 2t + \cos 2t$

$y(t) = -\sin 2t$

13(a) Sove $y'' - ty' + y = 1$

if $y(0)= 1, y'(0) =2$ (Raj SLET 2001)

Ans. $y = 2t +1$

(b) Solve $[t D^2 + (1 -2t) D - 2] y =0$

if $y(0) = 1, y'(0) =2$

Ans. $y = e^{2t}$

Objective Problems

Pick up the correct answer from the following

1. De Laplace was a

(a) German Mathematician (b) French Mathematican
(c) American Mathematican (d) British Mathematican

Ans. (b)

2. PAUL DIRAC (1902-1984), English Physicist, was awarded the Nobel Prize [jointly with----------- (1887-1961)] in 1933 for his work in quantum mechanics

(a) Laplace (b) Euler
(c) Erwin Schrödinger (d) C.V. Raman

Ans. (c)

3. Laplace transform of F(t); $t \geq o$ is defined as

(a) $\int_0^1 e^{-st} f(s)ds$ (b) $\int_0^{\infty} e^{-st} F(t)dt$

(c) $\int_{-\infty}^{\infty} e^{-st} F(t)ds$ (d) $\int_0^1 e^{-st} F(t)dt$

Ans. (b)

4. $L^{-1}\left\{\frac{1}{s^n}\right\}$ is possible only when n is

(a) Zero (b) –ve integer
(c) +ve integer (d) negative rational

Ans. (c)

5. Laplace transform of F(t) = 1 is

(a) 1 (b) $\frac{1}{s}$

(c) $\frac{1}{s^2}$ (d) o

Ans. (b)

6. $L\{e^t \cos t\}$ is equal to

(a) $\frac{s+1}{s^2+2s+2}$ (b) $\frac{s-1}{s^2-2s-2}$

(c) $\frac{s+1}{s^2-2s+2}$ (d) $\frac{s-1}{s^2-2s+2}$

Ans. (d)

7. L {e^{at} t^n} is equal to

(a) $\dfrac{\lfloor n}{(s+a)^n}$ (b) $\dfrac{\lfloor n}{(s-a)^n}$

(c) $\dfrac{\lfloor n}{(s+a)^{n+1}}$ (d) $\dfrac{\lfloor n}{(s-a)^{n+1}}$

Ans. (d)

8. L{F″(t)} is equal to

(a) s^2 f(s) – s F(o) – F′(o)

(b) s^2 f(s) + s F (o) – F′ (o)

(c) s^2 f(s) – sF(o) + F′ (o)

(d) s^2 f(s) – s F′(o) – F″ (o)

Ans. (a)

9. The inverse Laplace transform of $\dfrac{\lfloor n}{(s-a)^n}$, s > a is

(a) e^{at} t^n (b) e^t e^{nt}

(c) e^{at} t^{n-1} (d) e^{t+1} t^{na}

Ans. (c)

10. The inverse Laplace transform of $\dfrac{\lceil n}{s^n}$, s > a

(a) t^n (b) t^{n-1}

(c) t^{n+1} (d) nt

Ans. (b)

11. If L{F(t)} = f(s) , then L{e^{at} F(t)} is equal to

(a) f (s-a) (b) f(s+a)

(c) f(a-s) (d) f(s)

Ans. (a)

12. If L{F(t)} = f(s), then L{t^n F(t)} is equal to

(a) $f^n(s)$ (b) $(-1)^n$ f^n (s)

(c) $(-1)^n$ f′(s) (d) t^n f(s)

Ans. (b)

13. $L\left\{\dfrac{1-\cos at}{a^2}\right\}$ is equal to

(a) $\dfrac{1}{s^2+a^2}$ (b) $\dfrac{1}{s\left(s^2+a^2\right)}$

(c) $\dfrac{1}{a}$ (d) $\dfrac{1}{s^2+a^2}$

14. If $L\{F(t)\} = f(s)$, then $L\left[\int_0^t F(u)du\right]$ is equal to

(a) $s\,f(s)$ (b) $\frac{1}{s}\,f(s)$

(c) $\frac{1}{s}\,f\left(\frac{1}{s}\right)$ (d) $sf\left(\frac{1}{s}\right)$

Ans. (b)

15. The volue of $\int_0^\infty \frac{e^{-at} - e^{-bt}}{t} dt$ is equal to

(a) log a/b (b) log (a-b)

(c) log $(a^2 - b^2)$ (d) log (b/a)

Ans. (d)

16. $\int_0^\infty \frac{\sin t}{t} dt$ is equal to

(a) o (b) $\frac{\pi}{4}$

(c) π (d) $\pi/2$

Ans. (d)

17. The value of $\int_0^\infty \frac{e^{-t} - e^{-3t}}{t} dt$ is equal to

(a) log 3 (d) log (5/3)

(c) log 4 (d) log (1/3)

Ans (a)

18. The value of $\int_0^\infty te^{-2t} \sin t\, dt$ is equal to

(a) $\frac{4}{5}$ (b) $\frac{3}{7}$

(c) $\frac{4}{25}$ (d) $\frac{4}{125}$

Ans. (c)

19. $L\{F(t)\} = f(s)$, then $L\{F(at\}$ is equal to

(a) f (s/a) (b) $\frac{1}{a} f(s/a)$

(c) a f(s/a) (d) $\frac{1}{a} f(as)$

Ans. (b)

20. $L^{-1}\left\{\dfrac{s^2}{\left(s^2+4\right)^2}\right\}$ is equal to

(a) $\sin 2t + \frac{t}{2}\cos 2t$ (b) $\frac{1}{4}\sin 2t + \frac{t}{2}\cos 2t$

(c) $\frac{1}{4}\sin 2t + t\cos 2t$ (d) $\frac{1}{4}\sin 2t + \frac{t}{4}\sin 4t$

Ans. (b)

21. $L\{J_o(t)\}$ is equal to

(a) $\dfrac{1}{\sqrt{s^2+1}}$ (b) $\dfrac{-1}{\sqrt{s^2+1}}$

(c) $\dfrac{1}{\sqrt{(s^2-1)}}$ (d) $\dfrac{s}{\sqrt{s^2-1}}$

Ans. (a)

22. $L^{-1}\left[\dfrac{1}{(s-4)^2}\right]$ is equal to

(a) $\dfrac{t^3}{\underline{|3}}e^{3t}$ (b) $\dfrac{t^3}{6}e^{4t}$

(c) $\dfrac{t^3}{3}e^{4t}$ (d) $\dfrac{t^4}{\underline{|4}}e^{4t}$

Ans. (b)

23. $L^{-1}\left\{\tan^{-1}\dfrac{1}{s}\right\}$ is equal to

(a) $\sin t$ (b) $\dfrac{\sin t}{t}$

(c) $t \sin t$ (d) $t^2 \sin t$

Ans. (b)

24. If $L^{-1}\{f(s)\} = F(t)$ and $F(o)= O$, then $L^{-1}\{sf(s)\}$ is equal to

(a) $f(t)$ (b) $f''(t)$

(c) $F'(t)$ (d) $tF'(t)$

Ans. (c)

25. $L^{-1}\left\{\dfrac{e^{-\pi s}}{s^2+1}\right\}$ is equal to

(a) $e^{-t}\sin t$ (b) $e^t \sin t$

(c) $-\sin t \,.\, u(t-\pi)$ (d) $\sin t.\, u(t)$

Ans. (c)

26. $L^{-1}\left\{\dfrac{1}{s^2(s^2+1)}\right\}$ is equal to

(a) t + sin t
(b) t – sin t
(c) l – sin t
(d) t + cos t

Ans . (b)

27. $L^{-1}\left\{\dfrac{e^{-3s}}{s^3}\right\}$ is equal to

(a) $\dfrac{(t-3)^3}{\lfloor 3}H(t-3)$
(b) $\dfrac{(t-3)^3}{\lfloor 2}H(t-3)$
(c) $\dfrac{(t-3)^2}{\lfloor 2}H(t-3)$
(d) $\dfrac{(t-3)^3}{\lfloor 3}$

Ans. (c)

28. If H (t - a) is Heaviside unit step function, then F (t-a) H (t –a) is equal to

(a) L^{-1} {f(s)}
(b) L^{-1} {e^{-at} f(s)}
(c) L^{-1} {e^{-as} f(s)}
(d) L^{-1} {e^{as} f(s)}

Ans. (c)

29. If $\dfrac{d^2y}{dx^2}+y=o$, under the conditions y=1, $\dfrac{dy}{dx}=o$, when t = o, then y is equal to

(a) sin t
(b) cos t
(c) tan t
(d) cot t

Ans. (b)

30. If $L\{F(t)\} = f(s)$, then $L\left\{\dfrac{f(t)}{t}\right\}$ is equal to

(a) $\int_s^\infty f(x)dx$
(b) $\int_0^\infty f(s)ds$
(c) $\dfrac{1}{s}\int_s^\infty f(s)dx$
(d) $\dfrac{1}{s}\int_0^\infty f(s)ds$

Ans. (a)

31. $L\{\sin\sqrt{t}\}$ is equal to

(a) $\sqrt{\left(\dfrac{\pi}{s}\right)}\,e^{-1/4s}$
(b) $\dfrac{\sqrt{\pi}}{2s^{3/2}}e^{-1/4s}$

(c) $\frac{\sqrt{\pi}}{s^{3/2}} e^{-1/4s}$ (d) $\frac{1}{2}\sqrt{\left(\frac{\pi}{s}\right)} e^{-1/4s}$

Ans. (b)

32. $L\left\{\frac{\cos\sqrt{t}}{\sqrt{t}}\right\}$ is equal to

(a) $\left(\frac{\pi}{s}\right)^{1/2} e^{-1/4s}$ (b) $\frac{1}{2}\left(\frac{\pi}{2}\right)^{1/2} e^{-1/4.s}$

(c) $\frac{\sqrt{\pi}}{2s^{3/2}} e^{-1/4s}$ (d) $\frac{\sqrt{\pi}}{s^{3/2}} e^{-1/4s}$

Ans. (a)

33. $L^{-1}\left\{\frac{1}{(s-a)^3}\right\}$ is equal to

(a) $t^2 e^{at}$ (b) $\frac{1}{2} t^2 e^{at}$

(c) $\frac{1}{3} t^2 e^{at}$ (d) $\frac{1}{3} t\, e^{at}$

Ans. (b)

34. $L^{-1}\left\{\frac{1}{(s+a)^2}\right\}$ is equal to

(a) e^{at} (b) e^{-at}

(c) $t\, e^{-at}$ (d) te^{at}

Ans. (c)

35. $L\{ t^4 e^{-at}\}$ is equal to

(a) $\frac{\lfloor 4}{(s+a)^4}$ (b) $\frac{\lfloor 5}{(s+a)^5}$

(c) $\frac{\lfloor 4}{(s+a)^3}$ (d) $\frac{\lfloor 5}{(s-a)^5}$

Ans. (b)

36. $L^{-1}\left\{\frac{1}{s(s^2+1)}\right\}$ is equal to

(a) 1 + sin t (b) 1 – sin t

(c) 1 + cost (d) 1 – cost

Ans. (d)

37. The Laplace transform of the function

$$f(t) = \begin{cases} 1, & o \leq t < 2 \\ -1, & 2 \leq t < 4, \end{cases}$$

$f(t + 4) = f(t)$ is given as (U.P.T.U. 2009)

(a) $\dfrac{1-e^{-2s}}{s\left(1+e^{-2s}\right)}$ (b) $\dfrac{1+e^{-2s}}{s\left(1+e^{-2s}\right)}$

(c) O (d) $\dfrac{s+1}{s-1}$

Ans. (a)

38. The inverse Laplace transform of $\log\left(\dfrac{s+1}{s-1}\right)$ is given by

(U.P.T.U. 2009)

(a) $\dfrac{2}{t}$ cosh t (b) $\dfrac{2}{t}$sin ht

(c) 2t cost (d) 2t sin t

Ans. (b)

39. $L^{-1}\left\{\dfrac{s}{s^2-a^2}\right\}$ is equal to

(a) cos at (b) sin at

(c) cosh at (d) sin h at

Ans. (c)

40. $L^{-1}\left\{\dfrac{1}{s^2\left(s^2+1\right)}\right\}$ is equal to

(a) t + sin t (b) t - sin t

(c) l-sin t (d) t + cost

Ans. (b)

Select true or false answers in the following

1. Laplace transform of F(t) is defined for positive and negative value of t. (True or False)

Ans. False

2. $L^{-1}\left\{\dfrac{s}{s^2+a^2}\right\} = \cos at$

(True or False)

Ans. True

3. $L^{-1}\left\{\frac{1}{\left(s^2+a^2\right)^2}\right\} = \frac{1}{2a3}(\sin\ at - at\cos at)$

(True or False)

Ans. True

4. If L{F(t)} = f(s), then

$L\{t\,F(t)\} = \frac{d}{ds}\{f(s)\}$

Ans. False

5. $\frac{d^2}{ds^2}\left[L\{F(t)\} - L\left\{t^2F(t)\right\}\right] = o$

True or False

Ans. True

6. Match the items on the right hand side with those on the left hand side

(i) L {cos 4t}	(a) $\frac{2a\ (s-a)}{\left[(s-a)^2+a^2\right]^2}$
(ii) L {t sin 2t}	(b) $\frac{e^{-3t}}{t}\sin\ 2t$
(iii) L {t e^{at} sin at}	(c) $\frac{s}{s^2+16}$
(iv) $L^{-1}\left\{\cot^{-1}\left(\frac{s+3.}{2}\right)\right\}$	(d) $\frac{2s}{\left(s^2+4\right)^2}$

Ans. (i) -------- (c)
(ii)-------- (d)
(iii)--------(b)
(iv)--------(a)

UNIT - IV
FOURIER SERIES AND PARTIAL DIFFERENTIAL EQUATIONS

Chapter 7
Fourier Series

Introduction

Most of the events in nature and many other systems, being periodic in nature. In many engineering problems, especially in the study of periodic phenomena in conduction of heat, electro-dynamics and acoustic, it is necessary to express a function in a series of sines and cosines. Most of the single-valued functions which occur in applied mathematics can be expressed in the form

$$\frac{a_o}{2} + a_1\cos x + a_2\cos 2x + a_3\cos 3x + \ldots\ldots\ldots + b_1\sin x + b_2\sin 2x + b_3\sin 3x + \ldots\ldots$$

where $a_o, a_1, a_2, a_3, \ldots\ldots\ldots, b_1, b_2, b_3, \ldots\ldots\ldots$ are real constants and the series is known as Fourier series. The constants $a_o, a_1, a_2, a_3, \ldots\ldots, b_1, b_2, b_3, \ldots\ldots$ are called Fourier's Coefficients of the periodic function.

Euler's Formulae

The Fourier series for the function f(x) in the interval $\alpha < x < \alpha + 2\pi$ is given by

$$f(x) = \frac{a_o}{2} + \sum_{n=1}^{\infty} a_n \cos nx + \sum_{n=1}^{\infty} b_n \sin nx \qquad \text{(i)}$$

Determination of the Fourier's Constants:

To find a_o: assume that (i) can be integrated from x = c to x = c+2π, term by term, so that we have

$$\int_C^{C+2\pi} f(x)\,dx = \frac{a_o}{2}\int_C^{C+2\pi} 1\,dx + \int_C^{C+2\pi}\left(\sum_{n=1}^{\infty} a_n \cos nx\right)dx + \int_C^{C+2\pi}\left(\sum_{n=1}^{\infty} b_n \sin nx\right)dx$$

$$= \frac{a_o}{2}(x)_C^{C+2\pi} + 0 + 0$$

$$= \frac{a_o}{2}(C + 2\pi - C) = \pi a_o$$

or $$a_o = \frac{1}{\pi}\int_C^{C+2\pi} f(x)\,dx$$

To find a_n: multiplying each side of (i) by cosnx and integrate with respect to x between the limits x = C to x = C + 2π, so that we have

$$\int_C^{C+2\pi} f(x)\cos nx\,dx = \frac{a_o}{2}\int_C^{C+2\pi}\cos nx\,dx + \int_C^{C+2\pi}\left(\sum_{n=1}^{\infty} a_n \cos nx\right)\cos nx\,dx + \int_C^{C+2\pi}\left(\sum_{n=1}^{\infty} b_n \sin nx\right)\cos nx\,dx$$

$$= 0 + \int_C^{C+2\pi}\sum_{n=1}^{\infty} a_n \cos^2 nx\,dx + 0$$

$$= a_n \pi$$

or $$a_n = \frac{1}{\pi}\int_C^{C+2\pi} f(x)\cos nx\,dx$$

To find b_n multiply both side of (i) by sinnx and integrate with respect to x between the limits x = c to x = C+2π, so that we have

$$\int_C^{C+2\pi} f(x)\sin nx\,dx = \frac{a_o}{2}\int_C^{C+2\pi}\sin nx\,dx + \int_C^{C+2\pi}\left(\sum_{n=1}^{\infty} a_n \cos nx\right)\sin nx\,dx + \int_C^{C+2\pi}\left(\sum_{n=1}^{\infty} b_n \sin nx\right)\sin nx\,dx$$

$$= 0 + 0 + \int_C^{C+2\pi}\sum_{n=1}^{\infty} b_n \sin^2 nx\,dx = b_n\pi$$

$$b_n = \frac{1}{\pi}\int_C^{C+2\pi} f(x)\sin nx\,dx$$

The values $$a_o = \frac{1}{\pi}\int_C^{C+2\pi} f(x)\,dx$$

$$a_n = \frac{1}{\pi}\int_C^{C+2\pi} f(x)\cos nx$$

$$b_n = \frac{1}{\pi}\int_C^{C+2\pi} f(x)\sin nx\,dx$$

are called Euler's formulae

Euler's Formulae for Different Intervals

Case (i): If C = 0, then the interval for the above series (i) become 0<x<2π and the Euler's formulae reduce to

$$a_o = \frac{1}{\pi}\int_o^{2\pi} f(x)\,dx$$

$$a_n = \frac{1}{\pi}\int_o^{2\pi} f(x)\cos nx\,dx$$

$$b_n = \frac{1}{\pi}\int_o^{2\pi} f(x)\sin nx\,dx$$

Case (ii): If c = $-\pi$, then the above interval for the Fourier series become $-\pi<x<\pi$, then the Euler's formulae, reduce to

$$a_o = \frac{1}{\pi}\int_{-\pi}^{\pi} f(x)\, dx$$

$$a_n = \frac{1}{\pi}\int_{-\pi}^{\pi} f(x) \cos nx\, dx$$

$$b_n = \frac{1}{\pi}\int_{-\pi}^{\pi} f(x) \sin nx\, dx$$

Note: Periodic Functions

Consider any function of x as f(x) and let f(x) = f (x+T) = f (x + 2T) = ..., then the function f(x) is said to be periodic with its period T. This T is non-zero, smallest and positive real variable. For example consider f(x) = sin x, then f(x) = sin x = sin (x + 2π) = sin (x + 4π) = ... Hence, f(x) = sinx, is a periodic function with the period 2π. This is also called sinusoidal periodic function

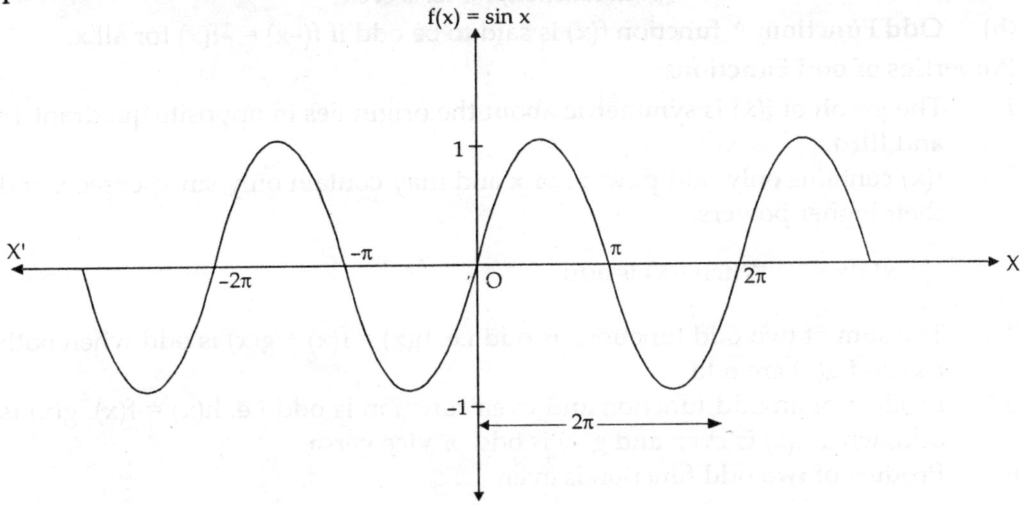

Even and odd Functions

(a) Even Function: A function f(x) is said to be even if f(–x) = f(x) for all x.

Properties of even functions:

1. The graph of f(x) is symmetrical about y axis.
2. f(x) contains only even powers of x and may contain only cos x, secx and their higher powers.
3. $\int_{-a}^{a} f(x)\,dx = 2\int_{0}^{a} f(x)\,dx$, when f(x) is even

4. the sum of two even functions is even i.e. h(x) = f(x) + g(x) is even when both f(x) and g(x) are even
5. Product of two even functions is even i.e. h(x) = f(x) g(x) is even when both f(x) and g(x) are even.
6. The Product of two odd function is an even function

For example: x^2, $2x^4$, cos x, $\cos^2 x$, $3x^2 + 5$, $x^4 + \cos 2x + 2$ are all even functions.

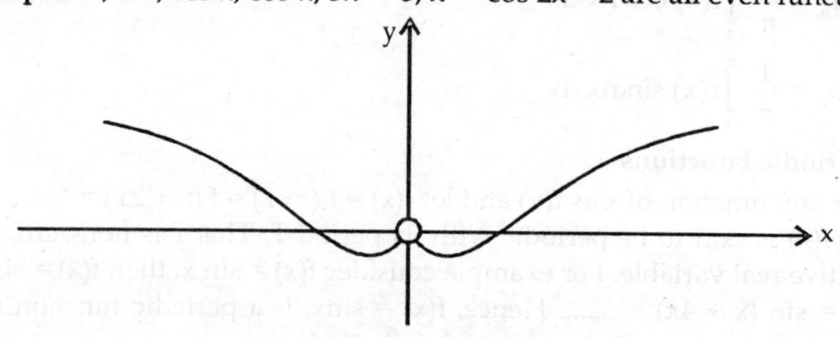

(Even function) cos nx is even

(b) Odd Function: A function f(x) is said to be odd if f(–x) = –f(x) for all x.

Properties of odd Functions:

1. The graph of f(x) is symmetric about the origin lies in opposite quadrant 1st and IIIrd.
2. f(x) contains only odd powers of x and may contain only sin x, cosec x and their higher powers.
3. $\int_{-a}^{a} f(x)\,dx = 0$, When f(x) is add.
4. The sum of two odd functions is odd i.e. h(x) = f(x) + g(x) is odd when both f(x) and g(x) are odd
5. Product of an odd function and even function is odd i.e. h(x) = f(x). g(x) is odd, when f(x) is even and g(x) is odd or vice versa
6. Product of two odd function is even

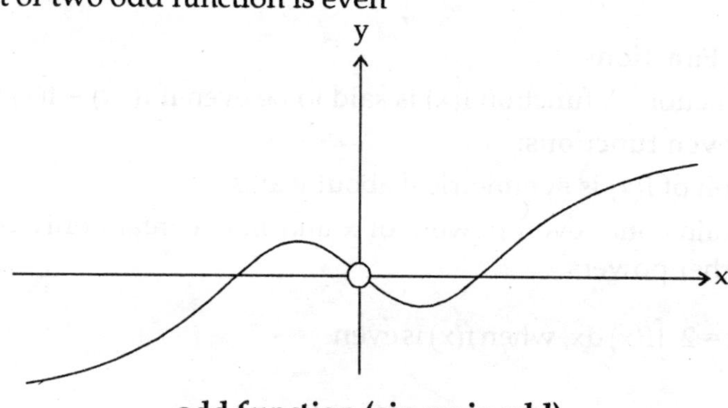

odd function (sin nx is odd)

For example: $2x^3$, sinx, tan x, $x^3 + x$ are odd functions.

Result: Most functions are neither even nor odd. But any function f(x) can be written as the arithmetic mean of an even and odd functions as

$$f(x) = \frac{1}{2}[f(x) + f(-x)] + \frac{1}{2}[f(x) - f(-x)]$$

Example 1. Find a Fourier series to represent $x-x^2$ from $x = -\pi$ to $x = \pi$ and hence deduce

$$\frac{\pi^2}{12} = \frac{1}{1^2} - \frac{1}{2^2} + \frac{1}{3^2} - \frac{1}{4^2} + \dots\dots\dots$$

Solution. Let $x - x^2 = \frac{a_o}{2} + \sum_{n=1}^{\infty} a_n \cos nx + \sum_{n=1}^{\infty} b_n \sin nx$ (1)

$$\text{Then } a_o = \frac{1}{\pi}\int_{-\pi}^{\pi}(x - x)^2\, dx = \frac{1}{\pi}\left[\frac{x^2}{2} - \frac{x^3}{3}\right]_{-\pi}^{\pi} = -\frac{2\pi^2}{3}$$

$$a_n = \frac{1}{\pi}\int_{-\pi}^{\pi}(x - x^2)\cos nx\, dx$$

$$= \frac{1}{\pi}\left[(x - x^2)\frac{\sin nx}{n} - (1 - 2x) - \left(-\frac{\cos nx}{n^2}\right) + (-2)\left(-\frac{\sin nx}{n^3}\right)\right]_{-\pi}^{\pi}$$

$$= \frac{-4(-1)^n}{n^2} \qquad \because \cos n\pi = (-1)^n$$

$$\therefore a_1 = 4/1^2, a_2 = -4/2^2, a_3 = 4/3^2, a_4 = -4/4^2 \text{ etc}$$

Finally

$$b_n = \frac{1}{\pi}\int_{-\pi}^{\pi}(x - x^2)\sin nx\, dx$$

$$= \frac{1}{\pi}\left[(x - x^2)\left(-\frac{\cos nx}{n}\right) - (1 - 2x)\left(-\frac{\sin nx}{n^2}\right) + (-2)\left(\frac{\cos nx}{n^3}\right)\right]_{-\pi}^{\pi}$$

$$= -2(-1)^n/n$$

$\therefore b_1 = 2/1, b_2 = -2/2, b_3 = 2/3, b_4 = -2/4$ etc

Substituting the values of a_o, a_n, b_n in (i) we get

$$x - x^2 = -\frac{\pi^2}{3} + 4\left[\frac{\cos x}{1^2} - \frac{\cos 2x}{2^2} + \frac{\cos 3x}{3^2} - \frac{\cos 4x}{4^2} + \dots\dots\right]$$

$$+ 2\left[\frac{\sin x}{1} - \frac{\sin 2x}{2} + \frac{\sin 3x}{3} - \frac{\sin 4x}{4} + \dots\dots\right]$$

Putting x = 0, in above we get

$$0 = -\frac{\pi^2}{3} + 4\left(\frac{1}{1^2} - \frac{1}{2^2} + \frac{1}{3^2} - \frac{1}{4^2} + \ldots\ldots\right)$$

$$\text{i.e. } \frac{1}{1^2} - \frac{1}{2^2} + \frac{1}{3^2} - \frac{1}{4^2} + \ldots\ldots\ldots = \frac{\pi^2}{12}$$

Example 2. Obtain the Fourier Series of $f(x) = \left(\frac{\pi - x}{2}\right)$ in the interval $(0, 2\pi)$ and hence deduce

$$\frac{\pi}{4} = 1 - \frac{1}{3} + \frac{1}{5} - \frac{1}{7} + \ldots\ldots\ldots \qquad \text{(U.P.T.U. 09)}$$

Solution Since we know

$$f(x) = \frac{a_o}{2} + \sum_{n=1}^{\infty} a_n \cos nx + \sum_{n=1}^{\infty} b_n \sin nx$$

$$\text{Let } \frac{\pi - x}{2} = \frac{a_o}{2} + \sum_{n=1}^{\infty} a_n \cos nx + \sum_{n=1}^{\infty} b_n \sin nx \qquad (1)$$

$$\text{Then } a_o = \frac{1}{\pi}\int_o^{2\pi} f(x)\,dx$$

$$= \frac{1}{\pi}\int_o^{2\pi}\left(\frac{\pi - x}{2}\right)dx = 0$$

$$a_n = \frac{1}{\pi}\int_o^{2\pi} f(x)\cos nx\,dx$$

$$= \frac{1}{\pi}\int_o^{2\pi}\left(\frac{\pi - x}{2}\right)\cos nx\,dx = 0$$

$$\text{Finally, } b_n = \frac{1}{\pi}\int_o^{2\pi} f(x)\sin nx\,dx$$

$$= \frac{1}{\pi}\int_o^{2\pi}\left(\frac{\pi - x}{2}\right)\sin nx\,dx = \frac{1}{n}$$

Substituting the values of a_o, a_n, b_n is (1) we get

$$f(x) = \frac{\pi - x}{2} = \sum_{n=1}^{\infty}\frac{1}{n}\sin nx$$

$$= \sin x + \frac{1}{2}\sin 2x + \frac{1}{3}\sin 3x + \dots\dots\dots$$

Put $x = \frac{\pi}{2}$, we have

$$\frac{\pi}{4} = 1 - \frac{1}{3} + \frac{1}{5} - \frac{1}{7} + \dots\dots\dots$$

Example 3. Given that $f(x) = x + x^2$ for $-\pi<x<\pi$, find the Fourier expression of f(x).

Deduce that $\frac{\pi^2}{6} = 1 + \frac{1}{2^2} + \frac{1}{3^2} + \frac{1}{4^2} + \dots$ (U.P.T.U. 2003)

Solution

$$\text{Let } x + x^2 = \frac{a_o}{2} + \sum_{n=1}^{\infty} a_n \cos nx + \sum_{n=1}^{\infty} b_n \sin nx \quad (1)$$

$$\text{Then } a_o = \frac{1}{\pi}\int_{-\pi}^{\pi} f(x)\,dx = \frac{1}{\pi}\int_{-\pi}^{\pi}(x + x^2)\,dx$$

$$= \frac{1}{\pi}\left[\frac{x^2}{2} + \frac{x^3}{3}\right]_{-\pi}^{\pi}$$

$$= \frac{1}{\pi}\left[\frac{\pi^2}{2} + \frac{\pi^3}{3} - \frac{\pi^2}{2} + \frac{\pi^3}{3}\right] = \frac{2\pi^2}{3}$$

$$a_n = \frac{1}{\pi}\int_{-\pi}^{\pi} f(x)\cos nx\,dx$$

$$= \frac{1}{\pi}\int_{-\pi}^{\pi}(x + x^2)\cos nx\,dx$$

$$= \frac{1}{\pi}\left[(x + x^2)\frac{\sin nx}{n} - (2x+1)\frac{(-\cos nx)}{n^2} + (2)\left(-\frac{\sin nx}{n^3}\right)\right]_{-\pi}^{\pi}$$

$$= \frac{1}{\pi}\left[(2\pi + 1)\frac{\cos n\pi}{n^2} - (-2\pi + 1)\frac{\cos(-n\pi)}{n^2}\right]$$

$$= \frac{1}{\pi}\left[4\pi\frac{\cos n\pi}{n^2}\right] = \frac{4(-1)^n}{n^2}$$

$$\text{Finally, } b_n = \frac{1}{\pi}\int_{-\pi}^{\pi} f(x)\sin nx\,dx = \frac{1}{\pi}\int_{-\pi}^{\pi}(x + x^2)\sin nx\,dx$$

$$= \frac{1}{\pi}\left[(x+x^2)\left(-\frac{\cos nx}{n}\right) - (2x+1)\left(-\frac{\sin nx}{n^2}\right) + 2\,\frac{\cos nx}{n^3}\right]_{-\pi}^{\pi}$$

$$= \frac{1}{\pi}\left[-(\pi+\pi^2)\,\frac{\cos n\pi}{n} + 2\,\frac{\cos n\pi}{n^3} + (-\pi+\pi^2)\,\frac{\cos n\pi}{n} - 2\,\frac{\cos n\pi}{n^3}\right]$$

$$= \frac{1}{\pi}\left[-\frac{2\pi}{n}\cos n\pi\right]$$

$$= -\frac{2}{n}(-1)^n$$

Substituting the values of a_o, a_n, b_n in (1) we get

$$x + x^2 = \frac{\pi^2}{3} + 4\left[-\cos x + \frac{1}{2^2}\cos 2x - \frac{1}{3^2}\cos 3x + ...\right] \qquad (2)$$

$$-2\left[-\sin x + \frac{1}{2}\sin 2x - \frac{1}{3}\sin 3x + ...\right]$$

putting $x = \pi$, (2) becomes

$$\pi + \pi^2 = \frac{\pi^2}{3} + 4\left[1 + \frac{1}{2^2} + \frac{1}{3^2} + \frac{1}{4^2} + ...\right] \qquad (3)$$

putting $x = -\pi$, (2) becomes

$$\pi + \pi^2 = \frac{\pi^2}{3} + 4\left[1 + \frac{1}{2^2} + \frac{1}{3^2} + \frac{1}{4^2} + ...\right] \qquad (4)$$

Adding (3) and (4), we have

$$2\pi^2 = \frac{2\pi^2}{3} + 8\left[1 + \frac{1}{2^2} + \frac{1}{3^2} + \frac{1}{4^2} + ...\right]$$

$$\frac{4\pi^2}{3} = 8\left[1 + \frac{1}{2^2} + \frac{1}{3^2} + \frac{1}{4^2} + ...\right]$$

$$\frac{\pi^2}{6} = 1 + \frac{1}{2^2} + \frac{1}{3^2} + \frac{1}{4^2} + ... = \sum_{n=1}^{\infty}\frac{1}{n^2}$$

Example 4. Expand $f(x) = x \sin x$ as a Fourier series for $0<x<2\pi$ (U.P.T.U. 2001)

Solution: Let $f(x) = \frac{a_o}{2} + \sum_{n=1}^{\infty} a_n \cos nx + \sum_{n=1}^{\infty} b_n \sin nx$ (1)

Then $a_o = \frac{1}{\pi}\int_o^{2\pi} x \sin x\, dx = \frac{1}{\pi}\left[x(-\cos x) - 1(-\sin x)\right]_o^{2\pi} = -2$

$$a_n = \frac{1}{\pi}\int_o^{2\pi} x \sin x \cos nx\, dx = \frac{1}{2\pi}\int_o^{2\pi} x\,(2\cos nx \sin x)\, dx$$

$$= \frac{1}{2\pi}\int_o^{2\pi} x\,[\sin(n+1)x - \sin(n-1)x]\, dx$$

$$= \frac{1}{2\pi}\left[x\left\{\frac{-\cos(n+1)x}{n+1} + \frac{\cos(n-1)x}{n-1}\right\} - 1\left\{-\frac{\sin(n+1)x}{(n+1)^2} + \frac{\sin(n-1)x}{(n-1)^2}\right\}\right]_o^{2\pi}$$

$$= \frac{1}{2\pi}\left[2\pi\left\{-\frac{\cos 2(n+1)\pi}{n+1} + \frac{\cos 2(n-1)\pi}{n-1}\right\}\right]$$

$$= \frac{2}{n^2-1},\ n \neq 1$$

When $n = 1$, $a_1 = \frac{1}{\pi}\int_o^{2\pi} x \sin x \cos x\, dx$

$$= \frac{1}{2\pi}\int_o^{2\pi} x \sin 2x\, dx$$

$$= \frac{1}{2\pi}\left[x\left(-\frac{\cos 2x}{2}\right) - 1\left(-\frac{\sin 2x}{4}\right)\right]_o^{2\pi}$$

$$= -\frac{1}{2}$$

Finally, $b_n = \frac{1}{\pi}\int_\pi^{2\pi} x \sin x \sin nx\, dx$

$$= \frac{1}{2\pi}\int_o^{2\pi} x\,[\cos(n-1)x - \cos(n+1)x]\, dx$$

$$= \frac{1}{2\pi}\left[x\left\{\frac{\sin(n-1)x}{n-1} - \frac{\sin(n+1)x}{n+1}\right\} - 1\left\{-\frac{\cos(n-1)x}{(n-1)^2} + \frac{\cos(n+1)x}{(n+1)^2}\right\}\right]_o^{2\pi}$$

$$= \frac{1}{2\pi}\left[\frac{\cos 2(n-1)\pi}{(n-1)^2} - \frac{\cos 2(n+1)\pi}{(n+1)^2} - \frac{1}{(n-1)^2} + \frac{1}{(n+1)^2}\right]$$

$= 0, \quad (n \neq 1)$

When n = 1, $b_1 = \frac{1}{\pi}\int_o^{2\pi} x \sin x \sin x \, dx$

$$= \frac{1}{2\pi}\int_o^{2\pi} x\,(1 - \cos 2x)\, dx$$

$$= \frac{1}{2\pi}\left[x\left(x - \frac{\sin 2x}{2}\right) - 1\left(\frac{x^2}{2} + \frac{\cos 2x}{4}\right)\right]_o^{2\pi}$$

$$= \pi$$

Substituting the values of a_o, a_n and b_n we get

$$x \sin x = -1 + \pi \sin x - \frac{1}{2}\cos x + \frac{2}{2^2-1}\cos 2x + \frac{2}{3^2-1}\cos 3x + ...$$

Example 5. Find the Fourier series expansion for

$$f(x) = x + \frac{x^2}{4}, -\pi \le x \le \pi \qquad \text{(U.P.T.U. 2009)}$$

Solution:- Let $f(x) = x + \frac{x^2}{4} = \frac{a_o}{2} + \sum_{n=1}^{\infty} a_n \cos nx + \sum_{n=1}^{\infty} b_n \sin nx \qquad (1)$

Then $a_o = \frac{1}{\pi}\int_{-\pi}^{\pi} f(x)\, dx$

$$= \frac{1}{\pi}\int_{-\pi}^{\pi}\left(x + \frac{x^2}{4}\right) dx$$

$$= \frac{1}{\pi}\left[\frac{x^2}{2} + \frac{x^3}{12}\right]_{-\pi}^{\pi} = \frac{\pi^2}{6}$$

$$a_n = \frac{1}{\pi}\int_{-\pi}^{\pi} f(x)\cos nx\, dx$$

$$= \frac{1}{\pi}\int_{-\pi}^{\pi}\left(x + \frac{x^2}{4}\right)\cos nx\, dx$$

$$= \frac{(-1)^n}{n^2}$$

Finally, $b_n = \frac{1}{\pi}\int_{-\pi}^{\pi} f(x)\sin nx\, dx$

$$= \frac{1}{\pi}\int_{-\pi}^{\pi}\left(x+\frac{x^2}{4}\right)\sin nx\,dx$$

$$= -\frac{2}{n}(-1)^n$$

Substituting the values of a_o, a_n and b_n in (1) we get

$$x+\frac{x^2}{4} = f(x) = \frac{\pi^2}{12}+\sum_{n=1}^{\infty}\frac{(-1)^n}{n^2}\cos nx - 2\sum_{n=1}^{\infty}\frac{(-1)^n}{n}\sin nx$$

Example 6: Obtain the Fourier series for $f(x) = e^{-x}$ in the interval $0<x<2\pi$ (S.V.T.U. 2007)

Solution: Let $e^{-x} = \frac{a_o}{2}+\sum_{n=1}^{\infty}a_n\cos nx+\sum_{n=1}^{\infty}b_n\sin nx$ (1)

Then $a_o = \frac{1}{\pi}\int_{0}^{2\pi}f(x)\,dx$

$$= \frac{1}{\pi}\int_{0}^{2\pi}e^{-x}\,dx = \frac{1}{\pi}\left[-e^{-x}\right]_0^{2\pi}$$

$$= \frac{1-e^{-2\pi}}{\pi}$$

and $a_n = \frac{1}{\pi}\int_{0}^{2\pi}f(x)\cos nx\,dx = \frac{1}{\pi}\int_{0}^{2\pi}e^{-x}\cos nx\,dx$

$$= \frac{1}{\pi(n^2+1)}\left[e^{-x}(-\cos nx+n\sin nx)\right]_0^{2\pi}$$

$$= \left(\frac{1-e^{-2\pi}}{\pi}\right)\frac{1}{n^2+1}$$

$$\therefore\ a_1 = \left(\frac{1-e^{-2\pi}}{\pi}\right)\frac{1}{2},\ a_2 = \left(\frac{1-e^{-2\pi}}{\pi}\right)\frac{1}{5}\ \text{etc}$$

Finally $b_n = \frac{1}{\pi}\int_{0}^{2\pi}f(x)\sin nx\,dx$

$$= \frac{1}{\pi}\int_{0}^{2\pi}e^{-x}\sin nx\,dx$$

$$= \frac{1}{\pi(n^2+1)}\left[e^{-x}(-\sin nx - n\cos nx)\right]_o^{2\pi}$$

$$= \left(\frac{1-e^{-2\pi}}{\pi}\right)\frac{n}{n^2+1}$$

$$\therefore\ b_1 = \frac{1-e^{-2\pi}}{\pi}\ \frac{1}{2},\ b_2 = \left(\frac{1-e^{-2\pi}}{\pi}\right)\frac{2}{5}\ \text{etc}$$

Substituting the values of a_o, a_n, b_n in (1), we get

$$e^{-x} = \frac{1-e^{-2\pi}}{\pi}\left\{\begin{matrix}\frac{1}{2}+\left(\frac{1}{2}\cos x+\frac{1}{5}\cos 2x+\frac{1}{10}\cos 3x+\ldots\ldots.\right)+ \\ \left(\frac{1}{2}\sin x+\frac{2}{5}\sin 2x+\frac{3}{10}\sin 3x+\ldots\ldots\right)\end{matrix}\right\}$$

Example 7. Obtain the Fourier Series expansion for the function $f(x) = x^2, -\pi<x<\pi$

Hence, deduce that

(a) $\frac{1}{1^2}+\frac{1}{2^2}+\frac{1}{3^2}+\frac{1}{4^2}+\ldots\ldots\ldots = \frac{\pi^2}{6}$

(b) $\frac{1}{1^2}-\frac{1}{2^2}+\frac{1}{3^2}-\frac{1}{4^2}+\ldots\ldots\ldots\ldots = \frac{\pi^2}{12}$

(c) $\frac{1}{1^2}+\frac{1}{3^2}+\frac{1}{5^2}+\ldots\ldots\ldots = \frac{\pi^2}{8}$ (UPTU, 2004)

Solution: We have $f(x) = x^2$

f(x) is an even function, therefore, f(x) contains only cosine terms. Hence $b_n = 0$

Let $f(x) = x^2 = \frac{a_o}{2} + \sum_{n=1}^{\infty} a_n \cos nx$ (1)

Then we have $a_o = \frac{1}{\pi}\int_{-\pi}^{\pi} f(x)\,dx = \frac{1}{\pi}\int_{-\pi}^{\pi} x^2\,dx$

$$= \frac{1}{\pi}\left[\frac{x^3}{3}\right]_{-\pi}^{\pi} = \frac{1}{3\pi}\left[\pi^3-(-\pi)^3\right]$$

$$= \frac{2\pi^3}{3\pi} = \frac{2\pi^2}{3}$$

and $a_n = \frac{1}{\pi}\int_{-\pi}^{\pi} f(x)\cos nx\,dx$

$$= \frac{2}{\pi}\int_{-\pi}^{\pi} x^2 \cos nx\,dx$$

$$= \frac{2}{\pi}\left[x^2\left(\frac{\sin nx}{n}\right) - 2x\left(\frac{-\cos nx}{n^2}\right) + 2\left(\frac{-\sin nx}{n^3}\right)\right]_o^{\pi}$$

$$= \frac{2}{\pi}\left[0 + \frac{2\pi}{n^2}\cos n\pi + 0\right]$$

$$= \frac{2}{\pi}\frac{2\pi}{n^2}(-1)^n = \frac{4}{n^2}(-1)^n$$

Substituting in (1), we get

i.e. $x^2 = \frac{\pi^2}{3} + 4\left[\sum_{n=1}^{\infty}\frac{(-1)^n}{n^2}\cos nx\right]$

$$= \frac{\pi^2}{3} + 4\left[\frac{(-1)^1}{1^2}\cos x + \frac{(-1)^2}{2^2}\cos 2x + \frac{(-1)^3}{3^2}\cos 3x + ...\right]$$

$$= \frac{\pi^2}{3} - 4\left[\frac{\cos x}{1^2} - \frac{\cos 2x}{2^2} + \frac{\cos 3x}{3^2} - ....\right]$$

or $x^2 = \frac{\pi^2}{3} - 4\left[\frac{\cos x}{1^2} - \frac{\cos 2x}{2^2} + \frac{\cos 3x}{3^2} - ....\right]$ (2)

At $x = \pi$ and $x = 0$, the function f(x) is continuous

Putting $x = \pi$, is (2), we get

$$\pi^2 = \frac{\pi^2}{3} - 4\left[\frac{\cos\pi}{1^2} - \frac{\cos 2\pi}{2^2} + \frac{\cos 3\pi}{3^2} - ....\right]$$

or $\pi^2 = \frac{\pi^2}{3} - 4\left[-\frac{1}{1^2} - \frac{1}{2^2} - \frac{1}{3^2} - \frac{1}{4^2} - ....\right]$

$$\pi^2 - \frac{\pi^2}{3} = 4\left[\frac{1}{1^2} + \frac{1}{2^2} + \frac{1}{3^2} + ...\right]$$

or $4\left[\frac{1}{1^2} + \frac{1}{2^2} + \frac{1}{3^2} + ...\right] = \frac{2\pi^2}{3}$

i.e. $\frac{1}{1^2} + \frac{1}{2^2} + \frac{1}{3^2} + ... = \frac{\pi^2}{6}$ (3)

putting x = 0, in (2), we get

$$0 = \frac{\pi^2}{3} - 4\left[\frac{\cos 0}{1^2} - \frac{\cos 0}{2^2} + \frac{\cos 0}{3^2} - \dots\dots\dots\right]$$

$$\text{or } 0 = \frac{\pi^2}{3} - 4\left[\frac{1}{1^2} - \frac{1}{2^2} + \frac{1}{3^2} - \frac{1}{4^2} + \dots\dots\right]$$

$$\text{or } 4\left[\frac{1}{1^2} - \frac{1}{2^2} + \frac{1}{3^2} - \frac{1}{4^2} + \dots\dots\right] = \frac{\pi^2}{3}$$

$$\text{or } \frac{1}{1^2} - \frac{1}{2^2} + \frac{1}{3^2} - \frac{1}{4^2} + \dots\dots = \frac{\pi^2}{12} \qquad (4)$$

Adding (3) and (4) we get

$$\frac{1}{1^2} + \frac{1}{3^2} + \frac{1}{5^2} + \dots\dots = \frac{\pi^2}{8}$$

Example 8. Obtain the Fourier series for

$f(x) = |x|$ in $-\pi < x < \pi$

Solution. we have $f(x) = |x|$

since $f(-x) = |-x| = |x| = f(x)$

$f(x)$ is an even function

Therefore f(x) contain only cosine terms and we have $b_n = 0$

$$\text{Let } f(x) = |x| = \frac{a_o}{2} + \sum_{n=1}^{\infty} a_n \cos nx$$

$$\text{we have } a_o = \frac{1}{\pi}\int_{-\pi}^{\pi} f(x)\,dx = \frac{2}{\pi}\int_{o}^{\pi} f(x)\,dx$$

$$= \frac{2}{\pi}\int_{o}^{\pi} |x|\,dx = \frac{2}{\pi}\int_{o}^{\pi} x\,dx$$

$$= \frac{2}{\pi}\left[\frac{x^2}{2}\right]_o^{\pi} = \frac{2}{\pi}\frac{(\pi^2 - o)}{2} = \pi$$

$$\text{and } a_n = \frac{1}{\pi}\int_{-\pi}^{\pi} |x| \cos nx\,dx = \frac{2}{\pi}\int_{o}^{\pi} x \cos nx\,dx$$

$$= \frac{2}{\pi}\left[x\left(\frac{\sin nx}{n}\right) - 1\left(-\frac{\cos nx}{n^2}\right)\right]_o^{\pi}$$

$$= \frac{2}{\pi}\left[0 + \frac{1}{n^2}(\cos nx)_o^\pi\right]$$

$$= \frac{2}{\pi}\frac{1}{n^2}(\cos n\pi - \cos o)$$

$$= \frac{2}{n^2\pi}\left[(-1)^n - 1\right]$$

Clearly $a_n = 0$, when n is even

and $a_n = \frac{-4}{\pi n^2}$, when n is odd

∴ The required Fourier series expansion is

$$|x| = \frac{\pi}{2} - \frac{4}{\pi}\sum_{n=1}^{\infty}\frac{\cos nx}{n^2} \quad \text{(n is odd)}$$

$$\text{i.e. } |x| = \frac{\pi}{2} - \frac{4}{\pi}\left[\cos x + \frac{\cos 3x}{3^2} + \frac{\cos 5x}{5^2} + \ldots\ldots\right]$$

Example 9. obtain the Fourier series of the function

$$f(x) = \begin{cases} 1\,, & -\pi<x<o \\ -1\,, & 0 \le x<\pi \end{cases}$$

Solution: The function f(x) is odd. Therefore, the Fourier series of f(x) contains only sine terms

$$b_n = \frac{1}{\pi}\int_o^\pi f(x)\ \sin nx\, dx$$

$$= \frac{1}{\pi}\int_o^\pi (-1)\sin nx\, dx$$

$$= \frac{1}{\pi}\left[\cos nx\right]_o^\pi$$

$$= \frac{1}{\pi}\left[\cos n\pi - \cos o\right] = \frac{1}{\pi}\left[(-1)^n - 1\right]$$

$$= \begin{cases} o & ; \quad \text{n is even} \\ -4/n\pi & ; \quad \text{n is odd} \end{cases}$$

∴ The Fourier series expansion of f(x) may be written as

$$f(x) = \frac{-4}{\pi}\left[\sin x + \frac{\sin 3x}{3} + \frac{\sin 5x}{5} + \ldots\ldots\right]$$

The graph of f(x) is given in figure as given below

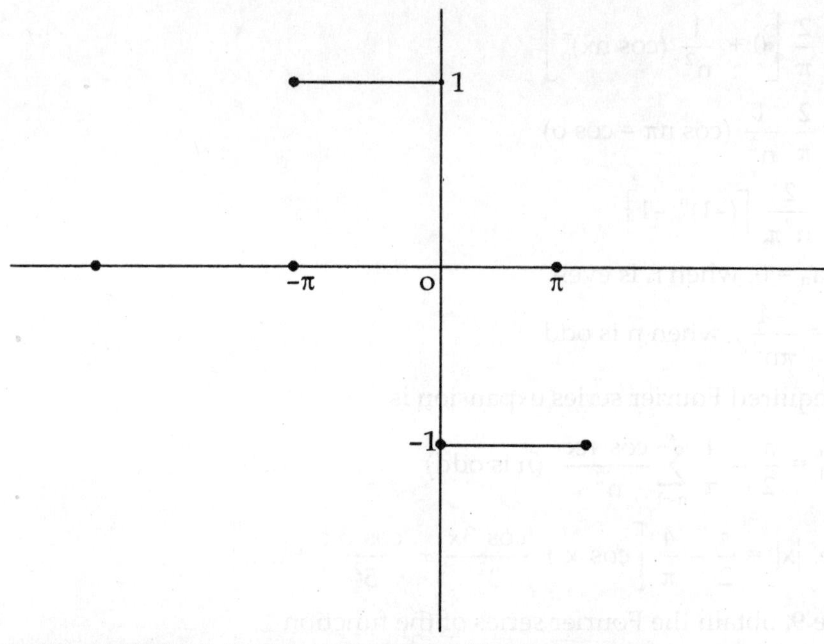

FUNCTIONS HAVING POINT OF DISCONTINUITY

In deriving Euler's formulae for the constants a_o, a_n and b_n we have assumed that f(x) is continuous in the given interval. In some cases f(x) may have a finite number of discontinuities. We can also express such functions as Foruier series. For example, consider a function f(x) defined as follows:

$$f(x) = f_1(x), \qquad c<x<x_o$$
$$= f_2(x), \qquad x_o<x<c+2\pi$$

Where x_o is a point of discontinuities for f(x) is the interval (c, c+2π) and $\lim_{x\to o-} f(x)$ i.e. $f(x_o - o)$ and $\lim_{x\to o+} f(x)$ i.e. $f(x_o + o)$ exist unequal and are finite. We determine the values of a_o, a_n and b_n can be computed as

$$a_o = \frac{1}{\pi}\left[\int_c^{x_o} f_1(x)\,dx + \int_{x_o}^{C+2\pi} f_2(x)\,dx\right]$$

$$a_n = \frac{1}{\pi\cdot}\left[\int_c^{x_o} f_1(x)\cos nx\,dx + \int_{x_o}^{c+2\pi} f_2(x)\cos nx\,dx\right]$$

$$\text{and } b_n = \frac{1}{\pi}\left[\int_c^{x_o} f_1(x)\sin nx\,dx + \int_{x_o}^{c+2\pi} f_2(x)\sin nx\,dx\right]$$

At the point of discontinuity i.e. at $x = x_o$ the Fourier series converges to

$$\frac{1}{2}\,[f(x_o + o) + f(x_o - o)] = \frac{1}{2}\,(FB + FC)$$

If f(x) satisfies Dirichlet's conditions and f(x) is expressed as

$$f(x) = \frac{a_o}{2} + \sum_{1}^{\infty} a_n \cos nx + \sum_{1} b_n \sin nx$$

in the interval {C, C+2π], then f(x) converges to $f(x_o)$ if x_o is a point of continuity of f(x) in the given interval

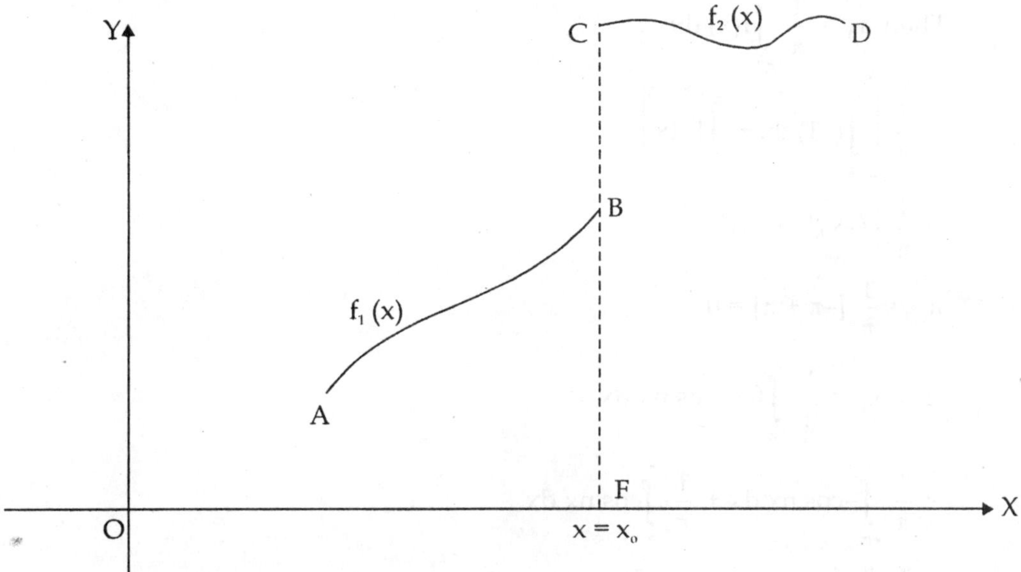

Thus from above we conclude that

(i) It may be seen from the graph, that at a point of finite discontinuity $x = x_o$ there is a finite jump equal to BC in the value of the function f(x) at $x = x_o$

(ii) A given function f(x) may be defined by different formulae in different regions. Such types of functions are quite common in Fourier series

(iii) At a point of discontinuity the sum of the series is equal to mean of the limits on the right and left.

FUNCTION DEFINED IN TWO OR MORE SUB-RANGE:

Example 10. Find Fourier series for the function defined by

$$f(x) = \begin{cases} -1 & \text{for } -\pi<x<o \\ 0 & \text{for } x = o \\ 1 & \text{for } 0<x<\pi \end{cases}$$

hence proved that

$$\frac{\pi}{4} = 1 - \frac{1}{3} + \frac{1}{5} - \frac{1}{7} + \frac{1}{9} + \ldots\ldots \qquad \text{(U.P.T.U. 2005)}$$

Solution: By Fourier series, we have

$$f(x) = \frac{a_o}{2} + \sum_{n=1}^{\infty} a_n \cos nx + \sum_{n=1}^{\infty} b_n \sin nx \qquad (1)$$

Then $a_o = \dfrac{1}{\pi} \displaystyle\int_{-\pi}^{\pi} f(x)\, dx$

$$= \frac{1}{\pi}\left[\int_{-\pi}^{o} (-1)\, dx + \int_{o}^{\pi} 1\, dx\right]$$

$$= \frac{1}{\pi}\left[(-x)_{-\pi}^{o} + (x)_{o}^{\pi}\right]$$

$$a_o = \frac{1}{\pi}[-\pi + \pi] = 0$$

and $a_n = \dfrac{1}{\pi} \displaystyle\int_{-\pi}^{\pi} f(x) \cos nx\, dx$

$$= \frac{1}{\pi}\int_{-\pi}^{o} -\cos nx\, dx + \frac{1}{\pi}\int_{o}^{\pi} \cos nx\, dx$$

$$= \frac{1}{\pi}\left\{-\frac{1}{n}[\sin nx]_{-\pi}^{o} + \frac{1}{n}[\sin nx]_{o}^{\pi}\right\}$$

$$= \frac{1}{\pi}[-o + o] = 0$$

Finally, $b_n = \dfrac{1}{\pi} \displaystyle\int_{-\pi}^{\pi} f(x) \sin nx\, dx$

$$= \frac{1}{\pi}\int_{-\pi}^{o} -\sin nx\, dx + \frac{1}{\pi}\int_{o}^{\pi} \sin nx\, dx$$

$$= \frac{1}{\pi}\left\{\frac{1}{n}[\cos nx]_{-\pi}^{o} - \frac{1}{n}[\cos nx]_{o}^{\pi}\right\}$$

$$= \frac{1}{n\pi}[1 - \cos n\pi - \cos n\pi + 1]$$

$$= \frac{1}{n\pi}[2 - 2\cos n\pi] = \frac{2}{n\pi}[1-(-1)^n]$$

$$\Rightarrow b_n = \frac{2}{n\pi}[1-(-1)^n],\ n = 1, 2, 3, \ldots\ldots..$$

If n is even $b_n = 0$

if n is odd $b_n = \frac{4}{n\pi}$

Hence from (1) we have

$$f(x) = \sum_{n=1}^{\infty} b_n \sin nx$$

$$= b_1 \sin x + b_2 \sin 2x + b_3 \sin 3x + b_4 \sin 4x + \ldots\ldots\ldots.$$

$$f(x) = \frac{4}{\pi}\sin x + \frac{4}{3\pi}\sin 3x + \frac{4}{5\pi}\sin 5x + \ldots\ldots. \quad (2)$$

The expansion (2) is required Fourier expansion

putting $x = \frac{\pi}{2}$ is (2) we get

$$1 = \frac{4}{\pi}\left[\sin\frac{\pi}{2} + \frac{1}{3}\sin\frac{3\pi}{2} + \frac{1}{5}\sin\frac{5\pi}{2} + \ldots\ldots.\right]$$

$$= \frac{4}{\pi}\left[1 + \frac{1}{3}(-1) + \frac{1}{5}(1) + \frac{1}{7}(-1) + \ldots..\right]$$

$$= \frac{4}{\pi}\left[1 - \frac{1}{3} + \frac{1}{5} - \frac{1}{7} + \ldots\ldots.\right]$$

$$= \frac{\pi}{4} = 1 - \frac{1}{3} + \frac{1}{5} - \frac{1}{7} + \ldots..$$

Example 11. Find the Fourier series for the function

$$f(t) = \begin{cases} -1 & \text{for} \quad -\pi<t<-\pi/2 \\ 0 & \text{for} \quad -\pi/2<t<\pi/2 \\ 1 & \text{for} \quad \frac{\pi}{2}<t<\pi \end{cases}$$

Solution. Let $f(t) = \frac{a_o}{2} + \sum_{n=1}^{\infty} a_n \cos nt + \sum_{n=1}^{\infty} b_n \sin nt$ (1)

$$\text{Then } a_o = \frac{1}{\pi}\left\{\int_{-\pi}^{-\pi/2}(-1)\,dt + \int_{-\pi/2}^{\pi/2} 0\,dt + \int_{\pi/2}^{\pi} 1\,dt\right\}$$

$$= \frac{1}{\pi}\left\{(-x)_{-\pi}^{-\pi/2} + [x]_{\pi/2}^{\pi}\right\}$$

$$= \frac{1}{\pi}\left(\frac{\pi}{2} - \pi + \pi - \frac{\pi}{2}\right)$$

$$= 0$$

$$a_n = \frac{1}{\pi}\left\{\int_{-\pi}^{-\pi/2}(-1)\cos nt\,dt + \int_{-\pi/2}^{\pi/2}(0)\cos nt\,dt + \int_{\pi/2}^{\pi}(1)\cos nt\,dt\right\}$$

$$= \frac{1}{\pi}\left\{\left[-\frac{\sin nt}{n}\right]_{-\pi}^{-\pi/2} + \left[\frac{\sin nt}{n}\right]_{\pi/2}^{\pi}\right\} = \frac{1}{n\pi}\left\{\sin\frac{n\pi}{2} - \sin\frac{n\pi}{2}\right\}$$

$$= 0$$

$$\text{and } b_n = \frac{1}{\pi}\left\{\int_{-\pi}^{-\pi/2}(-1)\sin nt\,dt + \int_{-\pi/2}^{\pi/2}(0)\sin nt\,dt + \int_{\pi/2}^{\pi}(1)\sin nt\,dt\right\}$$

$$= \frac{1}{\pi}\left\{\left[\frac{\cos nt}{n}\right]_{-\pi}^{-\pi/2} + \left[-\frac{\cos nt}{n}\right]_{\pi/2}^{\pi}\right\}$$

$$= \frac{2}{n\pi}\left(\cos\frac{n\pi}{2} - \cos n\pi\right)$$

$$\therefore\ b_1 = \frac{2}{\pi},\ b_2 = \frac{-2}{\pi},\ b_3 = \frac{2}{3\pi} \text{ etc.}$$

Substituting the values of a_o, a_n and b_n in (1) we get

$$f(t) = \frac{2}{\pi}\left(\sin t - \sin 2t + \frac{1}{3}\sin 3t + ...\right)$$

Example 12. Find the Fourier series for the periodic function

$$f(x) = \begin{cases} 0, & -\pi < x < 0 \\ x, & 0 < x < \pi \end{cases}$$

$$f(x + 2\pi) = f(x)$$

Solution. Let $f(x) = \dfrac{a_o}{2} + \sum_{n=1}^{\infty} a_n \cos nx + \sum_{n=1}^{\infty} b_n \sin nx$

Then $a_o = \frac{1}{\pi}\int_{-\pi}^{o} 0\, dx + \frac{1}{\pi}\int_{o}^{\pi} x\, dx = \frac{1}{\pi}\left[\frac{x^2}{2}\right]_o^{\pi}$

$= \frac{1}{\pi}\left(\frac{\pi^2}{2}\right) = \frac{\pi}{2}$

and $a_n = \frac{1}{\pi}\int_{o}^{\pi} x \cos nx\, dx$

$= \frac{1}{\pi}\left[x.\frac{\sin nx}{n} - (1)\left(-\frac{\cos nx}{n^2}\right)\right]_o^{\pi}$

$= \frac{1}{\pi}\left[\frac{(-1)^n}{n^2} - \frac{1}{n^2}\right] = \frac{-2}{n^2\pi}$, when n is odd

$= 0$, when n is even

$b_n = \frac{1}{\pi}\int_{o}^{\pi} x \sin nx\, dx$

$= \frac{1}{\pi}\left[x\left(-\frac{\cos nx}{n}\right) - (1)\left(-\frac{\sin nx}{n^2}\right)\right]_o^{\pi}$

$= \frac{1}{\pi}\left[-\pi\frac{(-1)^n}{n}\right] = -\frac{(-1)^n}{n}$

Substituting the values of a_o, a_n & b_n in (1) we get

$$f(x) = \frac{\pi}{4} - \frac{2}{\pi}\left[\frac{\cos x}{1^2} + \frac{\cos 3x}{3^2} + \frac{\cos 5x}{5^2} + \ldots\ldots\right] + \left[\frac{\sin x}{1} - \frac{\sin 2x}{2} + \frac{\sin 3x}{3} + \ldots\ldots\right]$$

DISCONTINUOUS FUNCTIONS

At a point of discontinuity, Fourier series gives the value of f(x) as the arithmetic mean of left and right limits

At the point of discontinuity, x = c

$$\text{At } x = c,\ f(x) = \frac{1}{2}\left[f(c-o) + f(c+o)\right]$$

Example 13. Obtain Fourier series of the function

$$f(x) = \begin{cases} x, & -\pi < x < o \\ -x, & o < x < \pi \end{cases}$$

and hence show that

$$1 - \frac{1}{4} + \frac{1}{9} - \frac{1}{16} + ... = \frac{\pi^2}{12}$$ (U.P.T.U. 2002, 2008)

Solution. Here f(x) is an even function so

$b_n = 0$

$$a_o = \frac{2}{\pi}\int_o^\pi f(x)\,dx$$

$$= \frac{2}{\pi}\int_o^\pi -x\,dx = -\frac{2}{\pi}\left[\frac{x^2}{2}\right]_o^\pi$$

$$= -\frac{2}{\pi}\left[\frac{\pi^2}{2}\right] = -\pi$$

$$a_n = \frac{2}{\pi}\int_o^\pi f(x)\cos nx\,dx$$

$$= \frac{2}{\pi}\int_o^\pi -x\cos nx\,dx$$

$$= -\frac{2}{\pi}\left[x\,\frac{\sin nx}{n} + \frac{\cos nx}{n^2}\right]_o^\pi$$

$$= -\frac{2}{\pi}\left[\frac{(-1)^n}{n^2} - \frac{1}{n^2}\right]$$

$$= \frac{2}{\pi n^2}[1-(-1)^n]$$

$$= \begin{cases} 0 & , \quad n \text{ is even} \\ \dfrac{4}{\pi n^2} & , \quad n \text{ is odd} \end{cases}$$

Thus, the Fourier series expansion is

$$f(x) = -\frac{\pi}{2} + \frac{4}{\pi}\left[\frac{\cos x}{1^2} + \frac{\cos 3x}{3^2} + \frac{\cos 5x}{5^2} + ...\right]$$

Example 14. Find the Fourier series of the function

$$f(x) = \begin{cases} x+\pi, & \text{for } o \le x \le \pi \\ -x-\pi, & \text{for } -\pi \le x < o \end{cases}$$

and $f(x + 2\pi) = f(x)$ (U.P.T.U., 2006)

Solution. Let the Fourier series expression be

$$f(x) = \frac{a_o}{2} + \sum_{n=1}^{\infty} a_n \cos nx + \sum_{n=1}^{\infty} b_n \sin nx$$

where $a_o = \frac{1}{\pi} \int_{-\pi}^{\pi} f(x)\, dx$

$$= \frac{1}{\pi} \int_{-\pi}^{o} (-x-\pi)\, dx + \frac{1}{\pi} \int_{o}^{\pi} (x+\pi)\, dx$$

$$= \frac{1}{\pi} \left[-\frac{x^2}{2} - \pi x \right]_{-\pi}^{o} + \frac{1}{\pi} \left[\frac{x^2}{2} + \pi x \right]_{o}^{\pi}$$

$$= \frac{1}{\pi} \left(\frac{x^2}{2} - \pi^2 \right) + \frac{1}{\pi} \left(\frac{\pi^2}{2} + \pi^2 \right) = \pi$$

and $a_n = \frac{1}{\pi} \int_{-\pi}^{\pi} f(x) \cos nx\, dx$

$$= \frac{1}{\pi} \int_{-\pi}^{o} (-x-\pi) \cos nx\, dx + \frac{1}{\pi} \int_{o}^{\pi} (x+\pi) \cos nx\, dx$$

$$= \frac{1}{\pi} \left[(-x-\pi) \frac{\sin nx}{n} - 1(-1) \left(\frac{-\cos nx}{n^2} \right) \right]_{-\pi}^{o} + \frac{1}{\pi} \left[(x+\pi) \frac{\sin nx}{n} - (1) \left(\frac{-\cos nx}{n^2} \right) \right]_{o}^{\pi}$$

$$= \frac{1}{\pi} \left[-\frac{1}{n^2} + \frac{(-1)^n}{n^2} \right] + \frac{1}{\pi} \left[\frac{(-1)^n}{n^2} - \frac{1}{n^2} \right]$$

$$= \frac{2}{n^2 \pi} \left[(-1)^n - 1 \right]$$

$$a_n = \begin{cases} -4/n^2\pi, & \text{if n is odd} \\ 0, & \text{if n is even} \end{cases}$$

$$b_n = \frac{1}{\pi} \int_{-\pi}^{\pi} f(x) \sin nx\, dx$$

$$= \frac{1}{\pi} \int_{-\pi}^{o} f(x) \sin nx\, dx + \frac{1}{\pi} \int_{o}^{\pi} f(x) \sin nx\, dx$$

$$= \frac{1}{\pi} \int_{-\pi}^{o} (-x-\pi) \sin nx\, dx + \frac{1}{\pi} \int_{o}^{\pi} (x+\pi) \sin nx\, dx$$

$$= \frac{1}{\pi}\left[(-x-\pi)\left(\frac{-\cos nx}{n}\right) - (-1)\left(\frac{-\sin nx}{n^2}\right)\right]_{-\pi}^{o} + \frac{1}{\pi}\left[(x+\pi)\left(\frac{-\cos nx}{n}\right) - (1)\left(\frac{-\sin nx}{n^2}\right)\right]_{o}^{\pi}$$

$$= \frac{1}{\pi}\left[\frac{\pi}{n}\right] + \frac{1}{\pi}\left[-\frac{2\pi}{n}(-1)^n + \frac{\pi}{n}\right]$$

$$= \frac{1}{n}\left[1-2(-1)^n+1\right] = \frac{2}{n}\left[1-(-1)^n\right]$$

$$b_n = \begin{cases} 4/n, & \text{if n is odd} \\ 0, & \text{if n is even} \end{cases}$$

Thus, the Fourier series is

$$f(x) = \frac{a_o}{2} + a_1 \cos x + a_2 \cos 2x + a_3 \cos 3x + \ldots\ldots\ldots + b_1 \sin x + b_2 \sin 2x + b_3 \sin 3x + \ldots\ldots\ldots$$

$$= \frac{\pi}{2} - \frac{4}{\pi}\left(\frac{\cos x}{1^2} + \frac{\cos 3x}{3^2} + \ldots\ldots\right) + 4\left(\frac{\sin x}{1} + \frac{\sin 3x}{3} + \ldots\ldots\right)$$

CHANGE OF INTERVAL AND FUNCTIONS HAVING ARBITRARY PERIOD

In many of the engineering problems (i.e. electrical engineering problems) the period of the function is not always 2π but it is different say 2C or T. This period must be converted to the length 2π. The independent variable x is also to be changed proportionally. Let function f(x) be defined in the interval (–C, C). Now we want to change the function to the period of 2π, so that we can use the formula of a_n, b_n as discussed in Euler's formula.

$\because$ 2C is the interval for variable x

$\therefore$ 1 is the interval for the variable $= \dfrac{x}{2C}$

$\therefore$ 2π is the interval for the variable $= \dfrac{x \cdot 2\pi}{2C}$

$$= \frac{\pi x}{C}$$

So put $z = \dfrac{\pi x}{C}$ or $x = \dfrac{zC}{\pi}$

Thus, the function f(x) of period 2C is transformed to the function $f\left(\dfrac{cz}{\pi}\right)$ or F(z) of period 2π. F(z) can be expanded in the Fourier series

$$F(z) = f\left(\frac{cz}{\pi}\right) = \frac{a_o}{2} + a_1 \cos z + a_2 \cos 2z + ... + b_1 \sin z + b_2 \sin 2z + ...$$

$$\text{Where } a_o = \frac{1}{\pi}\int_o^{2\pi} F(z)\,dz = \frac{1}{\pi}\int_o^{2\pi} f\left(\frac{cz}{\pi}\right) dz$$

$$= \frac{1}{c}\int_o^{2c} f(x)\, d\left(\frac{\pi x}{c}\right) = \frac{1}{c}\int_o^{2c} f(x)\,dx \text{ put } z = \frac{\pi x}{c}$$

$$\Rightarrow a_o = \frac{1}{c}\int_o^{2c} f(x)\,dx$$

$$a_n = \frac{1}{\pi}\int_o^{2\pi} F(z) \cos nz\, dz = \frac{1}{\pi}\int_o^{2\pi} f\left(\frac{cz}{\pi}\right) \cos nz\, dz$$

$$= \frac{1}{\pi}\int_o^{2C} f(x) \cos \frac{n\pi x}{c}\, d\left(\frac{\pi x}{c}\right) \text{ put } z = \frac{\pi x}{c}$$

$$= \frac{1}{c}\int_o^{2C} f(x) \cos \frac{n\pi x}{c}\, dx$$

$$a_n = \frac{1}{c}\int_o^{2C} f(x) \cos \frac{n\pi x}{c}\, dx$$

$$\text{Similarly } b_n = \frac{1}{c}\int_o^{2C} f(x) \sin \frac{n\pi x}{c}\, dx$$

Cor. Half range series interval (0, c)

Cosine series:

$$f(x) = \frac{a_o}{2} + a_1 \cos \frac{\pi x}{c} + a_2 \cos \frac{2\pi x}{c} + ...+ a_n \frac{\cos n\pi x}{c} +......$$

$$\text{Where } a_o = \frac{2}{c}\int_o^{C} f(x)\,dx,\ a_n = \frac{2}{c}\int_o^{C} f(x) \cos \frac{n\pi x}{c}\, dx$$

sine series:

$$f(x) = b_1 \sin \frac{\pi x}{c} + b_2 \sin \frac{2\pi x}{c} + ... + b_n \sin \frac{n\pi x}{c} + ...$$

$$\text{Where } b_n = \frac{2}{c}\int_o^{C} f(x) \sin \frac{n\pi x}{c}\, dx$$

HALF RANGE SERIES

Many a time it is required to obtain a Fourier expansion of a function f(x) for the range (0, C) which is half the period of the Fourier series. As it is immaterial whatever the function may be outside the range $0<x<c$, we extend the function to cover the range $-c<x<c$ so that the new function may be odd or even. The extension of the functions period being made in such a way that their graphs became either symmetrical to the axis of y or symmetrical to origin, and then the expansion contains either only the cosine terms along with a_o or only the sine terms. Thus, we may get the different forms of series for the same functions.

Sine Series: If we have to expand a function f(x) as a sine series in 0 to C, then we expand the function first from –C to C and then the make the reflection at the origin, so that f(x) = –f(–x) then the expanded function became odd and will give the required Fourier sine series

$$f(x) = \sum_{n=1}^{\infty} b_n \sin\frac{n\pi x}{C}$$

$$\text{Where } b_n = \frac{2}{C}\int_o^C f(x)\sin\frac{n\pi x}{C}\,dx$$

Cosine Series: If it is required to express f(x) as a cosine series in $0<x<C$, then first we expand the function from -C to C so that its reflection became about the axis of y i.e., the graph became symmetrical about the axis of y then the expanded Fourier Cosine series contains

$$f(x) = \frac{a_o}{2} + \sum_{n=1}^{\infty} a_n \cos\frac{n\pi x}{C}$$

$$\text{where } a_o = \frac{2}{C}\int_o^C f(x)\,dx$$

$$\text{and } a_n = \frac{2}{C}\int_o^C f(x)\cos\frac{n\pi x}{C}\,dx$$

Example 15. Expand for f(x) = k for
$0<x<2$ in a half range
(i) Sine series (ii) cosine series (U.P.T.U. 2007)

Solution. f(x) = k and C = 2

(i) $b_n = \frac{2}{C}\int_o^C f(x)\sin\frac{n\pi x}{C}$ in half range (0, C)

$$= \frac{2}{2}\int_0^C k \sin\frac{n\pi x}{2}\,dx$$

$$= k\,\frac{2}{n\pi}\left(-\cos\frac{n\pi x}{2}\right)_0^2 = \frac{2k}{n\pi}\left[-\cos n\pi + 1\right]$$

Half range sine series is

$$f(x) = \sum_{n=1}^{\infty} b_n \sin\frac{n\pi x}{2}$$

$$k = \sum_{n=1}^{\infty}\frac{2k}{n\pi}\left[1-\cos n\pi\right]\sin\frac{n\pi x}{2} = \frac{2k}{\pi}\sum_{n=1}^{\infty}[1-(-1)^n]\sin\frac{n\pi x}{2}$$

$$\Rightarrow k = \frac{2k}{\pi}\left[2\sin\frac{\pi x}{2} + 2\sin\frac{3\pi x}{2} + 2\sin\frac{5\pi x}{2} + \ldots\right]$$

(ii) $$a_o = \frac{2}{C}\int_0^C f(x)\,dx = \frac{2}{2}\int_0^2 k\,dx = k\,(x)_o^2 = 2k$$

$$a_n = \frac{2}{C}\int_0^C f(x)\cos\frac{n\pi x}{C}\,dx = \frac{2}{2}\int_0^2 k\cos\frac{n\pi x}{2}\,dx$$

$$= k\,\frac{2}{n\pi}\left[\sin\frac{n\pi x}{2}\right]_o^2 = \frac{2k}{n\pi}\sin n\pi$$

$$\Rightarrow a_n = 0$$

Therefore, from

$$f(x) = \frac{a_o}{2} + \Sigma a_n \cos\frac{n\pi x}{2} \text{ we have}$$

$$f(x) = k = \frac{1}{2}(2k) + \sum_{n=1}^{\infty} 0\cos\frac{n\pi x}{2}$$

$$\Rightarrow f(x) = k$$

Example 16. Find the Fourier half range cosine series of the function

$$f(t) = \begin{cases} 2t, & 0 < t < 1 \\ 2(2-t), & 1 < t < 2 \end{cases}$$ (U.P.T.U. 2001, 2006, 2007)

Solution. $f(t) = \dfrac{a_o}{2} + \displaystyle\sum_{n=1}^{\infty} a_n \cos\frac{n\pi t}{C}$

$$f(t) = \frac{a_o}{2} + \sum_{n=1}^{\infty} a_n \cos\frac{n\pi t}{2} \quad \text{As } C = 2 \tag{1}$$

Then $a_o = \frac{2}{C}\int_0^C f(t)\,dt$

$$= \frac{2}{2}\int_0^2 f(t)\,dt = \int_0^2 f(t)\,dt$$

$$= \int_0^1 2t\,dt + \int_1^2 2\,(2-t)\,dt$$

$$\Rightarrow a_o = [t^2]_0^1 + 2\left[2t - \frac{t^2}{2}\right]_1^2$$

$$= 1 + [(4t - t^2)]_1^2$$

$$\Rightarrow a_o = 1 + (8 - 4 - 4 + 1) = 2$$

$$\Rightarrow a_o = 2$$

and $a_n = \frac{2}{C}\int_0^C f(t)\cos\frac{n\pi t}{C}\,dt$

$$= \int_0^2 f(t)\cos\frac{n\pi t}{2}\,dt \text{ as } C = 2$$

$$= \int_0^1 2t\cos\frac{n\pi t}{2}\,dt + \int_1^2 2\,(2-t)\cos\frac{n\pi t}{2}\,dt$$

$$= \left[2t\left(\frac{2}{n\pi}\sin\frac{n\pi t}{2}\right) - 2\left(-\frac{4}{n^2\pi^2}\cos\frac{n\pi t}{2}\right)\right]_0^1 + \left[(4-2t)\left(\frac{2}{n\pi}\sin\frac{n\pi t}{2}\right) - (-2)\left(-\frac{4}{n^2\pi^2}\cos\frac{n\pi t}{2}\right)\right]_1^2$$

$$= \left[\frac{4}{n\pi}\sin\frac{n\pi}{2} + \frac{8}{n^2\pi^2}\cos\frac{n\pi}{2} - \frac{8}{n^2\pi^2}\right] + \left[0 - \frac{8}{n^2\pi^2}\cos n\pi - \frac{4}{n\pi}\sin\frac{n\pi}{2} + \frac{8}{n^2\pi^2}\cos\frac{n\pi}{2}\right]$$

$$= \frac{16}{n^2\pi^2}\cos\frac{n\pi}{2} - \frac{8}{n^2\pi^2}(1 + \cos n\pi)$$

$$\Rightarrow a_n = \begin{cases} 0, \text{ n is odd} \\ \frac{16}{n^2\pi^2}(-1)^{n/2} - \frac{16}{n^2\pi^2}, \text{ n is even} \end{cases} \quad \because \cos\frac{n\pi}{2} = (-1)^{n/2}$$

When n is even

Thus $a_2 = -\frac{32}{2^2\pi^2}$, $a_4 = 0$, $a_6 = \frac{-32}{6^2\pi^2}$, $a_8 = 0$, $a_{10} = \frac{-32}{10^2\pi^2}$ and so on

Hence from (i) the half range Fourier cosine series is

$$f(t) = \frac{2}{2} + \left(-\frac{8}{\pi^2}\right)\cos\frac{2\pi t}{2} + \left(\frac{-32}{36\pi^2}\right)\cos\frac{6\pi t}{2} + ...$$

$$\text{or } f(t) = 1 - \frac{8}{\pi^2}\cos\frac{2\pi t}{2} - \frac{8}{9\pi^2}\cos\frac{6\pi t}{2} + ...$$

$$\text{or } f(t) = 1 - \frac{8}{\pi^2}\cos\pi t - \frac{8}{9\pi^2}\cos 3\pi t + ...$$

Example 17. Obtain a Fourier cosine series expansion of the periodic function defined by

$$f(t) = \sin\left(\frac{\pi t}{l}\right),\ 0 < t < l \qquad \text{(U.P.T.U. 2001)}$$

Solution: We have

$$f(t) = \sin\left(\frac{\pi t}{l}\right),\ 0 < t < l$$

$$\text{Let } f(t) = \frac{a_o}{2} + \sum_{n=1}^{\infty} a_n \cos\frac{n\pi t}{l} + \sum_{n=1}^{\infty} b_n \sin\frac{n\pi t}{l} \qquad (1)$$

$$\text{Where } a_o = \frac{2}{l}\int_o^l \sin\frac{\pi t}{l}\,dt = \frac{2}{l}\left[-\frac{l}{\pi}\cos\frac{\pi t}{l}\right]_o^l$$

$$= -\frac{2}{\pi}(\cos\pi - \cos 0) = \frac{-2}{\pi}(-1-1) = \frac{4}{\pi}$$

$$a_n = \frac{2}{l}\int_o^l \sin\frac{\pi t}{l}\cos\frac{n\pi t}{l}\,dt$$

$$= \frac{1}{l}\int_o^l \left[\sin\left(\frac{\pi t + n\pi t}{l}\right) - \sin\left(\frac{n\pi t}{l} - \frac{\pi t}{l}\right)\right] dt$$

$$= \frac{1}{l}\int_o^l \sin(n+1)\frac{\pi t}{l}\,dt - \frac{1}{l}\int_o^l \sin(n-1)\frac{\pi t}{l}\,dt$$

$$= \frac{1}{l}\left[\frac{-l}{(n+1)\pi}\cos\frac{(n+1)\pi t}{l}\right]_o^l + \frac{1}{l}\left[\frac{1}{(n-1)\pi}\cos\frac{(n-1)\pi t}{l}\right]_o^l$$

$$= -\frac{1}{(n+1)\pi}[\cos(n+1)\pi - \cos o] + \frac{1}{(n-1)\pi}[\cos(n-1)\pi - \cos o]$$

$$= -\frac{1}{(n+1)\pi}\left[(-1)^{n+1}-1\right] + \frac{1}{(n-1)\pi}[(-1)^{n+1}-1]$$

$$= (-1)^{n+1}\left[-\frac{1}{(n+1)\pi} + \frac{1}{(n-1)\pi}\right] + \left[\frac{1}{(n+1)\pi} - \frac{1}{(n-1)\pi}\right]$$

$$= (-1)^{n+1}\frac{2}{(n^2-1)\pi} - \frac{2}{(n^2-1)\pi}$$

$$= \frac{2}{(n^2-1)\pi}\left[(-1)^{n+1}-1\right]$$

$$\Rightarrow a_n = \begin{cases} -\dfrac{4}{(n^2-1)} & ; \quad \text{When n is even} \\ 0 & ; \quad \text{When n is odd} \end{cases}$$

Here, we see that the above relation is not applicable to find a_1

$$\text{Thus } a_1 = \frac{2}{l}\int_o^l \sin\frac{\pi t}{l}\cos\frac{\pi t}{l}\,dt$$

$$= \frac{1}{l}\left(-\frac{l}{2\pi}\cos\frac{2\pi t}{l}\right)_o^l$$

$$= -\frac{1}{2\pi}(-\cos 2\pi + \cos o) = 0$$

Thus, we have

$$f(t) = \frac{a_o}{2} + a_1\cos\frac{\pi t}{l} + a_2\frac{\cos 2\pi t}{l} + a_3\cos\frac{3\pi t}{l} + a_4\cos\frac{4\pi t}{l} + \ldots..$$

$$= \frac{2}{\pi} - \frac{4}{\pi}\left[\frac{1}{3}\cos\frac{2\pi t}{l} + \frac{1}{15}\cos\frac{4\pi t}{l} + \frac{1}{35}\cos\frac{6\pi t}{l} + \ldots\ldots..\right]$$

Example 18. Obtain the half range series for the function $f(x) = x^2$ in the interval $o<x<3$ (U.P.T.U. 2002)

Solution. We know that half range sine series is given by $f(x) = \sum b_n \sin nx$

$$\text{Where } b_n = \frac{2}{C}\int_o^C f(x)\sin\frac{n\pi x}{C}\,dx \text{ in the half range } (0, C)$$

Here, we have half range $o<x<3$ and $f(x) = x^2$

$$b_n = \frac{2}{3}\int_o^3 x^2\sin\frac{n\pi x}{3}\,dx$$

$$= \frac{2}{3}\left[x^2 \frac{-3}{n\pi}\left(\cos\frac{n\pi x}{3}\right) + 2x\left(\frac{3}{x\pi}\right)\left(\frac{3}{n\pi}\right)\sin\frac{n\pi x}{3} - 2\left(\frac{3}{n\pi}\right)\left(\frac{3}{n\pi}\right)\left(\frac{3}{n\pi}\right)\cos\frac{n\pi x}{3}\right]_0^3$$

$$\Rightarrow b_n = \frac{2}{3}\left[\left\{\frac{-27}{n\pi}(-1)^n - \frac{54}{n^3\pi^3}(-1)^n\right\} + \frac{54}{n^3\pi^3}\right]$$

$$\text{or } b_n = \frac{2}{3}\left[\frac{54}{n^3\pi^3}\{1-(-1)^n\} - \frac{27}{n\pi}(-1)^n\right]$$

$$\Rightarrow b_n = \frac{2}{3}\left[\frac{108}{n^3\pi^3} + \frac{27}{n\pi}\right], \text{ when n is odd}$$

$$\text{and } b_n = -\frac{18}{n\pi}, \text{ when n is even}$$

Therefore, half range sine series is

$f(x) = b_1 \sin x + b_2 \sin 2x + b_3 \sin 3x + ..$

$$= \frac{2}{3}\left[\frac{27}{\pi}\left(\frac{\sin x}{1} + \frac{\sin 3x}{3} + \frac{\sin 5x}{5} + ...\right) - \frac{108}{\pi^3}\left(\frac{\sin x}{1^3} + \frac{\sin 3x}{3^3} + \frac{\sin 5x}{5^3} + ...\right) - \frac{18}{\pi}\left(\frac{\sin 2x}{2} + \frac{\sin 4x}{4} + ...\right)\right]$$

PRACTICAL HARMONIC ANALYSIS

As Fourier series

$$f(x) = \frac{a_o}{2} + \sum_{n=1}^{\infty} a_n \cos nx + \sum_{n=1}^{\infty} b_n \sin nx \qquad (1)$$

$$\left.\begin{aligned} \text{where } a_o &= \frac{1}{\pi}\int_0^{2\pi} f(x)\,dx \\ a_n &= \frac{1}{\pi}\int_0^{2\pi} f(x)\cos nx\,dx \\ b_n &= \frac{1}{\pi}\int_0^{2\pi} f(x)\sin nx\,dx \end{aligned}\right\} \qquad (2)$$

So far, the function has always been defined by an explicit function of an independent variable. In practice, however, the function is often given not by a formula but by a graph or by a table of corresponding values. In such cases, the integral (2) cannot be evaluated and instead, the following alternative forms of (2) are employed.

Since the mean value of a function

$y = f(x)$ over the range (a, b) is $\frac{1}{b-a}\int_a^b f(x)\,dx$

∴ The equation (2) gives

$$\left.\begin{aligned} a_o &= 2 \times \frac{1}{2\pi}\int_0^{2\pi} f(x)\,dx = 2\,[\text{mean value of } f(x) \text{ is } (o, 2\pi)] \\ a_n &= 2 \times \frac{1}{2\pi}\int_0^{2\pi} f(x)\cos nx\,dx = 2\,[\text{mean value of } f(x)\cos nx \text{ in } (0, 2\pi)] \\ b_n &= 2 \times \frac{1}{2\pi}\int_0^{2\pi} f(x)\sin nx\,dx = 2\,[\text{mean value of } f(x)\sin nx \text{ in } (o, 2\pi)] \end{aligned}\right\} \quad (3)$$

There are numerous other method of finding the value of a_o, a_n, b_n which constitute the field of harmonic analysis.
In (1) the term ($a_1 \cos x + b_1 \sin x$) is called the fundamental or first harmonic the term ($a_2 \cos 2x + b_2 \sin 2x$) the second harmonic and so on.

Example 19. The following table gives the variations of periodic current over a period

to sec	0	T/6	T/3	T/2	2T/3	5T/6	T
A amp	1.98	1.30	1.05	1.30	-0.88	-0.25	1.98

Show that there is a direct current part of 0.75 amp in the variable current and obtain the amplitude of the first harmonic (V.T.U. 2004)

Solution. Here length of the interval is T i.e. c = T/2

Then $A = \frac{a_o}{2} + a_1 \cos\frac{2\pi t}{T} + b_1 \sin\frac{2\pi t}{T} + a_2 \cos\frac{4\pi t}{T} + b_2 \sin\frac{4\pi t}{T} + ...$

The desired values are tabulated as follows:

t	2πt/T	Cos 2πt/T	sin 2πt/T	A	A cos 2πt/T	A sin 2πt/T
0	0	1.0	0.000	1.98	1.980	0.000
T/6	π/3	0.5	0.866	1.30	0.650	1.126
T/3	2π/3	-0.5	0.866	1.05	-0.525	0.909
T/2	π	-1.0	0.000	1.30	-1.300	0.000
2T/3	4π/3	-0.5	-0.866	-0.88	0.440	0.762
5T/6	5π/3	0.5	-0.866	0.25	-0.125	0.217
			Σ =	4.5	1.12	3.014

$\therefore\ a_o = 2\frac{1}{6}\,\Sigma A = \frac{1}{3}(4.5) = 1.5$

$a_1 = 2.\frac{1}{6}\,\Sigma A \cos\frac{2\pi t}{T} = \frac{1}{3}(1.12) = 0.373$

$b_1 = 2.\frac{1}{6}\,\Sigma A \sin\frac{2\pi t}{T} = \frac{1}{3}(3.014)$

$= 1.005$

Thus the direct current part in the variable current = $a_o/2 = 0.75$ and amplitude of the first harmonic = $\sqrt{a_1^2 + b_1^2} = \sqrt{(0.373)^2 + (1.005)^2}$

$= 1.072$

EXERCISE

1. Find a Fourier series to represent, $f(x) = \pi - x$ for $o<x<2\pi$

Ans. $2[\sin x + \frac{1}{2}\sin 2x + \frac{1}{3}\sin 3x + \ldots\ldots + \frac{1}{n}\sin nx + \ldots\ldots]$

2. Let $f(x) = x$ for $-\pi \le x \le \pi$. Write the Fourier series of f on $[-\pi, \pi]$

Ans. $f(x) = x = 2\sin x - \sin 2x + \frac{2}{3}\sin 3x - \frac{1}{2}\sin 4x + \frac{2}{5}\sin 5x +.$

3. Find the Fourier series expansion for the function $f(x) = x\cos x$, $-\pi<x<\pi$ (U.P.T.U. 2002)

Ans. $f(x) = -\frac{1}{2}\sin x + \frac{4}{2^2-1}\sin 2x - \frac{6\sin 3x}{3^2-1} + \ldots$

4. Expand the function $f(x) = x\sin x$ as a Fourier series in the interval $-\pi \le x \le \pi$. Hence deduce that

$$\frac{1}{1.3} - \frac{1}{3.5} + \frac{1}{5.7} - \frac{1}{7.9} + \ldots\ldots = \frac{\pi - 2}{4}$$ (U.P.T.U. 2002)

Ans. $x\sin x = 1 + 2\left[-\frac{1}{4}\cos x - \frac{1}{1.3}\cos 2x + \frac{1}{2.4}\cos 3x - \frac{1}{3.5}\cos 4x + \ldots\ldots\right]$

5. Find the Fourier series representing $f(x) = x$, $o<x<2\pi$ and sketch its graph from $x = -4\pi$ to $x = 4\pi$

Ans. $x = \pi - 2\left[\sin x + \frac{1}{2}\sin 2x + \frac{1}{3}\sin 3x + \ldots\ldots\right]$

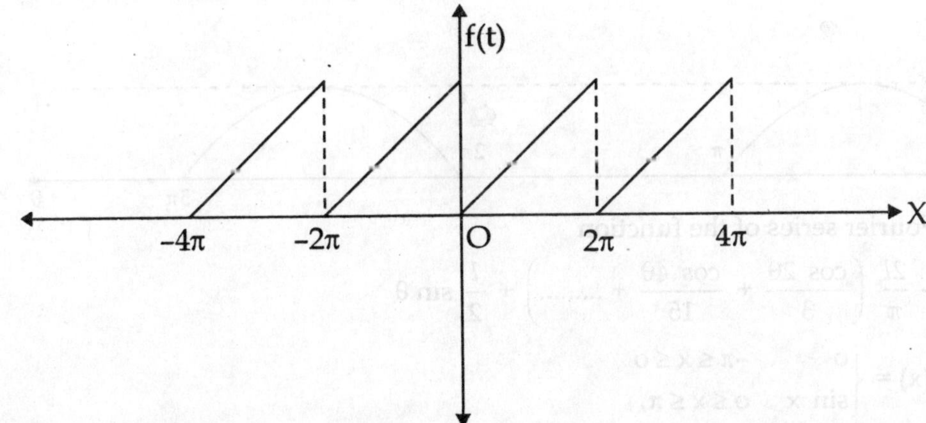

6. Find the Fourier series for f(x), if

$$f(x) = \begin{cases} -\pi & -\pi < x < o \\ x & o < x < \pi \end{cases}$$

hence deduce that $\frac{1}{1^2} + \frac{1}{3^2} + \frac{1}{5^2} + ... = \frac{\pi^2}{8}$

Ans. $f(x) = \frac{-\pi}{4} - \frac{2}{\pi}\left(\cos x + \frac{\cos 3x}{3^2} + \frac{\cos 5x}{5^2} + ...\right) + 3\sin x - \frac{\sin 2x}{2} + \frac{3\sin 3x}{3} - \frac{\sin 4x}{4} + ...$

7. Find the Fourier series of the function

$$f(x) = \begin{cases} -1 & \text{for } -\pi < x < o \\ 1 & \text{for } o < x < \pi \end{cases}$$

where $f(x+2\pi) = f(x)$ (UPTU 2005)

Ans. $\frac{4}{\pi}\left[\frac{1}{1}\sin x + \frac{1}{3}\sin 3x + \frac{1}{5}\sin 5x + \frac{1}{7}\sin 7x + ...\right]$

8. If $f(x) = \begin{cases} \pi x, & o < x < 1 \\ \pi(2-x), & 1 < x < 2 \end{cases}$

using half range cosine series, show that

$\frac{1}{1^4} + \frac{1}{3^4} + \frac{1}{5^4} + ... = \frac{\pi^4}{96}$ (U.P.T.U. 2007)

9. An alternating current after passing through a rectifier has the form

$i = l \sin\theta$, for $o < \theta < \pi$

$= 0$ for $\pi < \theta < 2\pi$

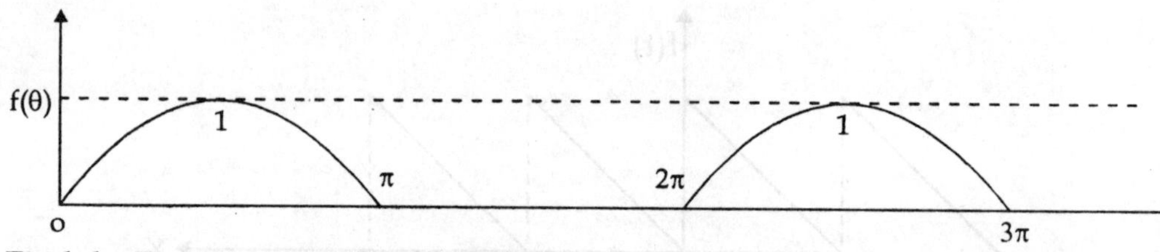

Find the Fourier series of the function

Ans. $\frac{l}{\pi} - \frac{2l}{\pi}\left(\frac{\cos 2\theta}{3} + \frac{\cos 4\theta}{15} + ...\right) + \frac{l}{2}\sin\theta$

10. If $f(x) = \begin{cases} o & -\pi \le x \le o \\ \sin x & o \le x \le \pi, \end{cases}$

Prove that $f(x) = \frac{1}{\pi} + \frac{\sin x}{2} - \frac{2}{\pi}\sum_{n=1}^{\infty}\frac{\cos 2nx}{4n^2-1}$

hence show that $\frac{1}{1.3} - \frac{1}{3.5} + \frac{1}{5.7} \ldots\ldots\ldots = \frac{1}{4}(\pi - 2)$

11. Obtain Fourier series for the function

$$f(x) = \begin{cases} \pi x \quad , & 0 \le x \le 1 \\ \pi(2-x) \, , & 1 \le x \le 2 \end{cases}$$ (U.P.T.U. 2001, V.T.U. 2004)

hence deduce that $\frac{1}{1^2} + \frac{1}{3^2} + \frac{1}{5^2} + \ldots\ldots\infty = \frac{\pi^2}{8}$

12. Express f(x) = x as a half range sine series in 0<x<2 (U.P.T.U. 2004)

Hint. The graph of f(x) = x in 0<x<2 is the line OA. Let us extend the function f(x) in the interval -2<x<0 (shown by the line BO) so that the new function is symmetrical about the origin and therefore, represents an odd function in (-2, 2) as shown is figure

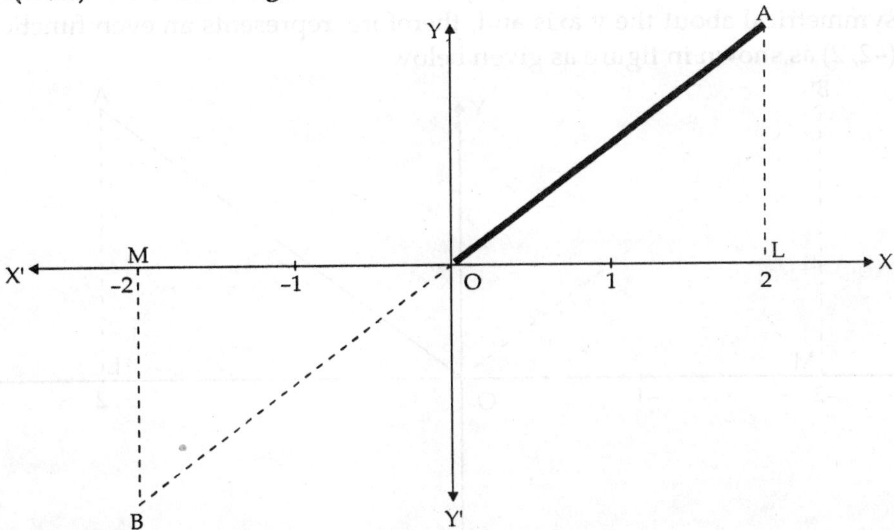

Hence, the Fourier series for f(x) over the full period (-2, 2) will contain only sine terms given by

$$f(x) = \sum_{n=1}^{\infty} b_n \sin \frac{n\pi x}{2}$$

where $b_n = \frac{2}{2} \int_0^2 f(x) \sin \frac{n\pi x}{2} \, dx$

$$= \int_0^2 x \sin \frac{n\pi x}{2} \, dx$$

$$= \left[-\frac{2x}{n\pi}\cos\frac{n\pi x}{2} + \frac{4}{n^2\pi^2}\sin\frac{n\pi x}{2} \right]_o^2$$

$$= \frac{-4(-1)^2}{n\pi}$$

Thus $b_1 = 4/\pi,\ b_2 = -4/2\pi,\ b_3 = 4/3\pi,\ b_4 = -4/4\pi$ etc

Hence the Fourier sine series for f(x) over the half range (0,2) is

$$f(x) = \frac{4}{\pi}\left(\sin\frac{\pi x}{2} - \frac{1}{2}\sin\frac{2\pi x}{2} + \frac{1}{3}\sin\frac{3\pi x}{2} - \frac{1}{4}\sin\frac{4\pi x}{2} + \ldots\ldots \right)$$

12. Express f(x) = x as a half range cosine series in o<x<2

Hint. The graph of f(x) = x in (0, 2) is the line OA. Let us extend the function f(x) in the interval (-2, 0) shown by line OB' so that the new function is symmetrical about the y axis and, therefore, represents an even function in (-2, 2) as shown in figure as given below

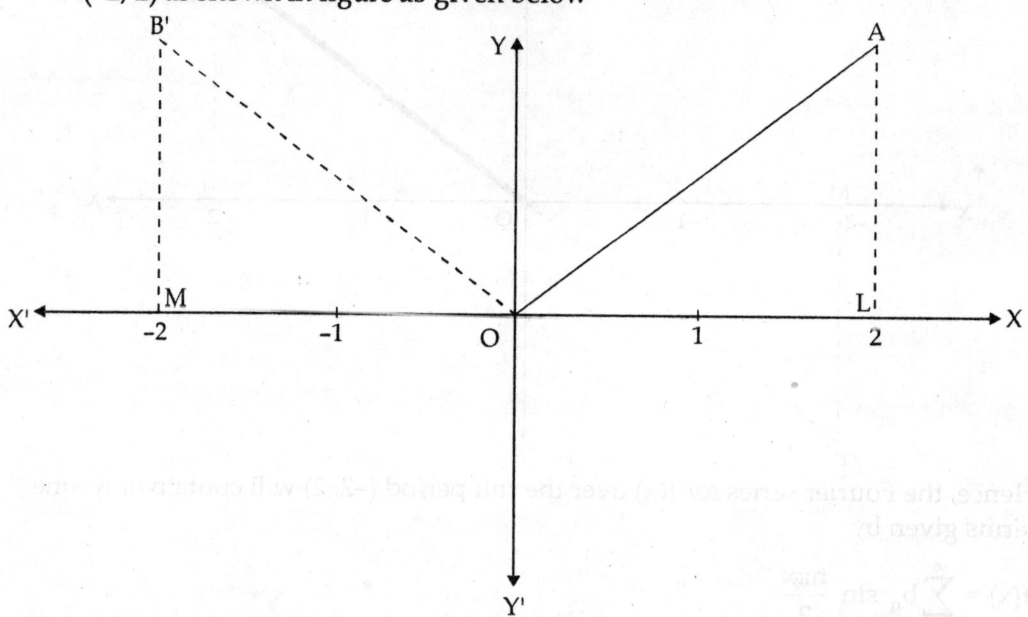

Hence the Fourier series for f(x) over the full period (-2, 2) will contain only cosine terms given by

$$f(x) = \frac{a_o}{2} + \sum_{n=1}^{\infty} a_n \cos\frac{n\pi x}{2}$$

where $a_o = \frac{2}{2}\int_o^2 f(x)\,dx = \int_o^2 x\,dx = 2$

and $a_n = \frac{2}{2}\int_o^2 f(x)\cos\frac{n\pi x}{2}\,dx = \int_o^2 x\cos\frac{n\pi x}{2}\,dx$

$= \left[\frac{2x}{n\pi}\sin\frac{n\pi x}{2} + \frac{4}{n^2\pi^2}\cos\frac{n\pi x}{2}\right]_o^2$

$= \frac{4}{n^2\pi^2}\left[(-1)^n - 1\right]$

Thus $a_1 = -8/\pi^2$, $a_2 = 0$, $a_3 = -8/3^2\pi^2$, $a_4 = 0$, $a_5 = -8/5^2\pi^2$, etc

Hence the desired Fourier series for f(x) over the half range (0, 2) is

$$f(x) = 1 - \frac{8}{\pi^2}\left[\frac{\cos \pi x/2}{1^2} + \frac{\cos {}^{3\pi x}\!/_{2}}{3^2} + \frac{\cos {}^{5\pi x}\!/_{2}}{5^2} + \ldots\ldots\right]$$

13. Find the half period series for f(x) given in the range (o, l) by the graph OPQ as shown in figure. (U.P.T.U. 2009)

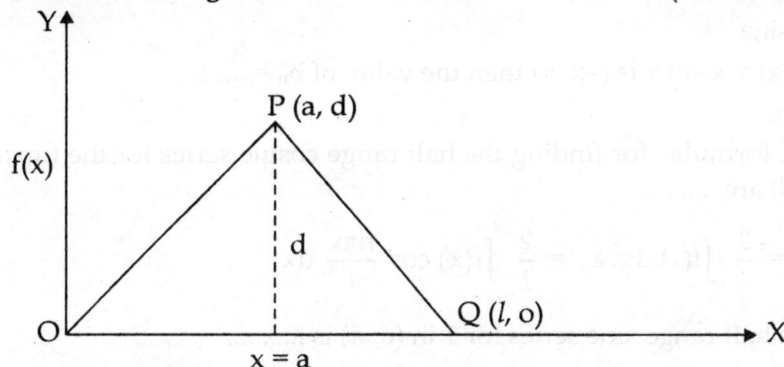

Hint. $f(x) = \begin{cases} \frac{xd}{a}, & o < x < a \\ \frac{d(l-x)}{l-a}, & a < x < l \end{cases}$

$f(x) = \sum_{n=1}^{\infty} b_n \sin\frac{n\pi x}{l}$

$b_n = \frac{2l^2 d}{a(l-a)\,n^2\pi^2}\sin\left(\frac{n\pi a}{l}\right)$

Therefore, the required half period series is given by

$$\frac{2l^2 d}{a(l-a)\pi^2}\left[\sin\frac{\pi a}{l}\sin\frac{\pi x}{l} + \frac{1}{2^2}\sin\frac{2\pi a}{l}\sin\frac{2\pi x}{l} + \frac{1}{3^2}\frac{\sin 3\pi a}{l}\sin\frac{3\pi x}{l} + \ldots\right]$$

Objective Problems

Fill up the Appropriate answers in the space provided

1. The period of cos 3x is x = ...

Ans. $2\pi/3$

2. If $f(x) = x^4$ in (-1, 1), then the Fourier coefficients b_n = ...

Ans. Zero

3. The period of a constant function is........

Ans. not defined

4. If $f(x) = \begin{cases} -1, & -1 < x < o \\ 1, & o < x < 1 \end{cases}$

Then f(t) is an..........

Ans. odd

5. The Fourier expansion of an even function f(x) is $(-\pi, \pi)$ has only............terms.

Ans. Cosine

6. If f(x) = x sin x is $(-\pi, \pi)$ then the value of b_n = ...

Ans. Zero

7. The formulae for finding the half range cosine series for the function f(x) in (o, l) are............

Ans. $a_o = \frac{2}{l}\int_o^l f(x)\,dx,\ a_n = \frac{2}{l}\int_o^l f(x)\cos\frac{n\pi x}{l}\,dx$

8. The half range sine series for 1 in (o, π) is ....

Ans. $\frac{4}{\pi}\left\{\sin x + \frac{\sin 3x}{3} + \frac{\sin 5x}{5} + ...\right\}$

9. The value of b_n in the Fourier series of $f(x) = |x|$ in $(-\pi, \pi)$ =........

Ans. Zero

10. If f(x) is defined in (o, l) then the period of f(x) to expand it as a half range sine series is.......

Ans. $2l$

11. Jacques Fourier was a ....

Ans. French mathematican and physicist

Indicate True/False for the statements made therein

12. (i) A function f(x) is even if $f(-x) = -f(x)$

(True/False)

Ans. False

(ii) A function f(x) is odd if $f(-x) = f(x)$

(True/False)

Ans. False

(iii) Most functions are neither even nor odd

(True/False)

Ans. True

(iv) A function f(x) can always be expressed as an arithematic mean of an even or odd function as

$$f(x) = \frac{1}{2}[f(x) + f(-x)] + \frac{1}{2}[f(x) - f(-x)]$$ (U.P.T.U. 2009)

True/False

Ans. True

13. f(x) is an odd function in $(-\pi, \pi)$, then the graph of f(x) is symmetric about the x axis.

(True/False)

Ans. False

14. A function f(x) defined in $(-\pi, \pi)$ can be expanded into Fourier series containing both sine and cosine terms

(True/False)

Ans. True

15. If $f(x) = x^2$ in $(-\pi, \pi)$, then the Fourier series of f(x) contains only sine terms (Ture/False)

Ans. False

16. The function $f(x) = \begin{cases} 1 - x \text{ in } -\pi<x<o \\ 1+x \text{ in } o<x<\pi, \end{cases}$ is an odd function

True/False

Ans. False

Chapter 8
Partial Differential Equations

Introduction

Partial differential equations arise, in cases when a dependent variable is a function of two or more independent variables. An ordinary differential equation can be formed by eliminating arbitrary Constants from a relation between two variables such as f (x, y) = 0, and in general the order of the differential equation is equal to the number of arbitrary constants eliminated. A partial differential equation, on the other hand, can be formed by eliminating not-arbitrary constants, but arbitrary functions, from a relation involving three or more variables, Provided such an elimination is possible, In many problems of science and engineering a dependent variable is connected implicity or explicitly with two or more independent variables. If z = z (x, y) is a dependent variable where x and y are the independent variables, then the first order partial derivative of z with respect to x and y are denoted by $p = \frac{\partial z}{\partial x}$ and $q = \frac{\partial z}{\partial y}$. The second order partial derivatives of z are given by

$$r = \frac{\partial^2 r}{\partial x^2}, s = \frac{\partial^2 z}{\partial x \partial y}, t = \frac{\partial^2 z}{\partial y^2}$$

Definition: An equation involving one or more partial derivatives of an unknown function of two or more independent variables is called a partial differential equation.

Example 1. $\frac{\partial^2 u}{\partial x^2} = a^2 \frac{\partial^2 u}{\partial t^2}$ is a partial differential equation.

Example 2. $\frac{\partial^2 u}{\partial x^2} + \frac{\partial^2 u}{\partial y^2} = 0$ is a partial differential equation.

Definition 2 The order of the highest derivative occuring is a partial differential equation is called the order of the equation.

Example 1. $x^2 \frac{\partial z}{\partial x} + y^2 \frac{\partial z}{\partial y} = z$ is a first order partial differential equation.

Example 2. The equation

$$\left(\frac{\partial^2 u}{\partial x^2}\right)^2 + u^2\left(\frac{\partial u}{\partial x}\right) = f(x, y)$$ is a second order partial differential equation.

Definition 3. The degree of a partial differential equation is the degree of the highest order partial derivative occuring in the equation.

Example The degree of the equation

$$\frac{\partial^2 u}{\partial x^2} + \frac{\partial^2 u}{\partial y^2} = 0$$ is one

Notation

If z is a function of two independent variables say x and y then, we shall use the following notation for the partial derivatives of z.

$$\frac{\partial z}{\partial x} = p,\ \frac{\partial z}{\partial y} = q,\ \frac{\partial^2 z}{\partial x^2} = r,\ \frac{\partial^2 z}{\partial x \partial y} = s,\ \frac{\partial^2 z}{\partial y^2} = t$$

i.e. $z_x = p$, $z_y = q$, $z_{xx} = r$, $z_{xy} = s$, $z_{yy} = t$

Formation of Partial Differential Equations

Partial differential equation can be formed in two ways- (1) eliminating arbitrary constants and (2) eliminating arbitrary functions

1. **By Elimination of Arbitrary constants:**

We can form partial differential equation by eliminating arbitrary constants from the given equations.

If the number of arbitrary constants is equal to the number of variables, in the given equation of a curve, we get a first order partial differential equation

Consider the equation

$$f(x, y, z, a, b) = 0 \tag{1}$$

Where a and b are arbitrary constants

Differentiating (1) partially with respect to x and y, we get

$$\frac{\partial f}{\partial x} + \frac{\partial f}{\partial z}\frac{\partial z}{\partial x} = 0 \text{ or } \frac{\partial f}{\partial x} + \frac{\partial f}{\partial z}p = 0 \tag{2}$$

and $$\frac{\partial f}{\partial y} + \frac{\partial f}{\partial z}\frac{\partial z}{\partial y} = 0 \text{ or } \frac{\partial f}{\partial y} + \frac{\partial f}{\partial z}q = 0 \tag{3}$$

Eliminating a and b from the equations (1), (2) and (3), we get an equation of the form

$$\phi\,(x, y, z, p, q) = 0$$

Which is the required partial differential equation of (1)

Example 1 Eliminate the constants a and b from the following equations

(A) $z = (x + a)(y + b)$

(B) $z = (x - a)^2 + (y - b)^2$

(C) $ax^2 + by^2 + z^2 = 1$

Solution (A) We have $z = (x+a)(y+b)$ (1)

Differentiating equation (1) partially with respect to x and y, we get

$$\frac{\partial z}{\partial x} = p = (y + b) \ \& \ \frac{\partial z}{\partial y} = q = (x + a)$$

Substituting in (1) we have z = pq which is the required differential equation

(B) The given equation is

$$z = (x - a)^2 + (y - b)^2 \quad (1)$$

Differentiating equation (1) partially with respect to x and y, we get

$$\frac{\partial z}{\partial x} = 2(x - a) \ \& \ \frac{\partial z}{\partial y} = 2(y-b)$$

On squaring and adding these equations, we get

$$\left(\frac{\partial z}{\partial x}\right)^2 + \left(\frac{\partial z}{\partial y}\right)^2 = 4\left[(x-a)^2 + (y-b)^2\right]$$

$= 4z$ using (1)

$\Rightarrow p^2 + q^2 = 4z$

Which is the required differential equation

(C) The given equation is

$$ax^2 + by^2 + z^2 = 1 \quad (1)$$

Differentiating equation (1) partially with respect to x and y, we get

$$2ax + 2z\frac{\partial z}{\partial x} = 0 \Rightarrow 2ax + 2zp = 0$$

$\Rightarrow ax = -zp$

$\Rightarrow a = -zp/x$

and $2by + 2z\dfrac{\partial z}{\partial y} = 0 \Rightarrow 2by + 2zq = 0$

$\Rightarrow by = -zq$

$\Rightarrow b = -zq/y$

putting the values of a and b in equation (1) we get

$$-\frac{z}{x}px^2 + \left(-\frac{z}{y}qy^2\right) + z^2 = 1$$

or $z(px + qy) = z^2 - 1$

Which is the required differential equation.

Example 2. Find the partial differential equation of all planes cutting of equal intercepts with x and y axes

Solution Let $\frac{x}{a} + \frac{y}{b} + \frac{z}{c} = 1$ (1)

be a equation of a plane making equal intercepts on x and y axis, so in this case a = b.

Differentiating (1) partially with respect to x and y, we get

$$\frac{1}{a} + \frac{1}{c}\frac{\partial z}{\partial x} = 0 \text{ i.e. } \frac{-c}{a} = p \quad (2)$$

$$\text{and } \frac{1}{b} + \frac{1}{c}\frac{\partial z}{\partial y} = 0 \text{ i.e. } \frac{-c}{a} = q \because a = b \quad (3)$$

From (2) and (3), we have

$p = q$

i.e. $p - q = 0$

Which is the required differential equation

Example 3 Find the partial differential equation of all spheres whose centres lie on the Z-axis (U.P.T.U. 2009)

Solution Let the equation of the sphere having its centre on z-axis be

$$x^2 + y^2 + (z-c)^2 = r^2 \quad (1)$$

Differentiating (1) partially with respect to x and y, we get

$$2x + 2(z-c)\frac{\partial z}{\partial x} = 0$$

or $x + (z - c)p = 0$

or $z - c = -x/p$ (2)

and $2y + 2(z-c)\frac{\partial z}{\partial y} = 0$

or $y + (z - c)q = 0$

or $z - c = -y/q$ (3)

From (2) and (3), we have

$$-\frac{x}{p} = -\frac{y}{q}$$

i.e. $xq - yp = 0$

which is the required differential equation

2. Formation of Partial Differential Equations by The Elimination of Arbitrary Function of Specific Function

When one Arbitrary Function is involved

In this case the resulting partial differential equation is a first order partial differential equation

Let the arbitrary function be of the form

$$z = f(u) \qquad (1)$$

where u is function of x, y, z

Differentiating (1) partially with respect to x and y, we get

$$\frac{\partial z}{\partial x} = \frac{\partial f}{\partial u}\frac{\partial u}{\partial x} + \frac{\partial f}{\partial u}\frac{\partial u}{\partial z}\frac{\partial z}{\partial x} \qquad (2)$$

and $$\frac{\partial z}{\partial y} = \frac{\partial f}{\partial u}\frac{\partial u}{\partial y} + \frac{\partial f}{\partial u}\frac{\partial u}{\partial z}\frac{\partial z}{\partial y} \qquad (3)$$

By eliminating the arbitrary function f, from (1), (2) and (3) we get a first order partial differential equation

Example 4 Form the partial differential equation by eliminating the arbitrary function f from the relation

$$z = y^2 + 2f\left(\frac{1}{x} + \log y\right)$$ (I.A.S. 2007, Bihar, P.C.S. 1995)

Solution We have

$$z = y^2 + 2f\left(\frac{1}{x} + \log y\right) \qquad (1)$$

Differentiating (1) partially with respect to x and y, We get

$$\frac{\partial z}{\partial x} = p = 2f'\left(\frac{1}{x} + \log y\right)\left(-\frac{1}{x^2}\right)$$

$$\text{or } -px^2 = 2f'\left(\frac{1}{x} + \log y\right) \qquad (2)$$

$$\text{and } \frac{\partial z}{\partial y} = q = 2y + 2f'\left(\frac{1}{x} + \log y\right)\left(\frac{1}{y}\right)$$

$$\text{or } qy - 2y^2 = 2f'\left(\frac{1}{x} + \log y\right) \qquad (3)$$

From (2) and (3) we have

$$-px^2 = qy - 2y^2$$

i.e. $x^2p + qy = 2y^2$

Which is the required partial differential equation.

Example 5. Eliminate the arbitrary function f from the equation

$$z = f\left(xy/z\right)$$

Solution. We have $z = f\left(\frac{xy}{z}\right)$ (1)

Differentiating (1) partially with respect to x and y, we get

$$\frac{\partial z}{\partial x} = p = f'(xy/z)\, y\left(\frac{z - xp}{z^2}\right) \quad (2)$$

$$\frac{\partial z}{\partial y} = q = f'(xy/z)\, x\left(\frac{z-yq}{z^2}\right) \quad (3)$$

Dividing (2) by (3) we have

$$\frac{p}{q} = \frac{y(z-xp)}{x(z-yq)}$$

$$\text{or } \frac{p}{q} = \frac{yz-xyp}{xz-xyq}$$

$$\text{or } px - qy = 0$$

Which is the required partial differential equation.

When Two Arbitrary Functions are Involved

When two arbitrary functions are to be eliminated from the given relation to form a partial differential equation, we differentiate twice or more number of times and eliminate the arbitrary functions from the relations obtained.

Example 6. Form a partial differential equation by eliminating the function f and F from

$$z = f(x + iy) + F(x - iy)$$

Solution. The given equation is

$$z = f(x + iy) + F(x-iy) \quad (1)$$

Differentiating (1) partially with respect to x and y, we get

$$\frac{\partial z}{\partial x} = f'(x+iy) + F'(x-iy) \quad (2)$$

and $\frac{\partial z}{\partial y} = i f'(x+iy) - i F'(x-iy)$ (3)

Differentiating equations (2) and (3) partially once again w.r.t. x and y, we get

$$\frac{\partial^2 z}{\partial x^2} = f''(x + iy) + F''(x - iy) \tag{4}$$

and $$\frac{\partial^2 z}{\partial y^2} = -f''(x + iy) - F''(x - iy) \tag{5}$$

Adding (4) and (5), we get

$$\frac{\partial^2 z}{\partial x^2} + \frac{\partial^2 z}{\partial y^2} = 0$$

Which is required partial differential equation of second order.

EXERCISE

1. Form Partial differential equations by eliminating the arbitrary constants from the following:

(i) $z = (x+a)(y+b)$ Ans. $pq = z$

(ii) $2z = (ax + y)^2 + b$ Ans. $q^2 = px + qy$

(iii) $z = ax + (1-a)y+b$ Ans. $p+q = 1$

(iv) $az + b = a^2x + y$ Ans. $pq = 1$

(v) $z = a \log\left(\frac{b(y-1)}{1-x}\right)$ Ans. $px + qy = p + q$

2. Form the partial differential equation by eliminating the arbitrary functions from

(i) $z = xy + f(x^2 + y^2)$ Ans. $py - qx = y^2 - x^2$

(ii) $z = x + y + f(xy)$ Ans. $px - qy = x - y$

(iii) $z = f(xy)$ Ans. $p+q = 0$

(iv) $z = f(x^2 + y^2)$ Ans. $xq - yp = 0$

(v) $z = f(x^2-y^2)$ Ans. $yp + xq = 0$

(vi) $f(x^2 + y^2 + z^2, z^2 - 2xy) = 0$ Ans $(p-q)z = y - x$

Hint. The given equation is $f(x^2 + y^2 + z^2, z^2 - 2xy) = 0$ (1)

Let $u = x^2 + y^2 + z^2$, $v = z^2 - 2xy$ (2)

so equation (1) becomes $f(u, v) = 0$ (3)

Differentiating (3) partially w.r.t x, we get

$$\frac{\partial f}{\partial u}\left(\frac{\partial u}{\partial x} + p\frac{\partial u}{\partial z}\right) + \frac{\partial f}{\partial v}\left(\frac{\partial v}{\partial x} + p\frac{\partial v}{\partial z}\right) = 0 \tag{4}$$

From (2) we have

$$\left.\begin{array}{l}\dfrac{\partial u}{\partial x} = 2x,\ \dfrac{\partial u}{\partial z} = 2z,\ \dfrac{\partial v}{\partial y} = -2x \\ \dfrac{\partial v}{\partial z} = 2z,\ \dfrac{\partial u}{\partial y} = 2y,\ \dfrac{\partial v}{\partial x} = -2y\end{array}\right\} \quad (5)$$

From (4) and (5) we have

$$\frac{\partial f/\partial u}{\partial f/\partial v} = -\left(\frac{\partial v}{\partial x} + p\frac{\partial v}{\partial z}\right) \Big/ \left(\frac{\partial u}{\partial x} + p\frac{\partial u}{\partial z}\right) \quad (6)$$

Similarly differentiating w.r.t. y, we get

$$\frac{\partial f/\partial u}{\partial f/\partial v} = -\frac{\left(\frac{\partial v}{\partial y} + q\,\partial v/\partial z\right)}{\left(\frac{\partial u}{\partial y} + q\frac{\partial u}{\partial z}\right)} \quad (7)$$

putting the values from (5) in equation (6) and (7), we get

$$\frac{\partial f/\partial u}{\partial f/\partial v} = -\frac{(-2y + 2pz)}{(2x + 2zp)}$$

$$\& \ \frac{\partial f/\partial u}{\partial f/\partial v} = -\left(\frac{-2x + 2qz}{2y + 2zq}\right)$$

$$\therefore \ \frac{pz - y}{pz + x} = \frac{qz - x}{qz + y}$$

$\Rightarrow pz\,(x + y) - qz\,(x + y) = y^2 - x^2$

or $(p - q)\,z = y - x$

3. Form a partial differential equation by eliminating the arbitrary function ϕ from $\phi\,(x + y + z,\ x^2 + y^2 - z^2) = 0$. What is the order of this partial differential equation? (U.P.P.C.S. 1993, Bihar P.C.S. 2007)

Hint. Given

$\phi\,(x+y+z,\ x^2 + y^2 - z^2) = 0$ (1)

Let $u = x + y + z$ & $v = x^2 + y^2 - z^2$ (2)

Then (1) becomes $f\,(u, v) = 0$ (3)

Differentiating (3) partially w.r.t. x, we get

$$\frac{\partial \phi}{\partial u}\left(\frac{\partial u}{\partial x} + p\frac{\partial u}{\partial z}\right) + \frac{\partial \phi}{\partial v}\left(\frac{\partial v}{\partial x} + p\frac{\partial v}{\partial z}\right) = 0 \quad (4)$$

From (2) $\left.\begin{aligned}\frac{\partial u}{\partial x} = 1, \frac{\partial u}{\partial z} = 1, \frac{\partial v}{\partial x} = 2x \\ \frac{\partial v}{\partial z} = -2z, \frac{\partial u}{\partial y} = 1, \frac{\partial v}{\partial y} = 2y\end{aligned}\right\}$ (5)

From (4) and (5) $\frac{\partial \phi}{\partial u}(1+p) + 2\frac{\partial \phi}{\partial v}(x - pz) = 0$

or $\frac{\partial \phi}{\partial u} \Big/ \frac{\partial \phi}{\partial v} = -2(x - pz)\Big/(1+p)$ (6)

Again differentiating (3) partially w.r.t y, we get

$$\frac{\partial \phi}{\partial u}\left(\frac{\partial u}{\partial y} + q\frac{\partial u}{\partial z}\right) + \frac{\partial \phi}{\partial v}\left(\frac{\partial v}{\partial y} + q\frac{\partial v}{\partial z}\right) = 0$$

or $\frac{\partial \phi}{\partial u}(1 + q) + 2\frac{\partial \phi}{\partial v}(y - zq) = 0$ using (5)

or $\frac{\partial \phi}{\partial u} \Big/ \frac{\partial \phi}{\partial v} = -2(y - qz)\Big/ 1+q$ (7)

From (6) and (7) by eliminating ϕ, we get

$$\frac{(x-pz)}{1+p} = \frac{y-qz}{1+q}$$

or $(1 + q)(x - pz) = (1 + p)(y - qz)$

or $(y + z)p - (x + z)q = x - y$

which is the desired partial differential equation of first order.

Solution of Partial Differential Equation by Direct integration:

Example 1. Solve $\frac{\partial^2 z}{\partial x^2} = xy$

Solution. Given $\frac{\partial^2 z}{\partial x^2} = xy$ (1)

treating y as constant and integrating (1) with respect to x, we get

$$\frac{\partial z}{\partial x} = y\frac{x^2}{2} + f(y) \text{ (say)} \quad (2)$$

Integrating (2) with respect to x keeping y as constant, we get

$z = y\,\frac{x^3}{6} + x\,f(y) + g(y)$

Hence the required solution is $z = \frac{x^3y}{6} + x\,f(y) + g(y)$

Example 2 Solve $\frac{x^2z}{\partial x\partial y} = \frac{1}{xy}$

Solution. Given $\frac{\partial^2 z}{\partial x\partial y} = \frac{1}{xy}$ (1)

treating y as constant and integrating (1) w.r.t x,
we get

$$\frac{\partial z}{\partial y} = \frac{1}{y}\log x + \phi\,(y) \qquad (2)$$

Now keeping x as constant and integrating (2) w.r.t y, we get

$$z = \log x \log y + \int \phi(y)\,dy + g(x)$$

or $z = \log x \log y + f(y) + g(x)$
which is the required solution

Example 3. Solve $\frac{\partial^3 z}{\partial x^2\partial y} = \cos\,(2x + 3y)$

Solution. we have $\frac{\partial^3 z}{\partial x^2\partial y} = \cos\,(2x + 3y)$ (1)

treating y as constant and integrating (1) w.r.t. x, we get

$$\frac{\partial^2 z}{\partial x\partial y} = \frac{1}{2}\sin\,(2x + 3y) + f(y)$$

Integrating w.r.t. x, we get

$$\frac{\partial z}{\partial y} = -\frac{1}{4}\cos\,(2x + 3y) + x\int f(y)dx + g(y)$$

$$= -\frac{1}{4}\cos\,(2x + 3y) + x\phi\,(y) + g(y)$$

Integrating w.r.t. y we get

$$z = -\frac{1}{12}\sin\,(2x + 3y) + x\int f(y)\,dy + \int g(y)\,dy$$

$$\Rightarrow z = -\frac{1}{12}\sin(2x+3y) + x\,\phi_1(y) + \phi_2(y)$$

Lagrange's Linear Equation

The partial differential equation of the form

$$Pp + Qq = R \tag{1}$$

where P, Q and R are functions of x, y, z is called Lagrange's linear partial differential equation. Lagrange's linear equation is a first order partial differential equation.

Method of solving Lagrange's equation

Equation (1) i.e. Lagrange's equation is obtain by eliminating arbitrary function from $\phi(u, v) = 0$ where u and v are functions of x, y and z

Differentiating

$$\phi(u, v) = 0 \tag{2}$$

partially with respect to x and y, we get

$$\frac{\partial\phi}{\partial u}\left(\frac{\partial u}{\partial x} + \frac{\partial u}{\partial z}p\right) + \frac{\partial\phi}{\partial v}\left(\frac{\partial v}{\partial x} + \frac{\partial v}{\partial z}p\right) = 0 \tag{3}$$

and $$\frac{\partial\phi}{\partial u}\left(\frac{\partial u}{\partial y} + \frac{\partial u}{\partial z}q\right) + \frac{\partial\phi}{\partial v}\left(\frac{\partial v}{\partial y} + \frac{\partial v}{\partial z}q\right) = 0 \tag{4}$$

Eliminating $\frac{\partial\phi}{\partial u}$ and $\frac{\partial\phi}{\partial v}$ from equation (3) and (4) we get

$$\begin{vmatrix} \frac{\partial u}{\partial x} + \frac{\partial u}{\partial z}p & \frac{\partial v}{\partial x} + \frac{\partial v}{\partial z}p \\ \frac{\partial u}{\partial y} + \frac{\partial u}{\partial z}q & \frac{\partial v}{\partial y} + \frac{\partial v}{\partial z}q \end{vmatrix} = 0$$

i.e. $$\left(\frac{\partial u}{\partial x} + \frac{\partial u}{\partial z}p\right)\left(\frac{\partial v}{\partial y} + \frac{\partial v}{\partial z}q\right) - \left(\frac{\partial u}{\partial y} + \frac{\partial u}{\partial z}q\right)\left(\frac{\partial v}{\partial x} + \frac{\partial v}{\partial z}p\right) = 0$$

or $$\left(\frac{\partial u}{\partial y}\frac{\partial v}{\partial z} - \frac{\partial u}{\partial z}\frac{\partial v}{\partial y}\right)p + \left(\frac{\partial u}{\partial z}\frac{\partial v}{\partial x} - \frac{\partial u}{\partial x}\frac{\partial v}{\partial z}\right)q = \frac{\partial u}{\partial x}\frac{\partial v}{\partial y} - \frac{\partial u}{\partial y}\frac{\partial v}{\partial x} \tag{5}$$

Comparing (1) and (5), we get

$$P = \frac{\partial u}{\partial x}\frac{\partial v}{\partial z} - \frac{\partial u}{\partial z}\frac{\partial v}{\partial y}$$

$$Q = \frac{\partial u}{\partial z}\frac{\partial v}{\partial x} - \frac{\partial u}{\partial x}\frac{\partial v}{\partial z}$$
$$R = \frac{\partial u}{\partial x}\frac{\partial v}{\partial y} - \frac{\partial u}{\partial y}\frac{\partial v}{\partial x}$$

Now let us suppose $u = C_1$, $v = C_2$ are two solutions of the Lagrange's equation $Pp + Qq = R$

Differentiating $u = C_1$ and $v = C_2$ partial with respect to x and y, we have

$$\frac{\partial u}{\partial x} dx + \frac{\partial u}{\partial y} dy + \frac{\partial u}{\partial z} dz = du = 0 \qquad (6)$$

and $$\frac{\partial v}{\partial x} dx + \frac{\partial v}{\partial y} dx + \frac{\partial v}{\partial z} dz = dv = 0 \qquad (7)$$

From (6) and (7) by cross-multiplication we have

$$\frac{dx}{\frac{\partial u}{\partial y}\frac{\partial v}{\partial z} - \frac{\partial u}{\partial z}\frac{\partial v}{\partial y}} = \frac{dy}{\frac{\partial u}{\partial z}\frac{\partial v}{\partial x} - \frac{\partial u}{\partial x}\frac{\partial v}{\partial z}} = \frac{dz}{\frac{\partial u}{\partial x}\frac{\partial v}{\partial y} - \frac{\partial u}{\partial y}\frac{\partial v}{\partial x}}$$

or $$\frac{dx}{P} = \frac{dy}{Q} = \frac{dz}{R}$$

The solutions of these equations are

$u = C_1$ and $v = C_2$

Hence $\phi(u, v) = 0$ is a solution of equation (1)

Working rule:

Step 1: Form auxiliary equations $\frac{dx}{P} = \frac{dy}{Q} = \frac{dz}{R}$

Step 2: Solve the above auxiliary equations Let the two solutions obtained be denoted by $u = C_1$ and $v = C_2$

Step 3: The required solution of the equation

$Pp + Qq = R$ is

$\phi(u, v) = 0$

Example 1. Solve $yzp + zxq = xy$

Solution. Given $yzp + zxq = xy$ (1)

Lagrange's auxiliary equations for (1) are

$$\frac{dx}{yz} = \frac{dy}{zx} = \frac{dz}{xy} \qquad (2)$$

Taking first two members, we have

$x\,dx - y\,dy = 0$

Integrating, we have

$x^2 - y^2 = C_1$

Similarly taking the first and the last members

we get

$x^2 - z^2 = C_2$

Therefore, the required solutions is

$f(x^2 - y^2, x^2 - z^2) = 0$

Example 2. Solve p tan x + q tan y = tan z (U.P.P.C.S. 1990)

Solution. Here Lagrange's auxiliary equations are

$$\frac{dx}{\tan x} = \frac{dy}{\tan y} = \frac{dz}{\tan z} \quad (1)$$

From first two fraction of (1) we get

$\cot x\,dx = \cot y\,dy$

Integrating, we get

$\log \sin x = \log \sin y + \log C_1$

or $\log\left(\sin x/\sin y\right) = \log C_1$

or $\sin x/\sin y = C_1$ (2)

Similarly, from the last two fractions,

we have

$\sin y/\sin z = C_2$ (3)

From (2) and (3) required general solution of the given equation is

$$\phi\left(\frac{\sin x}{\sin y}, \frac{\sin y}{\sin z}\right) = 0$$

Example 3. Find the general integral of

$(mz - ny)\,p + (nx - lz)\,q = ly - mx$ (I.A.S. 1977)

Solution. The auxiliary equations are

$$\frac{dx}{mz - ny} = \frac{dy}{nx - lz} = \frac{dz}{ly - mx} \quad (1)$$

Choosing x, y, z as multipliers, each fraction of (1)

$$= \frac{xdx + y\,dy + z\,dz}{x\,(mz - ny) + y\,(nx - lz) + n\,(ly - mx)}$$

$= \dfrac{x\,dx + y\,dy + z\,dz}{0}$

$\therefore x\,dx + y\,dy + z\,dz = 0$

Which on integration gives

$x^2 + y^2 + z^2 = C_1$ (2)

choosing l, m, n as multipliers, each fraction of (1)

$= \dfrac{l\,dx + m\,dy + n dz}{l\,(mz - ny) + m\,(nx - lz) + n\,(ly - mx)}$

$= \dfrac{l dx + m dy + n\,dz}{0}$

$\therefore l dx + m dy + n dz = 0$

which on integration, we have

$lx + my + nz = C_2$ (3)

$\therefore$ Required general solution is

$\phi\,(x^2 + y^2 + z^2, lx + my + nz) = 0$

where ϕ is an arbitrary function

Example 4. Find the general integral of $(y + zx)\,p - (x + yz)\,q = x^2 - y^2$

(U.P.P.C.S. 2002, Bihar P.C.S. 2002)

Solution. The Lagrange's auxiliary equations are

$$\frac{dx}{y+zx} = \frac{dy}{-(x+yz)} = \frac{dz}{x^2-y^2} \quad (1)$$

i.e. $\dfrac{y\,dx + x\,dy}{y^2-x^2} = \dfrac{dz}{x^2-y^2}$

or $d\,(xy) + dz = 0$

on integration, we get

$xy + z = C_1$ (2)

Again $\dfrac{dx + dy}{(y-x) - z\,(y-x)} = \dfrac{dz}{x^2 - y^2}$

or $\dfrac{dx + dy}{-(1-z)} = \dfrac{dz}{x+y}$

or $(x+y)\,(dx + dy) + (1-z)\,dz = 0$

or $d\left\{\dfrac{1}{2}\,(x+y)^2\right\} + (1-z)\,dz = 0$

Integrating above, we get

$$\frac{1}{2}(x+y)^2 + \left(z-\frac{z^2}{2}\right) = C_2$$

or $x^2 + y^2 + 2xy + 2z - z^2 = C_3$ where $2C_2 = C_3$

or $x^2 + y^2 - z^2 = C_4$ where $C_3 - 2C_1 = C_4$

Therefore, the general solution is

$f(x^2 + y^2 - z^2, xy + z) = 0$

where f is an arbitrary function

Example 5. Solve

$(2x^2 + y^2 + z^2 - 2yz - zx - xy)\,p + (x^2 + 2y^2 + z^2 - yz - 2zx - xy)q$

$= x^2 + y^2 + 2z^2 - yz - zx - 2xy$ (I.A.S. 1992)

Solution. The Lagrange's auxiliary equations are

$$\frac{dx}{2x^2 + y^2 + z^2 - 2yz - zx - xy} = \frac{dy}{x^2 + 2y^2 + z^2 - yz - 2zx - xy} = \frac{dz}{x^2 + y^2 + 2z^2 - yz - zx - 2xy}$$

$$\therefore \frac{dx - dy}{(x-y)(x+y+z)} = \frac{dy - dz}{(y-z)(x+y+z)} = \frac{dz - dx}{(z-x)(x+y+z)}$$

Taking first two fractions we have

$$\frac{dx - dy}{x - y} = \frac{dy - dz}{y - z}$$

Integrating, we get

$\log(x-y) = \log(y - z) + \log C_1$

or $\dfrac{x-y}{y-z} = C_1$

Similarly, we have

$$\frac{z-x}{y-z} = C_2$$

$\therefore$ The required solution is $\phi\left(\dfrac{x-y}{y-z}, \dfrac{z-x}{y-z}\right) = 0$

Example 6. Solve

$$(y+z+t)\frac{\partial t}{\partial x} + (z+x+t)\frac{\partial t}{\partial y} + (x+y+t)\frac{\partial t}{\partial z} = x+y+z \qquad \text{(I.A.S. 1995)}$$

Solution. The Lagrange's auxiliary equations are

$$\frac{dx}{y+z+t} = \frac{dy}{z+x+t} = \frac{dz}{x+y+t} = \frac{dt}{x+y+z} \qquad (1)$$

$$\therefore \frac{dx-dy}{-(x-y)} = \frac{dy-dz}{-(y-z)} = \frac{dz-dt}{-(z-t)} = \frac{dx+dy+dz+dt}{3(x+y+z+t)}$$

Taking first and Second fraction, we have

$$\frac{dx-dy}{x-y} = \frac{dy-dz}{y-z}$$

Integrating above we have

$\log (x-y) = \log (y-z) + \log C_1$

$$\frac{x-y}{y-z} = C_1 \qquad (2)$$

Similarly taking second and third fraction, we have

$$\frac{z-t}{y-z} = C_2 \qquad (3)$$

Again taking third and fourth fraction, we have

$$\frac{dz-dt}{z-t} + \frac{dx+dy+dz+dt}{3\,(x+y+z+t)} = 0$$

Integrating, above we have

$$\log (z-t) + \frac{1}{3} \log (x+y+z+t) = \log C_3$$

or $(z-t)\,(x+y+z+t)^{1/3} = C_3$

$\therefore$ The general integral is

$$f\left[\frac{x-y}{y-z}, \frac{z-t}{y-z}, (z-t)\,(x+y+z+t)^{1/3}\right] = 0$$

Example 7. Find the surface whose tangent planes cut off an intercept of constant length k from the axis of z. (IAS 1993)

Solution. Equation of the tangent plane at (x, y, z) is

$Z-z = p\,(X-x) + q\,(Y-y)$

Since k is the intercept on the axis of z

$\therefore$ when $X = 0 = Y$, $Z = k$

$\therefore k-z = p\,(-x) + q\,(-y)$

or $xp + yq = z-k$

The subsidiary equations are

$$\frac{dx}{x} = \frac{dy}{y} = \frac{dz}{z-k}$$

Taking the first two members, we have

$$\frac{dx}{x} = \frac{dy}{y}$$

$\therefore \log x = \log y - \log C_1$

or $\frac{x}{y} = C_1$

Again taking the first and last members.

we have

$$\frac{dx}{x} = \frac{dz}{z-k}$$

$\therefore \log x = \log (z-k) - \log C_2$

or $\frac{z-k}{x} = C_2$

Therefore, the general solution is $\phi\left(\frac{y}{x}, \frac{z-k}{x}\right) = 0$ which represents the required surface.

Example 8. Solve

$(z^2 - 2yz - y^2)p + (xy + xz)q = xy - xz$

If the solution of the above equation represents a sphere. What will be the coordinate of its centre. (Bihar P.C.S. 1999, Roorkee 1975)

Solution. Lagrange's auxiliary equations are

$$\frac{dx}{z^2 - 2yz - y^2} = \frac{dy}{x\,(y+z)} = \frac{dz}{x\,(y-z)} \qquad (1)$$

From the last two fractions of (1), we get

$$\frac{dy}{y+z} = \frac{dy}{y-z}$$

or $(y-z)\, dy = (y+z)\, dz$

or $ydy - (zdy + ydz) - zdz = 0$

or $ydy - d(yz) - zdz = 0$

Integrating, we get

$y^2 - 2yz - z^2 = C_1 \qquad (2)$

Again choosing x, y, z as multipliers, each fractions of (1)

$$= \frac{xdx + ydy + zdz}{x\,(z^2 - 2y\,z - y^2) + y\,(xy + xz) + z\,(xy - xz)}$$

$$= \frac{xdx + ydy + zdz}{0}$$

$xdx + ydy + zdz = 0$

Integrating we get

$x^2 + y^2 + z^2 = C_2$ (3)

From (2) and (3), the required general integral is

$\phi (y^2 - 2yz - z^2, x^2 + y^2 + z^2) = 0$ (4)

Where ϕ is an arbitrary function

From (4) we observe that if the solution represent a sphere, then co-ordinates of its centre must be (0, 0, 0) i.e. origin

NON-LINEAR PARTIAL DIFFERENTIAL EQUATIONS Those equations in which p and q occur other than in the first degree are called non-linear partial differential equations of the first order. In other words partial differential equations which contains p and q with powers higher than unity and the product of p and q are called non-linear partial differential equations.

Special Types of Equations

Standard I. Equations involving only p and q and no x, y, z That is, equation of the form f (p, q) = 0 (1)

i.e. equation which are independent of x, y, z.

Let the required solution be

$z = ax + by + c$ (2)

where a and b are connected by

$f (a, b) = 0$ (3)

Where a, b, c are constants,

Differentiating (1) partially with respect to x and y, we get

$$\frac{\partial z}{\partial x} = a \text{ and } \frac{\partial z}{\partial y} = b$$

Which when substituted in (3), gives (1)

From (3) we may find b in terms of a

i.e. $b = \phi (a)$ say

The required solution of (1) is

$z = ax + \phi (a) y + C$

Example 1 Solve $p^2 + q^2 = 1$

Solution. The equation is of the form f (p, q) = 0

The solution is given by

$z = ax + by + C$

where $a^2 + b^2 = 1$

or $b = \sqrt{(1-a^2)}$

Hence, the required solution is

$$z = ax + \sqrt{(1-a^2)}\, y + C$$

Equation Reducible to f (p, q) = 0

In some cases, we transform the equations into f(p, q) = 0 by making suitable substitutions

Example 2. Solve $(x + y)(p + q)^2 + (x - y)(p - q)^2 = 1$ (I.A.S. 1991)

Solution. putting $x + y = X^2$, $x - y = Y^2$

so that

$$p = \frac{\partial z}{\partial x} = \frac{\partial z}{\partial X}\frac{\partial X}{\partial x} + \frac{\partial z}{\partial Y}\frac{\partial Y}{\partial x} = \frac{1}{2X}\frac{\partial z}{\partial X} + \frac{1}{2Y}\frac{\partial z}{\partial Y}$$

$$\text{and } q = \frac{\partial z}{\partial y} = \frac{\partial z}{\partial X}\frac{\partial X}{\partial y} + \frac{\partial z}{\partial Y}\frac{\partial Y}{\partial y} = \frac{1}{2X}\frac{\partial Z}{\partial X} - \frac{1}{2Y}\frac{\partial Z}{\partial Y}$$

$$\therefore\ p + q = \frac{1}{X}\frac{\partial Z}{\partial X} \text{ and } p - q = \frac{1}{Y}\frac{\partial z}{\partial Y}$$

putting in the given equation, we have

$$\left(\frac{\partial z}{\partial X}\right)^2 + \left(\frac{\partial z}{\partial Y}\right)^2 = 1$$

Which is of the form of standard I

∴ The complete integral is given by

$$z = a\,X + b\,Y + c$$

Where $a^2 + b^2 = 1$ or $b^2 = \sqrt{(1-a^2)}$

∴ The required complete integral is

$$z = a\sqrt{(x+y)} + \sqrt{(1-a^2)}\,\sqrt{(x-y)} + c$$

Example 3. Find the complete integral of $(y - x)(qy - px) = (p - q)^2$ (I.A.S. 1992)

Solution. Let us put $X = x + y$ and $Y = xy$

$$\text{so that } p = \frac{\partial z}{\partial x} = \frac{\partial z}{\partial X}\frac{\partial X}{\partial x} + \frac{\partial z}{\partial Y}\frac{\partial Y}{\partial x} = \frac{\partial z}{\partial X} + y\frac{\partial z}{\partial Y}$$

$$\text{and } q = \frac{\partial z}{\partial y} = \frac{\partial z}{\partial X}\frac{\partial X}{\partial y} + \frac{\partial z}{\partial Y}\frac{\partial Y}{\partial y} = \frac{\partial z}{\partial X} + x\frac{\partial z}{\partial Y}$$

Substituting in the given equation we have

$$(y-x)(y-x)\frac{\partial z}{\partial X} = (y-x)^2\left(\frac{\partial z}{\partial Y}\right)^2$$

or $$\frac{\partial z}{\partial X} = \left(\frac{\partial z}{\partial Y}\right)^2$$

Which is of the form of standard I

$\therefore$ The complete integral is given by

$z = aX + bY + c$

where $a = b^2$

$\therefore$ The complete integral is

$z = b^2(x+y) + bxy + c$

Standard II Equations involving only p, q and z i.e. equations of the form f(z, p, q) = 0

Equations of the form $f(z, p, q) = 0$ (1)

Let us assume $z = f(x+ay)$ as a trial solution of given equation (1), where a is an arbitrary constant

$\therefore z = f(X)$ where $X = x + ay$

$$\therefore\ p = \frac{\partial z}{\partial x} = \frac{\partial z}{\partial X}\frac{\partial X}{\partial x} = \frac{\partial z}{\partial X}1 = \frac{\partial z}{\partial X}$$

and $$q = \frac{\partial z}{\partial y} = \frac{\partial z}{\partial X}\frac{\partial X}{\partial y} = a\frac{\partial z}{\partial X} = a\frac{dz}{dX}$$

$\therefore$ Equation (1) reduces to the form

$$f\left(z, \frac{dz}{dX}, a\frac{dz}{dX}\right) = 0$$

Which is an ordinary differential equation of order one. Integrating it we may get the complete integral

Example 4 Find the complete integral of

$z^2(p^2z^2 + q^2) = 1$ (I.A.S, 1997)

Solution Putting $z = f(x+ay) = f(X)$

where $X = x + ay$

so that $$p = \frac{\partial z}{\partial x} = \frac{dz}{dX}$$

and $$q = \frac{\partial z}{\partial y} = a\frac{dz}{dX}$$

The equation becomes

$$z^2\left[\left(\frac{dz}{dX}\right)^2 z^2 + a^2\left(\frac{dz}{dX}\right)^2\right] = 1$$

or $z^2 (z^2 + a^2)\left(\frac{dz}{dX}\right)^2 = 1$

or $z\sqrt{(z^2 + a^2)}\, dz = dX$

Integrating, we have

$$\frac{1}{3}(z^2 + a^2)^{3/2} = X + b$$

or $9 (x + ay + b)^2 = (z^2 + a^2)^3$

which is the required complete integral

Example 5. Solve $pq = x^m y^n z^l$ (I.A.S. 1989, 1994)

Solution. Putting $\frac{x^{m+1}}{m+1} = X,\ \frac{y^{n+1}}{n+1} = Y$

so that $p = \frac{\partial z}{\partial x} = \frac{\partial z}{\partial X}\frac{dX}{dx} = x^m \frac{\partial z}{\partial X}$

and $q = \frac{\partial z}{\partial y} = \frac{\partial z}{\partial Y}\frac{dY}{dy} = y^n \frac{\partial z}{\partial Y}$

Then the given equation reduce to

$$\frac{\partial z}{\partial X}\frac{\partial z}{\partial Y} = z^l \qquad (1)$$

Which is the form of standard II

$\therefore$ putting $z = f(X + aY) = f(u)$

where $u = X + aY$

$$\frac{\partial z}{\partial X} = \frac{dz}{du}\frac{\partial u}{\partial X} = \frac{dz}{du}$$

and $\frac{\partial z}{\partial Y} = \frac{dz}{du}\frac{\partial u}{\partial Y} = a\frac{dz}{du}$

Equation (1) becomes

$$a\left(\frac{dz}{du}\right)^2 = z^l$$

or $z^{-l/2}\, dz = \frac{du}{\sqrt{a}}$

Integrating, we have

$$\frac{z^{-\frac{l}{2}+1}}{-\frac{l}{2}+1} = \frac{u}{\sqrt{a}} + b$$

or $$\frac{z^{-\frac{l}{2}+1}}{-\frac{l}{2}+1} = \frac{1}{\sqrt{a}}\left(\frac{x^{m+1}}{m+1} + a\,\frac{y^{n+1}}{n+1}\right) + b$$

Standard III i.e. Equation of the form f (x, p) = F (y, q)

As a trial solution, let us put each side equal to a arbitrary constant

i.e. $f(x, p) = F(y, q) = a$

from which we obtain

$p = f_1(x, a)$ and $q = f_2(y, a)$

Now from $dz = pdx + qdy$

we have $dz = f_1(x, a)\,dx + f_2(y, a)\,dy$

$$z = \int f_1(x, a)\,dx + \int f_2(y, a)\,dy + b$$

Which is the complete integral

Example 6. Solve $p^2 + q^2 = x + y$

Solution. Let $p^2 - x = y - q^2 = a$

$\therefore\ p = \sqrt{(x+a)}$ and $q = \sqrt{(y-a)}$

putting in $dz = pdx + qdy$, we have

$$dz = \sqrt{(x+a)}\,dx + \sqrt{(y-a)}\,dy$$

Integrating, we get

$$z = \frac{2}{3}(x+a)^{3/2} + \frac{2}{3}(y-a)^{3/2} + b$$

Example 7. Solve $z(p^2 - q^2) = x-y$ (I.A.S. 1989, U.P.P.C.S. 1992)

Solution The given equation can be written as

$$\left(\sqrt{z}\,\frac{\partial z}{\partial x}\right)^2 - \left(\sqrt{z}\,\frac{\partial z}{\partial y}\right)^2 = x-y$$

putting $\sqrt{z}\,dz = dZ$, so that $Z = \frac{2}{3}z^{3/2}$

The equation becomes $\left(\frac{\partial Z}{\partial x}\right)^2 - \left(\frac{\partial Z}{\partial y}\right)^2 = x-y$

or $P^2 - Q^2 = x - y$

where $P = \frac{\partial Z}{\partial x}, Q = \frac{\partial Z}{\partial y}$

or $P^2 - x = Q^2 - y = a$

Which is of the form of standard III

$\therefore P^2 - x^2 = Q^2 - y = a$

$\therefore P = \sqrt{(x+a)}$ and $Q = \sqrt{(y+a)}$

Putting in $dZ = Pdx + Qdy$, we have

$dZ = \sqrt{(x+a)}\ dx + \sqrt{(y+a)}\ dy$

Integrating $Z = \frac{2}{3}(x+a)^{3/2} + \frac{2}{3}(y+a)^{3/2} + b$

or $z^{3/2} = (x+a)^{3/2} + (y+a)^{3/2} + c$

Standard IV Equation of the form Z = px + qy + f (p, q)
(Clairaut's form)

An equation of the form $z = px + qy + f(p, q)$, Which is linear in x and y is called Clairaut's equation

The complete solution of the Clairaut's equation is

$z = ax + by + f(a, b)$

i.e. the solution of Clairaut's equation is obtained putting p = a and q = b

Example 8 Solve $z = px + qy + c\sqrt{1+p^2+q^2}$

(I.A.S. 1989, Bihar P.C.S. 2007; U.P.P.C.S. 2005)

Solution. This is of the form of Standard IV

$\therefore$ The complete integral is

$z = ax + by + c\sqrt{(1+a^2+b^2)}$

EXERCISE

1. Solve the following partial differential equations

(a) $y^2p - xyq = x(z-2y)$

Ans. $\phi(x^2-y^2, zy-y^2) = 0$

(b) $xzp + yzq = xy$

Ans. $\phi(x/y, xy-z^2) = 0$

(c) $\left(\frac{y^2z}{x}\right) p + xz\,q = y^2$

Ans. $\phi(x^3 - y^3, x^2 - y^2) = 0$

(d) $z(xp - yq) = y^2 - x^2$

Ans. $f(xy, x^2 + y^2 + z^2) = 0$

(e) $p + 3q = 5z + \tan(y - 3x)$

Ans. $\phi[y-3x, e^{-5x}\{5z + \tan(y-3x)\}] = 0$

(f) $x(y+z)p - y(x^2 + z)q = z(x^2 - y^2)$

Ans. $f(x^2 + y^2 - 2z, xyz) = 0$

(g) $x^2(y - z)p + (z - x)y^2 q = z^2(x - y)$

Ans. $f\left(xyz, \frac{1}{x} + \frac{1}{y} + \frac{1}{z}\right) = 0$

(h) $(x+2z)p + (4zx - y)q = 2x^2 + y$ (Roorkee 1976)

Ans. $\phi(xy - z^2, x^2 - y - z) = 0$

(i) $px(z-2y^2) = (z-qy)(z-y^2-2x^3)$

Ans. $\phi\left(\frac{z}{x} + x^2 - \frac{y^2}{x}, \frac{y}{z}\right) = 0$

(j) $(x^2 - yz)p + (y^2 - xz)q = z^2 - xy$

Ans. $\phi\left(\frac{x-y}{y-z}, \frac{y-z}{z-x}\right) = 0$

2. Solve the following partial differential equations.

(a) $p = 2q^2 + 1$

Ans. $z = ax + \sqrt{(2b^2 + 1)}\, y + c$

(b) $x^2p^2 + y^2q^2 = z^2$ (Raj SLET 1997)

Ans. $\log z = a \log x + \sqrt{(1-a^2)} \log y + c$

(c) $(x^2 + y^2)(p^2 + q^2) = 1$

Ans. $z = \frac{a}{2}\log(x^2 + y^2) + \sqrt{(1-a^2)}\tan^{-1}\frac{y}{x} + C$

(d) $9(p^2 z + q^2) = 4$

Ans. $(z+a)^3 = (x + ay + b)^2$

(e) $z^2(p^2 z^2 + q^2) = 1$

Ans. $9(x + ay + b)^2 = (z^2 + a^2)^3$

(f) $p(1 + q^2) = q(z-a)$

Ans. $4(bz - ab - 1) = (x + by + c)^2$

(g) $p^2 = z^2(1 - pq)$

Ans. $\frac{1}{\sqrt{a}} \log\left[z\sqrt{a} + \sqrt{a(1+az^2)} + \sqrt{(1+az^2)}\right] = x + c$

(h) $p(1+q^2) = q(z-a)$

Ans. $4a(z-a) = 4 + (x+ay+b)^2$

(i) $q^2y^2 = z(z-px)$

Ans. $z^{2a^2/\{-1\pm\sqrt{(1+4a^2)}\}} = bxy^a$

(j) $p(1+q^2) = q(z-a)$

Ans. $4a(z-a) = 4 + (x+ay+b)^2$

(k) $z^2(p^2+q^2+1) = c^2$

Ans. $(1+a^2)(c-z^2) = (x+ay+b)^2$

3. Solve the following partial differential equations.

(a) $\sqrt{p} + \sqrt{q} = 2x$

Ans. $z = \frac{1}{6}(a+2x)^3 + a^2y + b$

(b) $yp = 2yx + \log q$

Ans. $az = ax^2 + a^2x + e^{ay} + ab$

(c) $z^2(p^2+q^2) = x^2 + y^2$

Ans. $z^2 = x\sqrt{(a+x^2)} + a\log\left\{x+\sqrt{(a+x^2)}\right\} + y\sqrt{(y^2-a)} - a\log\left\{y+\sqrt{(y^2-a)} + C\right\}$

(d) $p^2 - 2x^2 = q^2 - y$

Ans. $z = \frac{2}{3}x^3 + ax \pm \frac{2}{3}(y+a)^{3/2} + b$

(e) $p^2 + q^2 = z^2(x+y)$

Ans. $\log z = \frac{2}{3}(a+x)^{3/2} + \frac{2}{3}(y-a)^{3/2} + C$

(f) $z = px + qy + 2pq$

Ans. $z = ax + by + 2ab$

(g) $z = px + qy + p^2 + q^2$

Ans. $z = ax + by + a^2 + b^2$

(h) $z = px + qy + \log pq$

Ans. $z = ax + by + \log ab$

CHARPIT'S METHOD

Charpits method is a general method for solving equations with two independent variables. This method is generally applied to solve equations which cannot be reduced to any of the standard forms, let the given equation be

$$f(x, y, z, p, q) = 0 \tag{1}$$

If we are able to find another relation

$$F(x, y, z, p, q) = 0 \tag{2}$$

Then solving p and q from equations (1) and (2) and then, substituting in

$$dz = pdx + qdy \tag{3}$$

The equation (3) is integrable. The integral of equation (3) will give the complete solution of equation (1). Differentiating equations (1) and (2) with respect to x and y, we get

$$\frac{\partial f}{\partial x} + \frac{\partial f}{\partial z}p + \frac{\partial f}{\partial p}\frac{\partial p}{\partial x} + \frac{\partial f}{\partial q}\frac{\partial q}{\partial x} = 0 \tag{4}$$

$$\frac{\partial F}{\partial x} + \frac{\partial F}{\partial z}p + \frac{\partial F}{\partial p}\frac{\partial p}{\partial x} + \frac{\partial F}{\partial q}\frac{\partial q}{\partial x} = 0 \tag{5}$$

$$\frac{\partial f}{\partial y} + \frac{\partial f}{\partial z}q + \frac{\partial f}{\partial p}\frac{\partial p}{\partial y} + \frac{\partial f}{\partial q}\frac{\partial q}{\partial y} = 0 \tag{6}$$

$$\frac{\partial F}{\partial y} + \frac{\partial F}{\partial z}q + \frac{\partial F}{\partial p}\frac{\partial p}{\partial y} + \frac{\partial F}{\partial q}\frac{\partial q}{\partial y} = 0 \tag{7}$$

Eliminating $\frac{\partial p}{\partial x}$ from equations (4) and (5) (i.e. Multiplying (4) by $\frac{\partial F}{\partial p}$, (5) by $-\frac{\partial f}{\partial p}$ and adding) we get

$$\left(\frac{\partial f}{\partial x} + p\frac{\partial f}{\partial z}\right)\frac{\partial F}{\partial p} - \left(\frac{\partial F}{\partial x} + p\frac{\partial F}{\partial z}\right)\frac{\partial f}{\partial p} + \frac{\partial q}{\partial x}\left(\frac{\partial f}{\partial q}\frac{\partial F}{\partial p} - \frac{\partial F}{\partial q}\frac{\partial f}{\partial p}\right) = 0 \tag{8}$$

Similarly eliminating $\frac{\partial q}{\partial y}$ (i.e. multiplying equation (6) by $\frac{\partial F}{\partial q}$ and (7) by $-\frac{\partial F}{\partial q}$ and adding) from equation (6) and (7), we get

$$\left(\frac{\partial f}{\partial y} + q\frac{\partial f}{\partial z}\right)\frac{\partial F}{\partial q} - \left(\frac{\partial F}{\partial y} + q\frac{\partial F}{\partial z}\right)\frac{\partial f}{\partial q} + \frac{\partial p}{\partial y}\left(\frac{\partial f}{\partial p}\frac{\partial F}{\partial q} - \frac{\partial f}{\partial q}\cdot\frac{\partial F}{\partial p}\right) = 0 \tag{9}$$

since $\frac{\partial q}{\partial x} = \frac{\partial p}{\partial y}$

i.e. $\dfrac{\partial^2 z}{\partial x \partial y} = \dfrac{\partial^2 z}{\partial y \partial x}$

We find that the last terms of (8) and (9) cancel each other when added. Therefore, adding (8) and (9), we get

$$\left(\frac{\partial f}{\partial x} + p\frac{\partial f}{\partial z}\right)\frac{\partial F}{\partial p} + \left(\frac{\partial f}{\partial y} + q\frac{\partial f}{\partial z}\right)\frac{\partial F}{\partial q} + \left(-p\frac{\partial f}{\partial p} - q\frac{\partial f}{\partial q}\right)\frac{\partial F}{\partial x} + \left(\frac{-\partial f}{\partial q}\right)\frac{\partial F}{\partial y} = 0 \quad (10)$$

Equation (10) is a linear partial differential equation of first order with x, y, z, p, q as independent variables and F as the dependent variable.

$$\frac{dp}{\frac{\partial f}{\partial x} + p\frac{\partial f}{\partial z}} = \frac{dq}{\frac{\partial f}{\partial y} + q\frac{\partial f}{\partial z}} = \frac{dz}{-p\frac{\partial f}{\partial p} - q\frac{\partial f}{\partial q}} = \frac{dx}{\frac{-\partial f}{\partial p}} = \frac{dy}{-\frac{\partial f}{\partial q}} = \frac{dF}{0} \quad (11)$$

i.e. $\dfrac{dp}{f_x + pf_z} = \dfrac{dq}{f_y + qf_z} = \dfrac{dz}{-pf_p - qf_q} = \dfrac{dx}{-f_p} = \dfrac{dy}{-f_q} = \dfrac{dF}{0}$

we solve (11) for p and q substituting the values of p and q is (3) and integrating, we get the solution of (1)

Example 1. Solve $(p^2 + q^2)\, y = qz$

(U.P.T.U. 2006, U.P.P.C.S. 1990, Bihar P.C.S 2007, U.P.P.C.S, 1999)

Solution. Here, we have, $f(x, y, z, p, q) = (p^2 + q^2)\, y - qz = 0$ (1)

Now, the Charpit's auxiliary equations are

$$\frac{dx}{-f_p} = \frac{dy}{-f_q} = \frac{dz}{-pf_p - qf_q}\ \frac{dp}{f_x + pf_z} = \frac{dq}{f_y + qf_z}$$

which, here reduce to

$$\frac{dx}{-2py} = \frac{dy}{-(2qy - z)} = \frac{dz}{-p(2py) - q(2qy - z)} = \frac{dp}{0 + p(-q)} = \frac{dq}{(p^2 + q^2) + q(-q)}$$

From the last two fractions, we get

$$\frac{dp}{-pq} = \frac{dq}{p^2}$$

$\Rightarrow pdp + qdq = 0$

Integrating, we get

$p^2 + q^2 = a^2$ (say) (2)

Where a is an arbitrary constant

$\therefore$ From (1) we get $a^2 y - qz = 0$

$$\Rightarrow q = \frac{a^2 y}{z}$$

Therefore from (2) $p = \sqrt{(a^2 - q^2)}$

$= \sqrt{a^2 - (a^2y/z)^2}$

$= \dfrac{a\sqrt{(z^2 - a^2y^2)}}{z}$

Also, we know that

$dz = pdx + qdy$ (3)

putting p and q in equation (3), we get

$$dz = \frac{a}{z}\sqrt{(z^2 - a^2y^2)}\,dx + \frac{a^2y}{z}\,dy$$

$$\text{or } \frac{zdz - a^2ydy}{\sqrt{(z^2 - a^2y^2)}} = adx$$

putting $z^2 - a^2y^2 = t$, we get

$dt = a\,dx$

Integrating, we have

$t = \sqrt{(z^2 - a^2y^2)} = ax + b$

or $z^2 = a^2y^2 + (ax + b)^2$

which is the complete solution

Example 2. Solve $2zx - px^2 - 2q\,xy + pq = 0$

(I.A.S. 1991, 93, Bihar P.C.S. 2002, 1995)

Solution. Here, we have

$f(x, y, z, p, q) = 2zx - px^2 - 2qxy + pq = 0$ (1)

Now, the Charpit's auxiliary equations, are

$$\frac{dx}{-f_p} = \frac{dy}{-f_q} = \frac{dz}{-pf_p - qf_q} = \frac{dp}{f_x + pf_z} = \frac{dq}{f_y + qf_z}$$

which, here reduce to

$$\frac{dx}{-(-x^2 + q)} = \frac{dy}{-(-2xy + p)} = \frac{dz}{-p(-x^2 + q) - q(-2xy + p)}$$

$$= \frac{dp}{(2z - 2px - 2qy) + p(2x)} = \frac{dq}{-2qx + q(2x)}$$

From, the last fraction, we get

$dq = 0$

$\Rightarrow q = a$ (2)

putting in (1) we have

$2xz - px^2 - 2axy + pa = 0$

or $p = \dfrac{2x\,(z-ay)}{x^2 - a}$ (3)

Also, we know that

$dz = pdx + qdy$

putting the values of p & q from (2) & (3), we get

or $dz = \dfrac{2x(z-ay)}{x^2 - a}\,dx + ady$

or $\dfrac{dz - ady}{z - ay} = \dfrac{2x}{x^2 - a}\,dx$

Integrating above, we get

$\log (z - ay) = \log (x^2 - a) + \log b$

$\therefore\ z-ay = b\,(x^2 - a)$

or $z = ay + b\,(x^2 - a)$ is the complete integral. Where a & b are arbitrary constants.

Example 3. Solve $p = (qy + z)^2$ (U.P.P.C.S. 1992) (1)

Solution. Here, we have $f(x, y, z, p, q) = -p + (qy + z)^2 = 0$

Now, the Charpit's auxiliary equations, are

which, reduce to

$$\frac{dx}{-f_p} = \frac{dy}{-f_q} = \frac{dz}{-p\,f_p - qf_q} = \frac{dp}{f_x + pf_z} = \frac{dq}{f_y + qf_z}$$

$$\frac{dx}{-(-1)} = \frac{dy}{-2y(qy+z)} = \frac{dz}{-p(-1) - q2(qy+z)y} = \frac{dp}{2p\,(qy + z)} = \frac{dq}{4q\,(qy + z)}$$

Taking the second and fourth fractions, we get $\dfrac{dp}{p} + \dfrac{dy}{y} = 0$

Integrating, we have

$\log p + \log y = \log a$

or $p = \dfrac{a}{y}$

$\therefore$ From (1) $\dfrac{a}{y} = (qy + z)^2$

or $\sqrt{\left(\dfrac{a}{y}\right)} = qy + z$

$$\therefore q = \frac{\sqrt{a}}{y^{3/2}} - \frac{z}{y}$$

putting these values in dz = pdx + q dy, we get

$$dz = \frac{a}{y}\,dx + \left(\frac{\sqrt{a}}{y^{3/2}} - \frac{z}{y}\right) dy$$

$$\text{or } y dz + z dy = a\,dx + \frac{\sqrt{a}}{\sqrt{y}}\,dy$$

Integrating, we have

$$yz = ax + 2\sqrt{(ay)} + b$$

which is the complete integral

Example 4. Find a complete integral of

$p^2x + q^2y = z$

(IAS 2006, 2004 1976, 79; U.P.P.C.S 2004, 2002, 2001, 1997)

Solution. Here, we have

$f(x, y, z, p, q) = p^2x + q^2y - z = 0$ (1)

Now, the Charpits auxiliary equations are

$$\frac{dx}{-f_p} = \frac{dy}{-f_q} = \frac{dz}{-pf_p - qf_q} = \frac{dp}{f_x + pf_z} = \frac{dq}{f_y + qf_z}$$

which reduce to

$$\frac{dx}{-2px} = \frac{dy}{-2qy} = \frac{dz}{-2(p^2x + q^2y)} = \frac{dp}{-p^2 + p^2} = \frac{dq}{-q + q^2}$$

From which we have

$$\frac{p^2\,dx + 2px\,dp}{2p^3x + 2p^2x - 2p^3x} = \frac{q^2dy + 2qy\,dq}{2q^3y + 2q^2y - 2q^3y}$$

$$\text{i.e. } \frac{p^2\,dx + 2px\,dx}{2p^2x} = \frac{q^2\,dy + 2qy\,dq}{2q^2y}$$

$$\text{i.e. } \frac{d(p^2x)}{p^2x} = \frac{d(q^2y)}{q^2y}$$

Integrating, we have

$\log p^2x = \log q^2y + \log a$

$\Rightarrow p^2x = aq^2y$ (2)

where a is a constant

From (1) and (2) we have

$aq^2y + q^2y = z$

$$\therefore q = \left\{\frac{z}{(1+a)y}\right\}^{1/2}$$

From (1) $p = \left\{\frac{az}{(1+a)x}\right\}^{1/2}$

putting in $dz = pdx + qdy$, we have

$$dz = \left\{\frac{az}{(1+a)x}\right\}^{1/2} dx + \left\{\frac{z}{(1+a)y}\right\}^{1/2} dy$$

or $\sqrt{(1+a)}\ \frac{dz}{\sqrt{z}} = \sqrt{a}\ \frac{dx}{\sqrt{x}} + \frac{dy}{\sqrt{y}}$

Integrating above, we get

$(1+a)^{1/2}\ \sqrt{z} = \sqrt{(ax)} + \sqrt{y} + b$

which is the complete integral of (1)

Example 5. Solve $px + qy = z\,(1+pq)^{1/2}$

(I.A.S. 1992, Bihar P.C.S. 1993, 2005)

Solution. Here, we have

$f(x, y, z, p, q) = px + qy - z\,(1+pq)^{1/2} = 0$ (1)

Now, the Charpits auxiliary equations, are

$$\frac{dx}{-f_p} = \frac{dy}{-f_q} = \frac{dz}{-p\,f_p - q\,f_q} = \frac{dp}{f_x + p\,f_z} = \frac{dq}{f_y + q\,f_z}$$

Therefore Charpits auxiliary equations are

$$\text{................} = \frac{dp}{p - p\,(1+pq)^{1/2}} = \frac{dq}{q - q\,(1+pq)^{1/2}}$$

Taking the last two fractions, we have

$$\frac{dp}{p} = \frac{dq}{q}$$

Integrating above, we get

$\log p = \log q + \log a$

$\Rightarrow p = aq$ (2)

putting in (1) we have

$aqx + qy = z\,(1 + aq^2)^{1/2}$

or $q^2 (ax + y)^2 = z^2 (1 + aq)^2$

or $q^2 = \dfrac{z^2}{(ax+y)^2 - az^2}$

or $q = \dfrac{z}{\sqrt{\{(ax+y)^2 - az^2\}}}$ (3)

putting this value of q is (2) we have

$p = \dfrac{az}{\sqrt{\{(ax+y)^2 - az^2\}}}$ (4)

putting these values of p and q in dz = pdx + qdy, we get

$$dz = \frac{az\,dx + zdy}{\sqrt{\{(ax+y)^2 - az^2\}}}$$

or $\dfrac{dz}{z} = \dfrac{adx + dy}{\sqrt{\{(ax+y)^2 - az^2\}}}$

$\Rightarrow \dfrac{dz}{z} = \dfrac{\sqrt{a}\,du}{\sqrt{(au^2\ az^2)}}$

or $\dfrac{du}{dz} = \dfrac{1}{z}\sqrt{u^2 - z^2}$

Again putting u = vz

$\Rightarrow v + z\dfrac{dv}{dz} = \dfrac{1}{z}\sqrt{(v^2z^2 - z^2)}$

or $v + z\dfrac{dv}{dz} = \sqrt{(v^2 - 1)}$

or $z\dfrac{dv}{dz} = \sqrt{(v^2 - 1)} - v$

or $\dfrac{dz}{z} = \dfrac{dv}{\sqrt{(v^2 - 1)} - v}$

or $\dfrac{dz}{z} = -\left\{\sqrt{(v^2 - 1)} + v\right\} dv$

Integrating, above we have

$$\log z = -\left[\frac{v}{2}\sqrt{(v^2 - 1)} - \frac{1}{2}\log\left\{v + \sqrt{(v^2 - 1}\right\}\right] - \frac{v^2}{2} + b$$

or $-\log z + \frac{v^2}{2} + \frac{v}{2}\sqrt{(v^2-1)} - \frac{1}{2}\log\left\{v + \sqrt{(v^2-1)}\right\} = b$

which is the complete integral

where $v = \frac{u}{z} = \frac{ax+y}{z\sqrt{a}}$

Example 6. Find complete integral of $p^2 + q^2 - 2px - 2qy + 2xy = 0$

(I.A.S 2003, U.P.P.C.S. 1991)

Solution. Here, we have

$f(x, y, z, p, q) = p^2 + q^2 - 2px - 2qy + 2xy = 0$ (1)

Now, the Charpits auxiliary equations are

$$\frac{dx}{-f_p} = \frac{dy}{-f_q} = \frac{dz}{-pf_p - qf_q} = \frac{dp}{f_x + pf_z} = \frac{dq}{f_y + qf_z}$$

which here reduce to

$$\frac{dx}{-(2p-2x)} = \frac{dy}{-(2q-2y)} = \frac{dz}{p(2p-2x) - q(2q-2y)} = \frac{dp}{(-2p+2y) + p(o)} = \frac{dq}{(-2q+2x) + q(o)}$$

i.e. $\frac{dp}{-2p+2y} = \frac{dq}{-2q+2x} = \frac{dx}{2x-2p} = \frac{dy}{2y-2q}$

i.e. $\frac{dp}{-p+y} = \frac{dq}{-q+x} = \frac{dx}{x-p} = \frac{dy}{y-q}$

or $\frac{dp+dq}{x+y-p-q} = \frac{dx+dy}{x+y-p-q}$

or $dp + dq = dx + dy$

$\therefore (p-x) + (q-y) = a$ (2)

Equation (1) can be written as

$(p-x)^2 + (q-y)^2 = (x-y)^2$ (3)

putting the value of $(q-y)$ from (2) in equation (3) we have

$(p-x)^2 + [a - (p-x)]^2 = (x-y)^2$

or $2(p-x)^2 - 2a(p-x) + \{a^2 - (x-y)^2\} = 0$

$$\therefore p - x = \frac{2a \pm \sqrt{[4a^2 - 4\cdot 2\{a^2 - (x-y)^2\}}}{4}$$

$$= \frac{1}{2}\left[a + \sqrt{\{2(x-y)^2 - a^2\}}\right]$$

Taking positive sign only

$$p = x + \frac{1}{2}\left[a + \sqrt{\{2(x-y)^2 - a^2\}}\right]$$

$$\therefore \text{ From (2) } q - y = a - \frac{1}{2}\left[a + \sqrt{\{2(x-y)^2 - a^2\}}\right]$$

$$\text{or } q = y + \frac{1}{2}\left[a - \sqrt{\{2(x-y)^2 - a^2\}}\right]$$

putting these values of p and q in
$dz = p\,dx + q\,dy$ we get

$$dz = x\,dx + y\,dy + \frac{a}{2}(dx + dy) + \frac{1}{2}\sqrt{[2(x-y)^2 - a^2]}\,(dx - dy)$$

$$= xdx + y\,dy + \frac{a}{2}(dx + dy) + \frac{1}{\sqrt{2}}\sqrt{\left\{(x-y)^2 - \frac{a^2}{2}\right\}}\,(dx - dy)$$

Integrating we have

$$z = \frac{x^2}{2} + \frac{y^2}{2} + \frac{a}{2}(x+y) + \frac{1}{\sqrt{2}}\left[\frac{x-y}{2}\sqrt{\left\{(x-y)^2 - \frac{a^2}{2}\right\}} - \frac{a^2}{4}\log\left\{(x-y) + \sqrt{(x-y)^2 - \frac{a^2}{2}}\right\}\right]$$

$$\text{or } 2z = x^2 + y^2 + ax + ay + \frac{1}{\sqrt{2}}\left[(x-y)\sqrt{\left\{(x-y)^2 - \frac{a^2}{2}\right\}} - \frac{a^2}{2}\log\left\{(x-y) + \sqrt{\left\{(x-y)^2 - \frac{a^2}{2}\right\}}\right\}\right]$$

which is the complete integral

Example 7. Find a complete integral of $z^2(p^2 z^2 + q^2) = 1$ (I.A.S. 1997)

Solution. Here, we have

$f(x, y, z, p, q) = z^2(p^2 z^2 + q^2) - 1 = 0$ (1)

Now the Charpit's auxiliary equations are

$$\frac{dx}{-f_p} = \frac{dy}{-f_q} = \frac{dz}{-pf_p - qf_q} = \frac{dp}{f_x + pf_z} = \frac{dq}{f_y + qf_z}$$

$$\text{or } \frac{dx}{-2pz^4} = \frac{dy}{-2qz^2} = \frac{dz}{-2p^2z^2 - 2q^2z^2} = \frac{dp}{p(4p^2z^3 + 2zq^2)} = \frac{dq}{q(4p^2z^3 + 2zq^2)}$$

Taking first two fractions, we get

$$\frac{dp}{p} = \frac{dq}{q} \Rightarrow p = aq \qquad (2)$$

Solving equations (1) and (2) for p and q,
we have

$$p = \frac{q}{z\sqrt{a^2z^2 + 1}} \text{ and } q = \frac{1}{z\sqrt{a^2z^2 + 1}}$$

Therefore, $dz = pdx + q\,dy$

$$= \frac{adx}{z\sqrt{a^2z^2+1}} + \frac{dy}{z\sqrt{a^2z^2+1}}$$

$$= \frac{adx + dy}{z\sqrt{a^2z^2+1}}$$

or $adx + dy = z\sqrt{a^2z^2+1}\ dz$

Integrating, we get

$$ax + y = \int z\sqrt{(a^2\ z^2 + 1)}\ dz + b \qquad (3)$$

putting $a^2z^2 + 1 = t^2$

$\Rightarrow 2a^2z\ dz = 2t\ dt$

Thus, from equation (3) we have

$$ax + y = \int \frac{1}{a^2}\ t\ t\ dt + b$$

$$= \frac{1}{3a^2}\ t^3 + b$$

or $ax + y = \dfrac{1}{3\ a^2}\ (a^2z^2 + 1)^{3/2} + b$

which is the complete solution.

Example 8. Find a complete integral of

$p^2 + q^2 - 2px - 2qy + 1 = 0$ (I.A.S. 1999, U.P.P.C.S. 1998)

Solution. Here we have

$$f(x, y, z, p, q) = p^2 + q^2 - 2px - 2qy + 1 = 0 \qquad (1)$$

The Charpits auxiliary equations are

$$\frac{dx}{-(2p-2y)} = \frac{dy}{-(2q-2y)} = \frac{dz}{-p\,(2p-2x) - q\,(2q-2y)} = \frac{dp}{-2p} = \frac{dq}{-2q}$$

From the last two fractions, we have

$$\frac{dp}{-2p} = \frac{dq}{-2q} \Rightarrow \frac{dp}{p} = \frac{dq}{q}$$

on integrating we have

$\log p = \log q + \log a$

$$\Rightarrow p = a\,q \qquad (2)$$

Therefore, from equation (1) and (2), we have

$a^2q^2 + q^2 - 2aqx - 2qy + 1 = 0$

$(a^2 + 1) q^2 - (2ax + 2y) q + 1 = 0$

$$\text{or } q = \frac{2(ax + y) \pm \sqrt{4(ax + y)^2 - 4(a^2 + 1)}}{2(a^2 + 1)}$$

$$\text{Again } p = aq = a \, \frac{2(ax + y) \pm \sqrt{4(ax + y)^2 - 4(a^2 + 1)}}{2(a^2 + 1)}$$

putting these values in $dz = pdx + qdy$
we get

$$dz = \frac{(ax + y) \pm \sqrt{\{(ax + y)^2 - (a^2 + 1)\}}}{a^2 + 1} (adx + dy) \qquad (3)$$

putting $ax + y = v \Rightarrow adx + dy = dv$, we have

$$(a^2 + 1) dz = \left[v \pm \sqrt{v^2 - (a^2 + 1)} \right] dv$$

Integrating, we get

$$(a^2 + 1) z = \frac{1}{2} v^2 \pm \left[\frac{1}{2} v\sqrt{\{v^2 - (a^2 + 1)\}} \right] - \frac{1}{2} (a^2 + 1) \log \{v + \sqrt{v^2 - (a^2 + 1)}\} + b$$

$$\text{or } (a^2 + 1) z = \frac{1}{2} (ax + y)^2 \pm \left[\frac{1}{2} (ax + y) \sqrt{(ax + y)^2 - (a^2 + 1)} \right] - \frac{1}{2} (a^2 + 1) \log \{(ax + y) + \sqrt{(ax + y)^2 - (a^2 + 1)}\} + b$$

which is the complete integral

EXERCISE

1. Solve the following partial differential equations by Charpit's method.

(a) $px + qy = pq$

Ans. $az = \frac{1}{2} (y + ax)^2 + b$

(b) $pxy + pq + qy = yz$

Ans. $(z - ax) (y + a)^a = be^y$

(c) $(x^2 - y^2) pq - xy (p^2 - q^2) - 1 = 0$

Ans. $z = \frac{a}{2} \log (x^2 + y^2) + \frac{1}{a} \tan^{-1} \frac{y}{x} + b$

(d) $z^2 = pq\,xy$

Ans. $z = cx^{\sqrt{a}} \, y^{1/\sqrt{a}}$

(e) $(p^2 + q^2) x = pz$ (I.A.S. 1982)

Ans. $z^2 = a^2x + (ay + b)^2$

(f) $xp + 3yq = 2(z - x^2q^2)$

Ans. $az = ay + x(x^2b + a^2)$

(g) $16p^2z^2 + 9q^2z^2 + 4z^2 - 4 = 0$ (I.A.S. 1994)

Ans. $\sqrt{(16a^2+9)}\sqrt{1-z^2} + 2(ax+y) = b$

Linear Partial Differential Equations of second order with constant coefficients

Introduction

We shall restrict ourselves to consider only the linear partial differential equations of higher order with constant coefficients. These can be classified into the following two categories:

(a) homogeneous (b) non-homogeneous

An equation in which the derivatives involved are all of the same order, is called homogeneous

For example, $2\dfrac{\partial^3 z}{\partial x^3} - 3\dfrac{\partial^3 z}{\partial x^2 \partial y} + \dfrac{\partial^3 z}{\partial x \partial y^2} - 5\dfrac{\partial^3 z}{\partial y^3} = x - y$

is homogeneous of order three, whereas

$$4\frac{\partial^3 z}{\partial x^3} + 3\frac{\partial^2 z}{\partial x \partial y} + z = x e^{-y}$$

is certainly non-homogeneous

Homogeneous Linear partial differential Equation of Higher order with Constant Coefficients

A differential equation of the form

$$\frac{\partial^n z}{\partial x^n} + a_1\frac{\partial^n z}{\partial x^{n-1}\partial y} + a_2\frac{\partial^n z}{\partial x^{n-2}\partial y^2} + \dots\dots + a_n\frac{\partial^n z}{\partial y^n} = F(x, y) \qquad (1)$$

where $a_1, a_2, \dots\dots, a_n$ are constant's is called homogenous linear partial differential equation of nth order with constant coefficients

Let $D = \dfrac{\partial}{\partial x}$ and $D' = \dfrac{\partial}{\partial y}$, then equation (1) takes the form

$(D^n + a_1 D^{n-1} D' + a_2 D^{n-2} D'^2 + \dots\dots + a_2 D'^n) z = F(x, y)$

$\Rightarrow f(D, D') z = F(x, y)$ (2)

Where $f(D, D') = D^n + a_1 D^{n-1} D' + a_2 D^{n-2} D'^2 + \dots\dots + a_n D'^n$

is a homogenous function of D and D'

The method of solving equation (1) or (2) is similar to that for ordinary differential equations and complete solution = C.F. + P.I.

Determination of C.F.

The auxiliary equation is obtain by putting D = m and D' = 1 in f (D, D') = 0 There are two cases will aries.

(1) If $m_1, m_2, \ldots\ldots\ldots, m_n$ are the distinct roots of the auxiliary equation, then C.F of (1) is

$f_1 (y + m_1 x) + f_2 (y + m_2 x) + \ldots\ldots\ldots + f_n (y + m_n x)$

(2) When the auxiliary equation has equal (repeated) roots say m then C.F of (1) given by

$f_1 (y + mx) + x f_2 (y + mx) + \ldots\ldots + x^{r-1} f_r (y + mx)$

From above we conclude that

	Roots of A.E.	C.F.
1.	$m_1, m_2, m_3, \ldots\ldots\ldots\ldots$ (distinct roots)	$f_1 (y + m_1x) + f_2 (y + m_2x) + f_3 (y + m_3x) + \ldots\ldots\ldots\ldots\ldots\ldots\ldots\ldots$
2.	m_1, m_1, m_3 .. (Two equal roots)	$f_1 (y + m_1x) + x f_2 (y + m_1x) + f_3 (y + m_3x) + \ldots\ldots\ldots$
3.	m_1, m_1, m_1, .. (Three equal roots)	$f_1 (y + m_1x) + x f_2 (y + m_1x) + x^2 f_3 (y + m_1x) + \ldots\ldots\ldots$

Example 1. Solve 2r + 5s + 2t = 0

or $\frac{2\partial^2 z}{\partial x^2} + 5\frac{\partial^2 z}{\partial x \partial y} + 2\frac{\partial^2 z}{\partial y^2} = 0$

Solution. The given equation can be written as

$(2D^2 + 5D D' + 2D'^2) z = 0$

its auxiliary equation is putting D = m, D' = 1 i.e.

$2m^2 + 5m + 2 = 0$

or $(2m + 1) (m+2) = 0$

$\Rightarrow m = \frac{-1}{2}, -2$

Therefore C.F. = $f_1 (y - \frac{1}{2} x) + f_2 (y - 2x)$

or $z = f_1 (y - \frac{1}{2} x) + f_2 (y - 2x)$

when f's is are arbitrary

Example 2. Solve $(D^3 - 4D^2 D' + 4DD') Z = 0$

Solution. Its auxiliary equation is

$m^3 - 4m^2 + 4m = 0$

or $m (m^2 - 4m + 4) = 0$

or $m(m-2)^2 = 0$

or $m = 0, 2, 2$

Required solution is

$z = \phi_1(y + 0.x) + [\phi_2(y + 2x) + x f_3(y + 2x)]$

or $z = \phi_1(y) + \phi_2(y + 2x) + x f_3(y + 2x)$

where ϕ's are arbitrary

Example 3. solve $\dfrac{\partial^4 z}{\partial x^4} - \dfrac{\partial^4 z}{\partial y^4} = 0$

Solution. The given equation can be written as

$(D^4 - D'^4) z = 0$

its auxiliary equation is putting $D = m$, $D' = 1$

$m^4 - 1 = 0$

$\Rightarrow m = 1, -1, i, -i$

Therefore C.F $= \phi_1(y + x) + \phi_2(y - x) + \phi_3(y + ix) + \phi_4(y - ix)$

or $z = \phi_1(y + x) + \phi_2(y - x) + \phi_3(y + ix) + \phi_4(y - ix)$

The particular Integral

Consider the equation

$f(D, D') z = F(x, y)$

The particular integral may be written as

$$\frac{1}{f(D, D')} F(x, y)$$

and method of evaluating it is similar to those we have read in ordinary linear differential equations of higher orders with constant coefficients

Example 4. Solve the equation

$(D^2 - 2DD' + D'^2) z = 12xy$

Solution. Given equation is $(D^2 - 2DD' + D'^2) z = 12xy$ (1)

its auxiliary equation is

$m^2 - 2m + 1 = 0$ or $(m-1)^2 = 0$

or $m = 1, 1$

$\therefore$ C.F $= f_1(y + x) + x f_2(y + x)$

and P.I $= \dfrac{1}{D^2 - 2DD' + D'^2}(12xy)$

$= \dfrac{1}{(D - D')^2}(12xy)$

$$= \frac{1}{D^2}\left(1 - \frac{D'}{D}\right)^{-2} 12\,xy$$

$$= 12\frac{1}{D^2}\left[1 + \frac{2D'}{D} + 3\,\frac{D'^2}{D^2} + ...\right] xy$$

$$= 12\,\frac{1}{D^2}\left[xy + \frac{2}{D}\,D'\,(xy) + \frac{3}{D^2}\,D'^2\,(xy) + ...\right]$$

where $D = \frac{\partial}{\partial x}$, $D' = \frac{\partial}{\partial y}$

$$= 12\frac{1}{D^2}\left[xy + \frac{2}{D}\,x + \frac{3}{D^2}\,10\right]$$

$$= 12\,\frac{1}{D^2}\,(xy + x^2)$$

$$= 12\,\frac{1}{D}\left(\frac{1}{2}\,x^2y + \frac{x^3}{3}\right)$$

$$= 12\left(\frac{x^3}{6}\,y + \frac{1}{12}\,x^4\right)$$

$= x^4 + 2x^3\,y$

$\therefore$ Required solution is z = C.F + P.I

i.e. $z = f_1\,(y + x) + x\,f_2\,(y + x) + x^4 + 2x^3\,y$

where f_1 and f_2 are arbitrary functions

Example 5. Solve

$(D^2 + 3D\,D' + 2D'^2)\,z = x + y$ (I.A.S. 1986, 94)

Solution. The auxiliary equation is

$m^2 + 3m + 2 = 0 \Rightarrow m = -1, -2$

$\therefore$ C.F $= f_1\,(y - x) + f_2\,(y - 2x)$

and $\text{P.I} = \frac{1}{(D^2 + 3DD' + 2D'^2)}\,(x + y)$

$$= \frac{1}{D^2}\left(1 + \frac{3D'}{D} + \frac{2D'^2}{D^2}\right)^{-1}(x + y)$$

$$= \frac{1}{D^2}\left[1 - \left(\frac{3D'}{D} + \frac{2D'^2}{D^2}\right) + \left(\frac{3D'}{D} + \frac{2D'^2}{D^2}\right)^2 - ....\right](x + y)$$

$$= \frac{1}{D^2}\left[1 - \frac{3D'}{D} + ...\right](x + y)$$

$$= \frac{1}{D^2}\left[(x + y) - \frac{3}{D}(1)\right]$$

$$= \frac{1}{D^2}[(x + y) - 3x]$$

$$= \frac{1}{D^2}(y - 2x)$$

$$= \frac{1}{D}(yx - x^2)$$

$$= \frac{1}{2}yx^2 - \frac{x^3}{3}$$

$\therefore$ Required solution is z = C.I + P.I

i.e. $z = f_1(y - x) + f_2(y - 2x) + \frac{1}{2}y\,x^2 - \frac{1}{3}x^3$

where f' s are arbitrary functions

Example 6. Solve

$(2D^2 - 5DD' + 2D'^2)\,z = 24\,(y - x)$ (I.A.S. 1987)

Solution. The auxiliary equation is

$2m^2 - 5m + 2 = 0$

or $(2m - 1)(m-2) = 0$

$\Rightarrow m = \frac{1}{2}, 2$

C.F = $f_1(2y + x) + f_2(y + 2x)$

Now $P.I = \frac{1}{2D^2 - 5DD' + 2D'^2}\,24\,(y - x)$

$$= \frac{1}{2D^2}\left(1 - \frac{5D'}{2D} + \frac{D'^2}{D^2}\right)^{-1} 24\,(y - x)$$

$$= \frac{1}{2D^2}\left[1 + \frac{5D'}{2D} + ...\right] 24\,(y - x)$$

$$= \frac{1}{2D^2}\,24\,(y - x) + \frac{5}{4D^3}\,24$$

$$= 12\left(\frac{x^2}{2}y - \frac{x^3}{6}\right) + \frac{5}{4}\,24\,\frac{x^3}{6}$$

$= 6x^2 y + 3x^3$

Therefore z = C.F + P.I

or $z = f_1 (2y + x) + f_2 (y + 2x) + 6x^2 y + 3x^3$

Short Methods

Method I. When f(x, y) is function of (ax + by)

$$\frac{1}{f(D, D')} \phi^{(n)} (ax + by) = \frac{1}{f(a, b)} \phi (ax + by)$$

provided $f (a, b) \neq 0$

i.e. if $f (a, b) \neq 0$ then

$$\frac{1}{f(D, D')} \phi^{(n)} (ax + by) = \frac{1}{f(a, b)} \iint .... \int \phi(v)\, dv\, dv .... dv$$

where v = ax + by

Here after integrating ϕ (v) n times with respect to v, v is to be replaced by ax + by

Method II. When f(x, y) is a function of ax + by and f(a, b) = 0

$$\frac{1}{(D - mD')^n} \psi (y + mx) = \frac{x^n}{\underline{|n}} \psi (y + mx)$$

Example 7. Solve $(D^2 - 2DD' + D'^2) z = e^{x + 2y}$

Solution. Here, the auxiliary equation is

$m^2 - 2m + 1 = 0$

or $(m - 1)^2 = 0$ or m = 1, 1

$\therefore$ C.F = $f_1 (y + x) + x f_2 (y + x)$

where f_1 & f_2 is are arbitrary

and $P.I = \dfrac{1}{D^2 - 2DD' + D'^2} e^{x+2y}$

$= \dfrac{1}{(D - D')^2} e^{x+2y}$

$= \dfrac{1}{(1-2)^2} \iint e^v \, dv\, dv$

Applying method I

Here v = x + 2y, a = 1, b = 2, $f (a, b) \neq 0$

as $f (D, D') = (D - D')^2$

$= e^v = e^{x+2y}$

$\therefore$ Required solution is z = C.F + P.I

or $z = f_1 (y + x) + x f_2 (y + x) + e^{x + 2y}$

Example 8. Solve $(D^2 + D'^2) z = 30 (2x + y)$

Solution. Here, the auxiliary equation is

$m^2 + 1 = 0$

or $m = \pm i$

$\therefore$ C.F = $f_1(y + ix) + f_2(y - ix)$

where f_1 and f_2 are arbitrary

and $P.I = 30 \dfrac{1}{D^2 + D'^2}(2x + y)$

$= \dfrac{30}{2^2 + 1^2} \iint v\,dv\,dv$ where $v = 2x + y$

$= 6 \int \dfrac{1}{2} v^2 dv$

$= 3\left[\dfrac{v^3}{3}\right] = v^3 = (2x + y)^3$

$\therefore$ Required solution is z = C.F + P.I

or $z = f_1(y + ix) + f_2(y - ix) + (2x + y)^3$

Example 9. Solve $\dfrac{\partial^2 z}{\partial x^2} + 2\dfrac{\partial^2 z}{\partial x \partial y} + \dfrac{\partial^2 z}{\partial y^2} = 2x + 3y$

Solution. The given equation is

$(D^2 + 2DD' + D'^2)z = 2x + 3y$

Here, the auxiliary equation is

$m^2 + 2m + 1 = 0$

or $(m+1)^2 = 0$

or $m = -1, -1$

or C.F = $f_1(y - x) + x f_2(y - x)$

where f_1 and f_2 are arbitrary

$P.I = \dfrac{1}{D^2 + 2DD' + D'^2}(2x + 3y)$

$= \dfrac{1}{(D + D')^2}(2x + 3y)$

$= \dfrac{1}{(2 + 3)^2} \iint v\,dv\,dv$

where $v = 2x + 3y$

$= \dfrac{1}{25} \int \dfrac{v^2}{2} dv = \dfrac{1}{150} v^3 = \dfrac{1}{150}(2x + 3y)^3$

Therefore, required solution is

z = C.F + P.I

$= f_1(y - x) + x f_2(y - x) + \frac{1}{150}(2x + 3y)^3$

Example 10. Solve r – 2s + t = sin (2x + 3y)

or $\frac{\partial^2 z}{\partial x^2} - 2\frac{\partial^2 z}{\partial x \partial y} + \frac{\partial^2 z}{\partial y^2} = \sin(2x + 3y)$

Solution. The given equation is

$(D^2 - 2DD' + D'^2) z = \sin(2x + 3y)$

or $(D - D')^2 z = \sin(2x + 3y)$

Here, the auxiliary equation is

$(m - 1)^2 = 0$

$\Rightarrow m = 1, 1$

$\therefore$ C.F = $f_1(y + x) + x f_2(y + x)$

and P.I = $\frac{1}{(D - D')^2} \sin(2x + 3y)$

$= \frac{1}{(2 - 3)^2} \iint \sin v \, dv \, dv$ where v = 2x + 3y

$= -\int \cos v \, dv$

$= -\sin v$

$= -\sin(2x + 3y)$

$\therefore$ Required solution is z = C.F + P.I

i.e. $z = f_1(y + x) + x f_2(y + x) - \sin(2x + 3y)$

Example 11. Solve the P.D.E.

$\frac{\partial^2 z}{\partial x^2} - 2\frac{\partial^2 z}{\partial x \partial y} + \frac{\partial^2 z}{\partial y^2} = \sin x$ (U.P.T.U. 2004; U.P.T.U. 2009)

Solution. The given equation is

$(D^2 - 2DD' + D'^2) z = \sin x$

or $(D - D')^2 z = \sin x$

Here the auxiliary equation is

$(m - 1)^2 = 0$

$\Rightarrow m = 1, 1$

$\therefore$ C.F = $f_1(y + x) + x f_2(y + x)$

and P.I = $\frac{1}{D^2 - 2DD' + D'^2} \sin x$

$$= \frac{1}{(D - D')^2} \sin x$$

$$= \frac{1}{(1 - o)^2} \iint \sin v \, dv \, dv \text{ where } v = x$$

$$= - \int \cos v \, dv$$

$= - \sin v$

$= - \sin x$

Therefore complete solution is z = C.F + P.I

i.e. $z = f_1 (y + x) + x f_2 (y + x) - \sin x$

Example 12. Solve $(D^3 - 4D^2D' + 4DD'^2) z = 6 \sin (3x + 2y)$ (I.A.S 2006)

Solution. Here, the auxiliary equation is

$m^3 - 4m^2 + 4m = 0$

or $m (m^2 - 4m + 4) = 0$

or $m (m - 2)^2 = 0$

or $m = 0, 2, 2$

$\therefore$ C.F $= f_1 (y + ox) + [f_2 (y + 2x) + x f_3 (y + 2x)]$

$= f_1 (y) + f_2 (y + 2x) + x f_3 (y + 2x)$

and P.I $= \dfrac{1}{D^3 - 4D^2 D' + 4 DD'^2} 6 \sin (3x + 2y)$

$$= \frac{1}{3^3 - 4\, 3^2\, 2 + 4\, 3\, 2^2} 6 \iiint \sin v \, dv \, dv \, dv \text{ where } v = 3x + 2y$$

$$= \frac{1}{75 - 72} 6 \iiint (- \cos v) \, dv \, dv$$

$$= -2 \int \sin v \, dv$$

$= 2 \cos v$

$= 2 \cos (3x + 2y)$

$\therefore$ Required solution is z = C.F + P.I

$z = f_1 (y) + f_2 (y + 2x) + x f_3 (y + 2x) + 2 \cos (3x + 2y)$

Example 13. Solve $\dfrac{\partial^2 z}{\partial x^2} + \dfrac{\partial^2 z}{\partial y^2} = \cos mx \cos ny$

(I.A.S. 1996, U.P.P.C.S. 1994, 1996)

Solution. The given equation is

$(D^2 + D'^2) z = \cos mx \cos ny$

Here the auxiliary equation is

$m^2 + 1 = 0$

or $m = \pm i$

$\therefore$ C.F. = $f_1(y + ix) + f_2(y - ix)$

where f's are arbitrary

and P.I. = $\dfrac{1}{D^2 + D'^2} \cos mx \cos ny$

$$= \frac{1}{2} \frac{1}{D^2 + D'^2} \{2 \cos mx \cos ny\}$$

$$= \frac{1}{2} \frac{1}{D^2 + D'^2} [\cos (mx + ny) + \cos (mx - ny)]$$

$$= \frac{1}{2} \left[\frac{1}{D^2 + D'^2} \cos (mx + ny) + \frac{1}{D^2 + D'^2} \cos (mx - ny) \right]$$

$$= \frac{1}{2} \left[\frac{1}{m^2 + n^2} \iint \cos u \, du \, du + \frac{1}{m^2 + (-n)^2} \iint \cos v \, dv \, dv \right]$$

where $u = mx + ny$, $v = mx - ny$

$$= \frac{1}{2(m^2 + n^2)} \left[\int \sin u \, du + \int \sin v \, dv \right]$$

$$= \frac{1}{2(m^2 + n^2)} [-\cos u - \cos v]$$

$$= -\frac{1}{2(m^2 + n^2)} [\cos (mx + ny) + \cos (mx - ny)]$$

$$= -\frac{1}{2(m^2 + n^2)} [2 \cos mx \cos ny]$$

$$= -\frac{1}{2(m^2 + n^2)} [2 \cos mx \cos ny]$$

$$= -\frac{1}{m^2 + n^2} \cos mx \cos ny$$

$\therefore$ The required solution is z = C.F + P.I

or $z = f_1(y + ix) + f_2(y - ix) - \dfrac{1}{(m^2 + n^2)} \cos mx \cos ny$

Example 14. Solve $4r - 4s + t = 16 \log (x + 2y)$

Solution. Given that

$(4D^2 - 4DD' + D'^2) z = 16 \log (x + 2y)$

or $(2D - D')^2 z = 16 \log (x + 2y)$

its auxiliary equation is $(2m - 1)^2 = 0$

or $m = \frac{1}{2}, \frac{1}{2}$

$\therefore$ C.F. $= f_1 (y + \frac{1}{2} x) + x f_2 (y + \frac{1}{2} x)$

$= \phi_1 (2y + x) + x \phi_2 (2y + x)$

where ϕ's are arbitrary functions

and P.I. $= \frac{1}{(2D - D')^2} [16 \log (x + 2y)]$

$$= \frac{16}{(2D - D')^2} \log \left[2\left(y + \frac{1}{2} x\right)\right]$$

$$= \frac{16}{(2D - D')^2} \left[\log 2 + \log \left(y + \frac{1}{2} x\right)\right]$$

$$= \frac{16 \log 2}{4D^2\left(1 - \frac{D'}{2D}\right)} + \frac{16}{4\left(D - \frac{D'}{2}\right)^2} \log \left(y + \frac{1}{2} x\right)$$

$$= \frac{4}{D^2}\left(1 - \frac{D'}{2D}\right)^{-1} \log 2 + 4\left[\frac{x^2}{\lfloor 2} \log \left(y + \frac{1}{2} x\right)\right] \text{ using Method II}$$

$$= \frac{4}{D^2} (\log 2) + 2x^2 \log \left(y + \frac{1}{2} x\right)$$

$$= \frac{4}{D} x (\log 2) + 2x^2 \log \left(y + \frac{1}{2} x\right)$$

$$= 4 (\log 2) \frac{x^2}{2} + 2x^2 \log \left(y + \frac{1}{2} x\right)$$

$$= 2x^2 \left[(\log 2) + \log \left(y + \frac{1}{2} x\right)\right] = 2x^2 \log (2y + x)$$

Hence, the required solution is

$z = \phi_1 (2y + x) + x \phi_2 (2y + x) + 2x^2 \log (2y + x)$

Example 15. Solve $(D^2 - 6DD' + 9 D'^2) z = 6x + 2y$

Solution. Its auxiliary equation is $m^2 - 6m + 9 = 0$ or $(m - 3)^2 = 0$ or $m = 3, 3$

$\therefore$ C.F. $= f_1 (y + 3x) + x f_2 (y + 3x)$

and $\text{P.I.} = \dfrac{1}{(D^2 - 6DD' + 9D'^2)}(6x + 2y)$

$= \dfrac{1}{(D - 3D')^2}(6x + 2y)$

$= \dfrac{2}{(D - 3D')^2}(y + 3x)$

$= 2\dfrac{x^2}{\lfloor 2}(y + 3x) = x^2 (y + 3x)$

$\therefore$ Required solutions z = C.F. + P.I.

or $z = f_1(y + 3x) + xf_2(y + 3x) + x^2(y + 3x)$

Example 16. Solve $(D^2 - 2DD' + D'^2)\, z = x^3 + e^{x+2y}$

Solution. Its auxiliary equation is

$m^2 - 2m + 1 = 0$

$\Rightarrow m = 1, 1$

$\therefore$ C.F. $= f_1(y + x) + x\, f_2(y + x)$

where f's are arbitrary

and $\text{P.I.} = \dfrac{1}{D^2 - 2DD' + D^2}(x^3 + e^{x+2y})$

$= \dfrac{1}{(D - D')^2}x^3 + \dfrac{1}{(D - D')^2}e^{x+2y}$

$= \dfrac{1}{D^2}\left(1 - \dfrac{D'}{D}\right)^{-2} x^3 + \dfrac{1}{(1-2)^2}\iint e^v\, dv\, dv$ where v = x + 2y

$= \dfrac{1}{D^2}\left(1 + \dfrac{2D'}{D} + \ldots\ldots\ldots\ldots\right) x^3 + \int e^v\, dv$

$= \dfrac{1}{D^2}x^3 + e^v$

$= \dfrac{1}{D}\left(\dfrac{x^4}{4}\right) + e^{x+2y} = \dfrac{x^5}{20} + e^{x+2y}$

$\therefore$ Required solution is z = C.F. + P.I.

or $z = f_1(y + x) + x\, f_2(y + x) + \dfrac{1}{20}x^5 + e^{x+2y}$

Example 17. Solve $(D^2 - 5DD' + 4\, D')^2\, z = \sin(4x + y)$

Solution. Its auxiliary equation is

$m^2 - 5m + 4 = 0$

$\Rightarrow m = 1, 4$

$\therefore$ C.F. $= f_1 (y + x) + f_2 (y + 4x)$

where f's are arbitrary

and P.I. $= \dfrac{1}{D^2 - 5DD' + 4D'^2} \sin (4x + y)$

$= \dfrac{1}{(D - 4D')(D - D')} \sin (4x + y)$

$= \dfrac{1}{(D - 4D')} \left[\dfrac{1}{D - D'} \sin (4x + y)\right]$

$= \dfrac{1}{(D - 4D')} \left[\dfrac{1}{4 - 1} \int \sin v \, dv\right]$ using method I, where $v = 4x + y$

$= \dfrac{1}{3(D - 4D')} (-\cos v) = -\dfrac{1}{3}\left[\dfrac{1}{D - 4D'} \cos (4x + y)\right]$

$= -\dfrac{1}{3}\left[\dfrac{x}{\underline{|1}} \cos (4x + y)\right]$ using method II

$= -\dfrac{1}{3} x \cos (4x + y)$

$\therefore$ The required solution is z = C.F. + P.I.

or $z = f_1 (y + x) + f_2 (y + 4x) - \dfrac{1}{3} x \cos (4x + y)$

Example 18 : Solve $(D^3 - 4D^2D' + 4DD'^2) z = 4 \sin (2x + y)$ (Bihar, P.C.S. 1999)

Solution. Its auxiliary equation is

$m^3 - 4m^2 + 4m = 0$

$\Rightarrow m (m^2 - 4m + 4) = 0$

$\Rightarrow m (m - 2)^2 = 0$

$\Rightarrow m = 0, 2, 2$

$\therefore$ C.F $= f_1 (y) + f_2 (y + 2x) + x f_3 (y + 2x)$

where f_1, f_2 and f_3 are arbitrary.

and P I $= \dfrac{1}{D^3 - 4D^2D' + 4DD'^2} [4 \sin (2x + y)]$

$= 4 \dfrac{1}{D(D^2 - 4D' + 4D'^2)} \sin (2x + y)$

$$= 4\frac{1}{D(D-2D')^2}\sin(2x+y) = 4\frac{1}{(D-2D')^2}\left[\frac{1}{D}\sin(2x+y)\right]$$

$$= \frac{4}{D(D-2D')^2}\left[-\frac{1}{2}\cos(2x+y)\right]$$

$$= \frac{-2}{(D-2D')^2}\cos(2x+y)$$

$$= -2\frac{x^2}{\lfloor 2}\cos(2x+y) \qquad \text{by IInd method}$$

$$= -x^2\cos(2x+y)$$

$\therefore$ Required solution is z = C. F + P.I

or $z = f_1(y) + f_2(y + 2x) + xf_2(y+2x) - x^2 \cos(2x+y)$

General Method of finding P.I.

Let the given equation be f (D, D') z = F (x, y)

where f (D, D') is a homogeneous function of D and D'.

we take $\frac{1}{D-mD'}F(x,y)$ as $\int F(x, c-mx)dx$

where c is to be replaced by y + mx after integration

Example 19 : Solve $(D^2 - DD' - 2D'^2)\, z = (y - 1)e^x$

(I.A.S. 2004, Bihar, P.C.S 1993)

or

$(D+D')(D-2D')\, z = (y-1)e^x$

Solution. It auxiliary equation is (m+1) (m-2) = 0

$\Rightarrow$ m = -1, 2

$\therefore$ C.F = $f_1(y-x) + f_2(y+2x)$

and P.I. $= \frac{1}{(D+D')(D-2D')}(y-1)e^x$

$$=\frac{1}{(D+D')}\left[\frac{1}{(D-2D')}(y-1)e^x\right]$$

$$=\frac{1}{(D+D')}\left[\int(c-2x-1)e^x dx\right]$$

$\because$ we take $\frac{1}{D-mD'}F(x,y)$ as $\int F(x,c-mx)dx$ and $c = y + 2x$

$$=\frac{1}{(D+D')}\left[(c-2x-1)e^x - \int(-2)e^x dx\right]$$

$$=\frac{1}{(D+D')}\left[(c-2x-1)e^x + 2e^x\right]$$

$$=\frac{1}{D+D'}(c-2x+1)e^x$$

$$=\frac{1}{(D+D')}\{(y+2x)-2x+1\}e^x, \text{ replacing c by y+2x}$$

$$=\frac{1}{(D+D')}(y+1)e^x = \int(C'+x+1)e^x dx$$

$\because$ we use $\frac{1}{D-mD'}F(x,y)$ as $\int F(x,c-mx)dx$ and $C' = y - x$

$=(C' + x + 1)e^x - e^x = (C' + x)e^x$ where $c' = y - x$

$= \{(y-x) + x\}e^x = ye^x$

Therefore required solution is z = C. F + P.I

or $z = f_1(y - x) + f_2(y + 2x) + y\,e^x$

Example 20 Solve $r + s - 6t = y \cos x$ (I.A.S. 1992)

OR

$(D^2 + DD' - 6D'^2)\,z = y \cos x$

Solution . Given equation is $(D^2 + DD' - 6D'^2)\,z = y \cos x$

(Bihar P.C.S 2002, U.P.P.C.S. 1999)

or $(D + 3D')(D - 2D')\,z = y \cos x$

its auxiliary equation is $(m + 3)(m - 2) = 0$

$\Rightarrow \quad m = 2, -3$

$\therefore$ C, F $= f_1(y + 2x) + f_2(y - 3x)$

and $\text{P.I} = \frac{1}{(D+3D')(D-2D')}y\cos x$

$= \frac{1}{(D+3D')}\left[\frac{1}{D-2D'} y\cos x\right]$, where c is to be replaced by y + 2x after integration

$= \frac{1}{(D+3D')}\left[\int (C-2x)\cos x\,dx\right],$

$= \frac{1}{D+3D'}\left[(c-2x)\sin x - \int(-2)\sin x\,dx\right]$ integration by parts

$= \frac{1}{D+3D'}\left[(c-2x)\sin x + 2(-\cos x)\right]$

$= \frac{1}{D+3D'}\left[\{(y+2x)-2x\}\sin x - 2\cos x\right]$, replacing c by y + 2x

$= \frac{1}{D+3D'}(y\sin x - 2\cos x)$

$= \int\{(c'+3x)\sin x - 2\cos x\}dx$,, where c' is to be replaced by y – 3x after integration

$= -(c' + 3x)\cos x + \int 3\cos x\,dx - 2\sin x$

$= -(y - 3x + 3x)\cos x + 3\sin x - 2\sin x$ (replacing c' by y – 3x)

= sin x – y cos x, on simplifying

∴ Required solution is z = C. F + P.I.

or $z = f_1(y+2x) + f_2(y-3x) + \sin x - y\cos x$

Example 22 Solve $r - t = \tan^3 x \tan y - \tan x \tan^3 y$ (U.P.P.C.S. 1992)

Solution Give equation is

$(D^2 - D'^2)z = \tan^3 x \tan y - \tan x \tan^3 y$

or $(D + D')(D - D')z = \tan^3 x \tan y - \tan x \tan^3 y$

Its auxiliary equation is (m +1) (m–1) = 0

⇒ m = –1, 1

∴ $C.F = f_1(y - x) + f_2(y + x)$

and $P.I. = \frac{1}{(D+D')(D-D')}\left[\tan^3 x \tan y - \tan x \tan^3 y\right]$

$= \frac{1}{(D+D')}\left[\frac{1}{(D-D')}(\tan^3 x \tan y - \tan x \tan^3 y)\right]$

$= \frac{1}{(D+D')}\left[\int \tan^3 x \tan(c-x) - \tan x \tan^3(c-x)\,dx\right]$

where c is to be replaced by y + x after integration

$$=\frac{1}{(D+D')}\int\left[\tan x\left(\sec^2 x-1\right)\tan(c-x)-\tan x\tan(c-x)\left\{\sec^2(c-x)-1\right\}\right]dx$$

$$=\frac{1}{(D+D')}\left[\tan(c-x).\frac{\tan^2 x}{2}+\int\sec^2(c-x)\left\{\frac{1}{2}\tan^2 x\right\}dx-\tan x\left\{-\frac{1}{2}\tan^2(c-x)\right\}\right.$$
$$\left.+\int\sec^2 x\left\{-\frac{1}{2}\tan^2(c-x)\right\}dx\right]$$

$$=\frac{1}{2(D+D')}\left[\tan(c-x)\tan^2 x+\int\sec^2 x(c-x)\tan^2 x dx+\tan x\tan^2(c-x)\right.$$
$$\left.-\int\sec^2 x\tan^2(c-x)dx\right]$$

$$=\frac{1}{2(D+D')}\left[\tan(c-x)\tan^2 x+\int\left\{1+\tan^2(c-x)\right\}\tan^2 x dx+\tan x\tan^2(c-x)\right.$$
$$\left.-\int\left(1+\tan^2 x\right)\tan^2(c-x)dx\right]$$

$$=\frac{1}{2(D+D')}\left[\tan(c-x)\tan^2 x+\tan x\tan^2(c-x)+\int\left\{\tan^2 x-\tan^2(c-x)\right\}dx\right]$$

$$=\frac{1}{2(D+D')}\left[\tan(c-x)\tan^2 x+\tan x\tan^2(c-x)+\int\left\{\left(\sec^2 x-1\right)-\left\{sce^2(c-x)-1\right\}\right\}dx\right]$$

$$=\frac{1}{2(D+D')}\left[\tan(c-x)\tan^2 x+\tan x\tan^2(c-x)+\tan x+\tan(c-x)\right]$$

$$=\frac{1}{2(D+D')}\left[\tan y\tan^2 x+\tan x\tan^2 y+\tan x+\tan y\right]$$ replacing c by y + x

$$=\frac{1}{2(D+D')}\left[\tan y\left(\tan^2 x+1\right)+\tan x(\tan^2 y+1)\right]$$

$$=\frac{1}{2(D+D')}\left[\tan y\sec^2 x+\tan x\sec^2 y\right]$$

$=\frac{1}{2}\int\left[\tan(c'+x)\sec^2 x+\tan x\sec^2(c'+x)\right]dx$ where c' is to be replaced by y-x after integration

$$=\frac{1}{2}\left[\left\{\tan(c'+x)\sec^2 x dx+\int\tan x\sec^2(c'+x)dx\right\}\right]$$

$=\frac{1}{2}\left[\left\{\tan(c'+x)\tan x-\int\sec^2(c'+x)\tan x dx\right\}+\int\tan x\sec^2(c'+x)dx\right]$ integrating only first integral by parts

$= \frac{1}{2}\tan x \tan(c' + x)$

$= \frac{1}{2}$ tan x tan y, replacing c' by y – x

∴ Required solution is z = C.F.+ P.I.

or $z = f_1(y - x) + f_2(y + x) + \frac{1}{2}\tan x \tan y$

EXERCISE

Solve the following Partial differential Equations

1. $(D^3 - 6D^2D' + 12DD'^2 - 8D'^3)z = 0$

Ans. $z = f_1(y + 2x) + xf_2(y + 2x) + x^2 f_3(y + 2x)$

2. $\frac{\partial^3 z}{\partial x^3} - 3\frac{\partial^3 z}{\partial x^2 \partial y} + 4\frac{\partial^3 z}{\partial y^3} = e^{x+2y}$ (U.P.T.U. 2007)

3. $\frac{\partial^2 z}{\partial x^2} - 4\frac{\partial^2 z}{\partial x \partial y} + 4\frac{\partial^2 z}{\partial y^2} = e^{2x+y}$

Ans. $z = f_1(y + 2x) + x f_2(y + 2x) + \frac{x^2}{2}e^{2x+y}$

4. $\frac{\partial^2 z}{\partial x^2} - 2\frac{\partial^2 z}{\partial x \partial y} + \frac{\partial^2 z}{\partial y^2} = \sin(2x + 3y)$

Ans. $z = f_1(y - x) + x f_2(y - x) - \frac{1}{25}\sin(2x + 3y)$

5. $\frac{\partial^2 z}{\partial x^2} - 3\frac{\partial^2 z}{\partial x \partial y} + 2\frac{\partial^2 z}{\partial y^2} = e^{2x-y} + e^{x+y} + \cos(x + 2y)$

Ans. $z = f_1(y + x) + f_2(y + 2x) + \frac{1}{12}e^{2x-y} - xe^{x+y} - \frac{1}{3}\cos(x + 2y)$ (U.P.T.U. 2006)

6. $\frac{\partial^3 z}{\partial x^3} - \frac{\partial^3 z}{\partial y^3} = x^3 y^3$ (I.A.S. 1997, U.P.P.C.S. 1998)

Ans. $z = f_1(y + x) + f_2(y + wx) + f_3(y + w^2 x) + \frac{1}{120}x^6 y^3 + \frac{1}{10080}x^9$

Hint. A. F is $m^3 - 1$ or $m = 1, w, w^2$, where w and w^2 are Complex cube root of unity.

7. $(D - D')(D + 2D')z = (y + x)e^x$

Ans. $z = f_1(y + x) + f_2(y - 2x) + y e^x$

8. $(D^2 + DD' - 6D'^2)z = x^2 \sin(x + y)$

Ans. $z = f_1(y + 2x) + f_2(y - 3x) + \frac{1}{4}\left(x^2 - \frac{13}{8}\right)\sin(y+x) - \frac{3}{8}x\cos(x+y)$

9. $\frac{\partial^2 z}{\partial x^2} - \frac{\partial^2 z}{\partial x \partial y} = \sin x \cos 2y$ (U.P.T.U. 2008)

Ans. $z = f_1(y) + f_2(y + x) + \frac{1}{2}\sin(x+2y) - \frac{1}{6}\sin(x+2y)$

Classification of Partial Differential Equations

Generally, we use the following types of partial differential equations:

Wave equation $\frac{\partial^2 u}{\partial t^2} = c^2 \frac{\partial^2 u}{\partial x^2}$

one-dimensional heat flow

$$\frac{\partial u}{\partial t} = c^2 \frac{\partial^2 u}{\partial x^2}$$

Two -dimensional heat flow

$$\frac{\partial^2 u}{\partial x^2} + \frac{\partial^2 u}{\partial y^2} = 0$$

and

Radio equation

$$-\frac{\partial v}{\partial x} = L\frac{\partial I}{\partial t}, -\frac{\partial I}{\partial x} = c\frac{\partial v}{\partial t}$$

Now, let us consider second order partial differential equation

$$A\frac{\partial^2 u}{\partial x^2} + B\frac{\partial^2 u}{\partial x \partial y} + C\frac{\partial^2 u}{\partial y^2} + F(x,y,u,p,q) = 0 \qquad \text{(i)}$$

Where A, B and C are constants or functions of x and y.

Then equation (i) is

(a) elliptic if $B^2 - 4AC < 0$

(b) Parabolic if $B^2 - 4AC = 0$

(c) hyperbolic if $B^2 - 4AC > 0$

Example 1 Show that the equation

$\frac{\partial^2 u}{\partial t^2} = c^2 \frac{\partial^2 u}{\partial x^2}$ is hyperbolic (U.P.T.U. 2003)

Solution

$$\frac{\partial^2 u}{\partial t^2} - c^2 \frac{\partial^2 u}{\partial x^2} = 0$$

Here A =1, B= 0, C = – c^2

Now B^2 – 4AC = 0–4.1. (– c^2) = 4 c^2

⇒ $B^2 - 4AC = (2c)^2$ is positive always

⇒ $B^2 - 4AC > 0$

∴ The given equation is hyperbolic.

Example 2. Characterize the following P.D.E into elliptic parabolic, and hyperbolic equations.

$$A\frac{\partial^2 u}{\partial x^2} + 2B\frac{\partial^2 u}{\partial x \partial y} + C\frac{\partial^2 u}{\partial y^2} = F(x, y, u, u_x, u_y)$$ (U.P.T.U. 2007)

Here A, B and C may be functions of x and y.

Solution We have

$$A\frac{\partial^2 u}{\partial x^2} + 2B\frac{\partial^2 u}{\partial x \partial y} + C\frac{\partial^2 u}{\partial y^2} \quad F(x, y, u, u_x, u_y)$$

For elliptic equations, we have $B^2 - 4AC < 0$, Here given that

$B^2 - 4AC < 0$

Here given that

B = 2B, A= A, C = C

Therefore $(2B)^2 - 4AC < 0$

or $B^2 - 4AC < 0$

For parabolic equation, we have

$B^2 - 4AC = 0$

⇒ $(2B)^2 - 4AC = 0$

⇒ $B^2 - AC = 0$

and, for hyperbolic equations

⇒ $B^2 - 4AC > 0$

i.e. $(2B)^2 - 4AC > 0$

⇒ $B^2 - 4AC > 0$

These are the conditions for the elliptic, parabolic and hyperbolic equations.

Example 3 Match the column for the items of the lefts side to that of right side.

A second order P.D.E in the function 'u' of two independent variables x,y given with usual symbols

$Au_{xx} + Bu_{xy} + Cu_{yy} + F(u) = 0$, Then

(i) Hyperbolic	(a) $B^2 - 4AC = 0$
(ii) Parabolic	(b) $B^2 - 4AC < 0$
(iii) Elliptic	(c) B2 – 4AC > 0
(iv) Not Classified	(d) A = B = C =0

(U.P.T.U. 2009)

Solution

(i) (a) $B^2 - 4AC > 0$
(ii) (b) $B^2 - 4AC = 0$
(iii) (c) $B^2 - 4AC < 0$
(iv) (d) $A = B = C = 0$

Example 4. Classify the equation

$$\left(1-x^2\right)\frac{\partial^2 z}{\partial x^2} - 2xy\frac{\partial^2 z}{\partial x \partial y} + \left(1-y^2\right)\frac{\partial^2 z}{\partial y^2} + x\frac{\partial z}{\partial x} + 3x^2y\frac{\partial z}{\partial y} - 2z = 0$$

Solution

Here $A = 1 - x^2$, $B = -2xy$, $C = 1-y^2$

Now $B^2 - 4AC = 4x^2y^2 - 4(1-x^2)(1-y^2)$

$= 4x^2y^2 - 4 + 4x^2 + 4y^2 - 4x^2y^2$

$= 4(x^2 + y^2 - 1)$

$\Rightarrow \quad B^2 - 4AC = 4(x^2 + y^2 - 1)$

if $x^2 + y^2 > 1 \Rightarrow B^2 - 4AC > 0$ (Hyperbolic)

if $x^2 + y^2 = 1 \Rightarrow B^2 - 4AC = 0$ (Parabolic)

and if $x^2 + y^2 < 1 \Rightarrow B^2 - 4AC < 0$ (Elliptic)

Choose the correct answer from the following parts:

1. The Fourier coefficient a_n of a function $f(x)$ in the interval $(0, 2\pi)$ is:

(a) $\frac{1}{\pi}\int_{-\pi}^{\pi} f(x) \cos nx\, dx$ (b) $\frac{1}{\pi}\int_{0}^{2\pi} f(x) \cos nx\, dx$

(c) $\frac{1}{\pi}\int_{0}^{\pi} f(x) \cos nx\, dx$ (d) $\frac{1}{\pi}\int_{0}^{2\pi} f(x)\, dx$

Ans (b)

2. Te Fourier coefficient b_n of a function $f(x)$ in the interval $(0, 2\pi)$ is:

(a) $\frac{1}{\pi}\int_{0}^{\pi} f(x) \sin nx\, dx$ (b) $\frac{1}{\pi}\int_{-\pi}^{\pi} f(x) \sin nx\, dx$

(c) $\frac{1}{\pi}\int_{0}^{2\pi} f(x)\, dx$ (d) $\frac{1}{\pi}\int_{0}^{2\pi} f(x) \sin nx\, dx$

Ans (d)

3. The value of $1 + \frac{1}{2^2} + \frac{1}{3^2} + \frac{1}{4^2} + \ldots\ldots\ldots\ldots$ is equal to:

(a) $\frac{\pi}{6}$ (b) $\frac{\pi^2}{8}$

(c) $\frac{\pi^2}{6}$ (d) $\frac{\pi^2}{12}$

Ans (c)

4. The value of $1 - \frac{1}{2^2} + \frac{1}{3^2} - \frac{1}{4^2} + \ldots\ldots$ is equal to:

(a) $\frac{\pi^2}{12}$ (b) $\frac{\pi}{12}$

(c) $\frac{\pi^2}{8}$ (d) $\frac{\pi^2}{6}$

Ans (a)

5. The value of $1 + \frac{1}{3} - \frac{1}{5} - \frac{1}{7} + \frac{1}{9} - \ldots\ldots$ is equal to:

(a) $\frac{\pi}{2}$ (b) $\frac{\pi}{4}$

(c) $\frac{\pi}{8}$ (d) $\frac{\pi^2}{2}$

Ans (b)

6. The value of $1 + \frac{1}{3^2} + \frac{1}{5^2} + \ldots\ldots$ is equal to:

(a) $\frac{\pi^2}{8}$ (b) $\frac{\pi^2}{6}$

(c) $\frac{\pi}{4}$ (d) $\frac{\pi^2}{2}$

Ans (a)

7. The value of $\frac{1}{1.3} - \frac{1}{3.5} + \frac{1}{5.7} - \frac{1}{7.9} + \ldots\ldots\ldots$ is equal to:

(a) $\frac{\pi}{12}$ (b) $\frac{\pi}{4} - 1$

(c) $\frac{\pi}{4} - \frac{1}{2}$ (d) $2\pi - 1$

Ans (c)

8. The half range sine series for the function $f(x)$ in $(0, \pi)$ is:

(a) $\frac{1}{\pi}\int_0^\pi f(x) \sin nx\, dx$ (b) $\frac{2}{\pi}\int_0^\pi f(x) \sin nx\, dx$

(c) $\int_0^\pi f(x) \sin nx\, dx$ (d) All the above.

Ans (b)

9. The value of $\int_{-1}^{1} [f(x)]^2\, dx$ is equal to:

(a) $\frac{1}{2}\left[a_0^2 + \sum_{n=0}^{\infty}(a_n^2 + b_n^2)\right]$ (b) $\Sigma\,(a_n^2 + b_n^2)$

(c) $l\left\{\frac{a_0^2}{2} + \sum_{n=0}^{\infty}(a_n^2 + b_n^2)\right\}$ (d) $l\left\{a_0 + \sum_{n=0}^{\infty}(a_n + b_n)\right\}$

Ans (c)

10. The value of $\frac{1}{1^4} + \frac{1}{3^4} + \frac{1}{5^4} + \ldots\ldots\ldots$ is equal to:

(a) $\frac{\pi^2}{16}$ (b) $\frac{\pi^4}{16}$

(c) $\frac{\pi^4}{96}$ (d) $\frac{\pi^4}{12}$

Ans (c)

11. The general solution of $\frac{\partial^2 z}{\partial x \partial y} = xy^2$ is:

(a) $z = \frac{x^2y^3}{6} + f(y) + \phi(x)$ (b) $z = \frac{x^2y^3}{4} + f(y) + \phi(x)$

(c) $z = \frac{x^3y^2}{6} + f(y) + \phi(x)$ (d) $z = \frac{xy^2}{4} + f(y) + \phi(x)$

Ans (a)

12. The general solution of $\frac{\partial^2 z}{\partial x \partial y} = e^y \cos x$ is:

(a) $z = e^x \sin y + f(y) + \phi(x)$ (b) $z = e^y \sin x + f(y) + \phi(x)$

(c) $z = e^x \sin x + f(y) + \phi(x)$ (d) $z = e^y \cos y + f(x) + \phi(y)$

Ans (b)

13. The general solution of $\frac{\partial z}{\partial y} = a \frac{\partial z}{\partial x}$ is:

(a) $z = f(y + ax)$ (b) $z = f(x^2 + y^2)$

(c) $z = f(x + ay)$ (d) $z = f(ax + by)$

Ans (c)

14. The general solution of $(D^4 - D'^4)\, z = 0$ is:

(a) $z = f_1(x^2 + y^2) + f_2(x^2 - y^2)$

(b) $z = f_1(x + y) + f_2(y - x) + f_3(y + ix) + f_4(y - ix)$

(c) $z = f_1(x^2 - y) + f_2(x - y^2) + f_3(y + ix) + f_4(y - ix)$

(d) $z = f_1(x^2 - y^2) + f_2(x - y^2) + f_3(y^2 + ix) + f_4(y - ix)$

Ans (b)

15. The general solution of $(D^3 - 6D^2D' + 11DD'^2 - 6D'^3)\, z = 0$ is:

(a) $z = f_1(y - x) + f_2(y - 2x) + f_3(y - 3x)$

(b) $z = f_1(y + x) + f_2(y - 2x) + f_3(y - 3x)$

(c) $z = f_1(y + x) + f_2(y + 2x) + f_3(y + 3x)$

(d) $z = f_1(x) + f_2(y + x) + f_3(x^2)$

Ans (c)

16. The P.I. of $(D^2 - 2DD' + D'^2)\, z = e^{x + 2y}$ is:

(a) $e^{x + 3y}$ (b) $e^{x + y}$

(c) $2e^{x + 2y}$ (d) $e^{x + 2y}$

Ans (d)

17. The P.I. of $(D^2 + 2DD' + D'^2)\, z = e^{2x + 3y}$ is:

(a) e^{2x+3y} (b) $\frac{1}{5}e^{2x+3y}$

(c) $\frac{1}{25}e^{2x+3y}$ (d) $\frac{1}{15}e^{2x+3y}$

Ans (c)

18. The general solution of $(D^2 + D'^2)\, z = \cos mx \cos ny$ is:

(a) $z = f_1(y - ix) + f_2(y + ix) - \frac{1}{(m^2 - n^2)} \cos mx \cos ny$

(b) $z = f_1(x + y) + f_2(y - x) - \frac{1}{(m^2 - n^2)} \cos mx \cos ny$

(c) $z = f_1(y - ix) + f_2(y + ix) - \frac{1}{(m^2 - n^2)} \cos mx \cos ny$

(d) $z = f_1(y + x) + f_2(y - x) - \frac{1}{(m^2 + n^2)} \cos mx \cos ny$

Ans (a)

19. The particular integral of $(4r - 4s + t) = 16 \log (x + 2y)$ is:

(a) $2x \log (x + 2y)$ (b) $2x^2 \log (2x + y)$

(c) $2y^2 \log (x + 2y)$ (d) $2x^2 \log (x + 2y)$

Ans (d)

20. The P.I. of $(r + s - 6t) = y\cos x$ is:

(a) $-y \cos x + \sin x$ (b) $y \cos x + \sin x$

(c) $\cos x + y \sin x$ (d) $x \cos x + \sin x$

Ans (a)

21. The general solution of $r = a^2 t$ is:

(a) $z = f_1 (y + ax) + f_2 (y - x)$

(b) $z = f_1 (y + x) + f_2 (y - ax)$

(c) $z = f_1 (y + ax) + f_2 (y - ax)$

(d) $z = f_1 (y + x) + f_2 (y^2 - ax)$

Ans (c)

22. The general solution of $(D^2 + DD' + D' - 1)\,z = 0$ is:

(a) $z = e^{-x} f_1(y) + e^{x} f_2(y - x)$

(b) $z = e^{x} f_1(y) + e^{-x} f_2(y + x)$

(c) $z = e^{-x} f_1(y) + e^{-x} f_2(y^2 - x)$

(d) $z = e^{x} f_1(y) + e^{-x} f_2(y + x^2)$

Ans (a)

23. The P.I. of $(D - 3D' - 2)^2\,z = 2e^{2x} \tan(y + 3x)$ is:

(a) $xe^{2x} \tan(y + 3x)$

(b) $\frac{e^{2x}}{x} \tan(y + 3x)$

(c) $x^2e^{2x} \tan(y + 3x)$

(d) $\frac{e^{2x}}{x^2} \tan(y + 3x)$

Ans (c)

24. The P.I. of $(D^2 - 5DD' + 4D'^2)\,z = \sin(4x + y)$ is:

(a) $\frac{x}{3} \cos(y + 4x)$ (b) $x \cos(y + 4x)$

(c) $\cos(y + 4x)$ (d) $-\frac{x}{3} \cos(y + 4x)$

Ans (d)

25. The P.I. of $(D^2 + DD' - 6D'^2)\,z = \cos(2x + y)$ is:

(a) $\cos(2x + y)$ (b) $\sin(2x + y)$

(c) $\frac{1}{25} \cos(2x + y)$ (d) $\frac{1}{25} \sin(2x + y)$

Ans (c)

26. The general solution of $(D^2 - 5DD' + 6D'^2) = e^{3x - 2y}$ is:

(a) $z = f_1(y + 2x) + f_2(y + 3x) + \frac{1}{63} e^{3x - 2y}$

(b) $z = f_1(3x + y) + y f_2(2x + y) + \frac{1}{63} e^{3x-2y}$

(c) $z = f_1(2y + x) + f_2(y + 3x) + \frac{1}{63} e^{3x - 2y}$

(d) $z = f_1(3y + 2x) + f_2(2y + 3x) + \frac{1}{63}e^{3x - 2y}$

Ans (a)

27. The general solution of $\frac{\partial z}{\partial x} + \frac{\partial z}{\partial y} = 0$ is:

(a) $z = f(x - y)$ (b) $z = f(x + y)$

(c) $z = af(xy)$ (d) $z = f(x + ay)$

Ans (a)

28. The general solution of $r + t + 2s = 0$ is:

(a) $z = f_1(y + x) + x f_2(y + x)$

(b) $z = f_1(y - x) + xf_2(y - x)$

(c) $z = f_1(xy) + xf_2(xy)$

(d) $z = f_1(x\,y) + xf_2(xy)$

Ans (b)

29. The general solution of $(D^3 - 3D^2D' + 3DD'^2 - D'^3)\, z = 0$ is:

(a) $z = f_1(y + x) + xf_2(y + x) + x^2f_3(y + x)$

(b) $z = f_1(y - x) + xf_2(y - x) + x^2f_3(y - x)$

(c) $z = f_1(y + x) + yf_2(y + x) + y^2f_3(y + x)$

(d) $z = f_1(x + y) + xf(y - x) + x^2f(y - x)$

Ans (a)

30. The general solution of $(D^2 - 2aDD' + a^2 D'^2)\, z = f(y + ax)$ is:

(a) $z = \phi_1(y - x) + x\phi_2(y - x) + \frac{x^2}{2} f(y + ax)$

(b) $z = \phi_1(y + x) + x\phi_2(y + x) + \frac{x^2}{2} f(y + ax)$

(c) $z = \phi_1(y + x) + y\phi_2(y + x) + \frac{y^2}{2} f(y + ax)$

(d) $z = f_1(xy) + xf_2(yx) + \frac{x^2}{2} f(y + ax)$

Ans (b)

31. The general solution of $\frac{\partial z}{\partial x} + \frac{\partial z}{\partial y} = \sin x$ is:

(a) $z = f(y + x) + \cos x$ (b) $z = f(y + x) - \cos x$

(c) $z = f(y - x) + \cos x$ (d) $z = f(y - x) - \cos x$

Ans (d)

32. The P.I. of $(D^2 - DD' - 2D'^2)\, z = (y - 1)\, e^x$:

(a) xe^x (b) ye^y

(c) x^2e^y (d) ye^x

Ans (d)

33. The P.I. of $(r + s - 6t) = y \cos x$ is:

(a) $\cos x - y \sin x$ (b) $\cos x - y \cos x$

(c) $\sin x - y \cos x$ (d) $\sin x + y \cos x$.

Ans (c)

34. The general solution of $(D^2 - D'^2 + D - D')\, z = 0$ is:

(a) $z = f_1(y + x) + e^{-x} f_2(y - x)$ (b) $z = f_1(y - x) + e^{-x} f_2(y - x)$

(c) $z = f_1(y - x) + e^{-x} f_2(y + x)$ (d) $z = f_1(y + x) + e^{x} f_2(y - x)$

Ans (a)

35. The general solution of $(D^2 - a^2 D'^2 + 2abD + 2a^2bD')\, z = 0$ is:

(a) $z = f_1(y + ax) + e^{-2abx} f_2(y + ax)$ (b) $z = f_1(y + ax) + e^{-2abx} f_2(y - ax)$

(c) $z = f_1(y - ax) + e^{-2abx} f_2(y + ax)$ (d) $z = f_1(y - ax) + e^{2abx} f_2(xy)$

Ans (c)

36. The P.I. of $(D^2 - 4DD' + D - 1)\, z = e^{3x - 2y}$ is:

(a) $e^{3x - 2y}$ (b) $e^{3x + 2y}$

(c) $\frac{1}{35} e^{3x - 2y}$ (d) $\frac{1}{35} e^{3x + 2y}$

Ans (c)

37. The general solution of $r = 6x$ is:

(a) $z = x^3 + xf_1(y) + f_2(y)$ (b) $z = x^3 + yf_1(y) + f_2(y)$

(c) $z = y^3 + xf_1(y) + f_2(y)$ (d) $z = 3x^3 + xf_1(y) + f_2(y)$

Ans (a)

38. The general solution of $t = \sin(xy)$ is:

(a) $z = -\frac{1}{x^2}\sin(xy) + yf_1(x) + f_2(x)$ (b) $z = -\sin(xy) + f_1(x) + f_2(y)$

(c) $z = \frac{1}{x^2}\sin(xy) + yf_1(x) + f_2(x)$ (d) $z = \frac{2}{x}\sin(xy) + f_1(x) + f_2(y)$

Ans (a)

39. The general solution of $\log s = x + y$ is:

(a) $z = \log(x + y) + f_1(y) + f_2(x)$ (b) $z = e^{x+y} + f_1(y) + f_2(x)$

(c) $z = \log x + \log y + f_1(y) + f_2(x)$ (d) $z = e^{x-y} + f_1(y) + f_2(x)$

Ans (b)

40. The general solution of $xr + 2p = 0$ is:

(a) $z = -(1/x)f_1(x) + f_2(x)$ (b) $z = -(1/y)f_1(x) + f_2(x)$

(c) $z = -(1/x)f_1(y) + f_2(y)$ (d) $z = -(1/x)f_1(xy) + f_2(y)$

Ans (c)

Fill in the blanks in the following problems:

1. The solution of $\frac{\partial^2 z}{\partial y^2} = \sin(xy)$ is...

Ans. $-x^{-2}\sin(xy) + yf(x) + \phi(x)$

2. The order of the partial differential equation obtained by eliminating f from $z = f(x^2 + y^2)$ is...

Ans. First order

3. The solution of $\frac{\partial^3 z}{\partial x^3} = 0$ is...

Ans. $z = f_1(y) + xf_2(y) + x^3 f_3(y)$

4. The P.I. of $(D^2 - D'^2)z = \cos(x + y)$ is....

Ans. $\frac{x^2}{2}\sin(x + y)$

5. The solution $\frac{\partial^2 z}{\partial x^2} - \frac{\partial^2 z}{\partial y^2} = 0$ is.....

Ans. $z = f_1(y + x) + f_2(y - x)$

6. P.I. of $(2D^2 - 3DD' + D'^2)z = e^{x+2y}$ is...

Ans. $-\frac{x}{2}e^{x+2y}$

7. By eliminating a and b from $(x - a)^2 + (y - b)^2 + z^2 = C^2$, the partial differential equation formed is...

Ans. $z^2 (p + q + 1) = C^2$

8. Eliminating the constants a and b from $z = (x^2 + a)(y^2 + b)$, the partial differential equation formed is...

Ans. $\frac{\partial^2 z}{\partial x \partial y} = 4xy$

9. The particular integral of $(D^2 + DD')z = \sin(x + y)$ is....

Ans. $-\frac{1}{2}\sin(x + y)$

10. The equation $\frac{\partial^2 z}{\partial x^2} + 2xy\left(\frac{\partial z}{\partial x}\right)^2 + \frac{\partial z}{\partial y} = 10$, is of order.......and degree.....

Ans. Order two, degree one

11. The solution of $x\frac{\partial z}{\partial x} = 2x + y$ is.......

Ans. $z = 2x + y \log x + f(xy)$

12. By eliminating a and b from $z = a(x + y) + b$, the partial differential equation formed is..

Ans. $\frac{\partial z}{\partial x} = \frac{\partial z}{\partial y}$

Select True 'or' False answers in the following:

13. $u\frac{\partial z}{\partial x} + \frac{\partial^2 u}{\partial x^2} = 0$ is a linear partial differential eqation. (False)

14. $x\frac{\partial u}{\partial x} + t\frac{\partial u}{\partial t} = 2u$ is an ordinary differential equation. (False)

15. The solution of $(D^2 - DD' + D' - 1)z = 0$ is $e^x f(y) + e^{-x} f_2(x + y)$. (True)

16. $x\frac{\partial u}{\partial x} + y\frac{\partial u}{\partial y^2} = u^2$ is a non-linear partial differential equation. (False)

17. $u = x^2 - y^2$ is a solution of $\frac{\partial^2 u}{\partial x^2} + \frac{\partial^2 u}{\partial y^2} = 0$. (True)

18. $u = e^{-t}\sin x$ is a solution of $\frac{\partial^2 u}{\partial t^2} + \frac{\partial^2 u}{\partial x^2} = 0$. (True)

19. Those equation in which p and q occur other than in the first degree are called non-linear partial differential equation of the first degree. (True)

20. The equation of the form

$$\frac{\partial^n z}{\partial x^n} + a_1 \frac{\partial^n z}{\partial x^{n-1}\partial y} + a_2 \frac{\partial^n z}{\partial x^{n-2}\partial y^2} + ...+a_n \frac{\partial^n z}{\partial y^n} = F(x,y)$$

in which a_1, a_2,....a_n are constats is called a homogeneous linear partial differential equation of nth order with constant coefficients. (True)

21. Match the following:

(i)	$z = ax + by$, forms the p.d.e:	(A)	$z = xf(y) + g(y)$
(ii)	The solution of $\frac{\partial^2 z}{\partial x^2} = 0$	(B)	$\frac{\partial^2 z}{\partial x^2} + \frac{\partial^2 z}{\partial y^2} = 0$
(iii)	The solution of $\frac{\partial^2 z}{\partial x^2} = \cos x$ is:	(C)	$z = px + qy$
(iv)	$z = f_1(x + iy) + f_2(x - iy)$ forms the p.d.e.:	(D)	$z = -\cos x + x f(y) + g(y)$.

21. (i)–(C), (ii)–(A), (iii)–(D), (iv)–(B).

UNIT - V
APPLICATIONS OF PARTIAL DIFFERENTIAL EQUATIONS

Chapter 9

Applications of Partial Differential Equations

INTRODUCTION

The problems related to fluid mechanics, solid state physics, heat transfer, electromagnetic theory, Wave equation and other areas of physics and engineering are governed by partial differential equations subject to certain given conditions, called boundary conditions. The process to find all solutions of a partial differential equation under given conditions is known as a boundary value problem. The method of solution of such equations differ from that used in the case of ordinary differential equations. Method of separation of variables is a powerful tool to solve such boundary value problem when partial differential equation is linear with homogenous boundary conditions. Most of the problems involving linear partial differential equations Can be solved by the method of separation of variables discussed below.

METHOD OF SEPARATION OF VARIABLES

It involves a solution which breaks up into a product of functions each of which contains only one of the variables. The following example explain this method.

Example 1 Apply the method of separation of variables to solve

$$\frac{\partial^2 z}{\partial x^2} - 2\frac{\partial^2 z}{\partial x} + \frac{\partial z}{\partial x} = 0 \qquad \text{(U.P.T.U. 2005, 09)}$$

Solution Assume the trial solution z = X(x) Y(y) (i)

where X is a function of x alone and Y that of y alone, substituting this value of z in the given equation we have

$$X''.\, Y - 2X'.Y + XY' = 0 \text{ where } X' = \frac{dX}{dx},\ Y' = \frac{dY}{dy}\ .\ \text{etc}$$

separating the variables, we get

$$\frac{X'' - 2X'}{X} = -\frac{Y'}{Y} \qquad \text{(ii)}$$

since x and y are independent variables, therefore, (ii) can only be true if each side is equal to the same constant, K(say), so we have

$$\frac{X'' - 2X'}{X} = -\frac{Y'}{Y} = K$$

Therefore, $\frac{X''-2X'}{X} = K$ i.e $X''-2X' - kX = 0$ (iii)

and $-\frac{Y'}{Y} = k$ i.e. $Y' + kY = 0$ (iv)

To solve the ordinary linear equation (iii) the auxiliary equation is
$m^2 - 2m - k = 0$

$$\Rightarrow m = 1 \pm \sqrt{(1+k)}$$

$\therefore$ The solution of (iii) is $X = C_1 e^{\{1+\sqrt{(1+k)}\}x} + c_2 e^{\{1-\sqrt{(1+k)}\}x}$ and the solution of (iv) is $Y = C_2 e^{-ky}$

Substituting these values of X and Y in (i), we get

$$z = \left\{c_1 e^{\{1+\sqrt{(1+k)}\}x} + c_2 e^{\{1-\sqrt{(1+k)}\}x}\right\}.c_3 e^{-ky}$$

$$\text{i.e. } z = \left\{a e^{\{1+\sqrt{(1+k)}\}x} + b e^{\{1-\sqrt{(1+k)}\}x}\right\} e^{-ky}$$

where $a = c_1 c_3$ and $b = c_2 c_3$
which is the required complete solution.

Example 2 Using the method of separation of variables, solve $\frac{\partial u}{\partial x} = 2\frac{\partial u}{\partial t} + u$, where $u(x, 0) = 6\,e^{-3x}$

(U.P.T.U. 2006)

Solution Assume the solution $u(x, t) = X(x)\,T(t)$ (i)

Substituting in the given equation, we have

$$X'T = 2XT' + XT$$

or $(X' - X)\,T = 2\,XT'$

or $\frac{X'-X}{2X} = \frac{T'}{T} = k(\text{say})$

$\therefore X' - X - 2kX = 0$

or $\frac{X'}{X} = 1 + 2k$ (ii)

and $\frac{T'}{T} = k$ (iii)

solving (ii), $\log X = (1+2k)x + \log c$

or $X = ce^{(1+2k)x}$

From (iii), $\log T = kt + \log c'$

or $T = c'e^{kt}$

Thus $u(x,t) = XT$

$= cc'e^{(1+2k)x} e^{kt}$

Now $6e^{-3x} = u(x, 0) = cc' e^{(1+2k)x}$ (iv)

$\therefore cc' = 6$ and $1+2k = -3$ or $k = -2$

Substituting these values in (iv) we get

$u = 6e^{-3x} e^{-2t}$ i.e. $u = 6e^{-(3x + 2t)}$ which is the required solution

Example 3 Solve by the method of separation of variables, $\frac{\partial u}{\partial x} = 2\frac{\partial u}{\partial y} + u$, where

$u(x, 0) = 3e^{-5x} - 2e^{-3x}$

Solution Assume the trial solution $u = X(x)\, Y(y)$ (i)

where X is the function of x alone and Y that of y alone.

substituting this value of u in the given equation we have

$$X'Y = 2XY' + XY$$

$$\frac{X'}{X} = \frac{2Y'}{Y} + 1 = k(\text{say}) \qquad \text{(ii)}$$

Now $\frac{X'}{X} = k \Rightarrow \frac{1}{X}\frac{dX}{dx} = k$

$$\Rightarrow \frac{dX}{X} = k\,dx$$

on integrating, we get

$$\log_e X = kx + \log_e C_1$$

$$\Rightarrow \quad X = c_1 e^{kx}$$

And taking last two terms of equation (ii), we have

$$\frac{2}{Y}\frac{dY}{dy} + 1 = k$$

$$\Rightarrow \frac{2}{Y}\frac{dY}{dy} = K - 1$$

$$\Rightarrow \frac{dY}{Y} = \frac{(k-1)}{2} dy$$

on integrating, $\log_e Y = \frac{(k-1)}{2} y + \log c_2$

$$\Rightarrow Y = c_2 e^{(k-1)\frac{y}{2}}$$

From (i), we get

$u = c_1 c_2 e^{kx} e^{(k-1)y/2}$

From (i), we get

$u = c_1 c_2 e^{kx} e^{(k-1)y/2}$

$$\Rightarrow u = \sum_{n=1}^{\infty} b_n e^{k_n x} e^{(k_n-1)y/2} \qquad \text{(iii)}$$

$b_n = c_1c_2$ and $k = K_n$

which is the most general solution of given equation, putting y =0 and $u = 3e^{-5x} - 2e^{-3x}$ in equation (iii) we get

$$3e^{-5x} - 2e^{-3x} = \sum_{n=1}^{\infty} b_n e^{k_n x} = b_1 e^{k_1 x} + b_2 e^{k_2 x}$$

comparing the terms on both sides, we get

$b_1 = 3, k_1 = -5, b_2 = -2, k_2 = -3$

Hence the required solution of given equation is from (iii), we have

$u = 3e^{-5x} - e^{-3y} + (-2)e^{-3x} e^{-2y}$

$\Rightarrow u = 3e^{-(5x+3y)} - 2e^{-(3x+2y)}$

Example 4.

Use the method of separation of variables to solve the equation.

$\frac{\partial^2 v}{\partial x^2} = \frac{\partial v}{\partial t}$ given that v =0 when $t \to \infty$, as well as v =0 at x =0 and x = l.

Solution. Assume the trial solution v = XT (i)

where X is a function of x alone and Y that of y given

$$\Rightarrow \frac{\partial v}{\partial t} = X\frac{dT}{dt} \quad \text{and} \quad \frac{\partial^2 v}{\partial x^2} = T\frac{d^2X}{dx^2}$$

Substituting these values in the given differential equation, we get

$$T\frac{d^2X}{dx^2} = X\frac{dT}{dt} \text{ or } \frac{1}{T}\frac{dT}{dt} = \frac{1}{X}\frac{d^2X}{dx^2} = -p^2 \text{ (say)}$$

$$\Rightarrow \frac{1}{T}\frac{dT}{dt} = -p^2 \text{ or } \frac{dT}{dt} + p^2T = 0 \qquad \text{(ii)}$$

$$\text{and } \frac{1}{X}\frac{d^2X}{dx^2} = -p^2 \text{ or } \frac{d^2X}{dx^2} + p^2X = 0 \qquad \text{(iii)}$$

Solving (ii) and (iii) we get

$T = c_1 e^{-p^2 t}$ and $X = c_2 \cos px + c_3 \sin px$

Substituting these values of X and T in (i) we get

$$v = c_1 e^{-p^2 t} (c_2 \cos pt + c_3 \sin pt) \qquad \text{(iv)}$$

putting $x = 0$, $v = 0$ in (ii), we get

$0 = c_1 e^{-p^2t} c_2 \Rightarrow \quad c_2 = 0$ since $c_1 \neq 0$

$\Rightarrow \qquad v = c_1 e^{-p^2t} c_3 \sin px \qquad$ (v)

putting $x = l$, $v = o$ in (v), we get

$c_1 c_3 e^{--p^2t} \sin pl = 0$

$\Rightarrow \quad \sin pl = 0 = \sin n\pi$

$\Rightarrow \quad p = n\pi/l$, n is any integer

$$\therefore \quad v = c_1 c_3 e^{-(n^2\pi^2 t)/l^2 \sin\left(\frac{n\pi x}{l}\right)}$$

$$= b_n \, e^{-n^2\pi^2 t/l^2} \sin\left(\frac{n\pi x}{l}\right)$$

where $b_n = c_1 c_3$

The most general solution is $v = \sum_{n=1}^{\infty} b_n e^{-n^2\pi^2 t/l^2} \sin\left(\frac{n\pi x}{l}\right)$

PARTIAL DIFFERENTIAL EQUATIONS OF ENGINEERING

A number of problems in engineering give rise to the following well known partial differential equations.

(i) Wave equation $\frac{\partial^2 y}{\partial t^2} = c^2 \frac{\partial^2 y}{\partial x^2}$

(ii) One dimensional heat flow equation $\frac{\partial u}{\partial t} = c^2 \frac{\partial^2 u}{\partial x^2}$

(iii) Two dimensional heat flow equation which in steady state becomes the two dimensional Laplace's equation $\frac{\partial^2 u}{\partial x^2} + \frac{\partial^2 u}{\partial y^2} = 0$

(iv) Transmission line equations

(v) Vibrating membrane. Two dimensional wave equation.

(vi) Laplace's equation in three dimensions.

Besides these, the partial differential equations frequently occur in the theory of Elasticity and Hydraulics.

Starting with the method of separation of variables, we find their solutions subject to specific boundary conditions and the combination of such solution gives the desired solution. Quite often a certain condition is not applicable. In such cases, the most general solution is written as the sum of the particular solutions already found and the constants are determined using Fourier series so as to satisfy the remaining conditions.

Vibrations of a stretched string (one dimensional wave Equation)

Consider a tightly stretched elastic string of length l and fixed ends A and B and subjected to constant tension T as shown in figure. The tension T will be considered to be large as compared to the weight of the string so that the effects of gravity are negligible.

Let the string be released from rest and allowed to vibrate. We shall study the subsequent motion of the string, with no external forces acting on it, assuming that each point of the string makes small vibrations at right angles to the equilibrium position AB, of the string entirely in one plane.

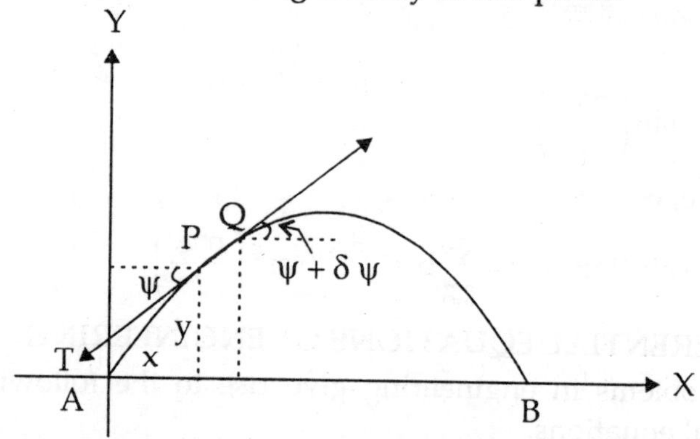

Taking the end A as the origin, AB as the x-axis and AY perpendicular to it as the y-axis, so that the motion takes place entirely in the xy -plane. Above figure shown the string in the position APB at times t. Consider the motion of the element PQ of the string between its points P(x, y) and Q (x + δx, y + δy), where the tangents make angles ψ and ψ + δψ with the x axis. Clearly the element is moving upwards with the acceleration $\frac{\partial^2 y}{\partial t^2}$. Also the vertical component of the force acting on this element

$= T \sin(\psi + \delta\psi) - T \sin \psi$

$\approx T(\psi + \delta\psi - \psi) \because \sin\psi = \psi$, as ψ is very small

$= T\,\delta\psi$ (approximately)

The acceleration of the elements in the QY direction is $\frac{\partial^2 y}{\partial t^2}$. If the length of PQ is δs, then the mass of PQ is m.δs.

Hence, by Newton's second law, the equation of motion becomes

$$m\delta s \frac{\partial^2 y}{\partial t^2} = T\delta\psi$$

or $\frac{\partial^2 y}{\partial t^2} = \frac{T}{m}\frac{\delta\psi}{\delta s}$

As Q →P, δs →0. Therefore, taking limit as δs →0, the above equation becomes

$$\frac{\partial^2 y}{\partial t^2} = \frac{T}{m}.\frac{\delta\psi}{\delta s}$$

Where $\frac{\delta\psi}{\delta s}$ = Curvature at P of the deflection curve

$$\frac{\delta\psi}{\delta s} = \frac{\frac{\partial^2 y}{\partial x^2}}{\left[1+\left(\frac{\partial y}{\partial x}\right)^2\right]^{3/2}}$$ using formula for the radius of curvature

$= \frac{\partial^2 y}{\partial x^2}$, approximately, since $\left(\frac{\partial y}{\partial x}\right)^2$ is negligible because $\frac{\partial y}{\partial x}$ is small

$$\therefore \frac{\partial^2 y}{\partial t^2} = \frac{T}{m}\frac{\partial^2 y}{\partial x^2}$$

Putting $\frac{T}{m} = C^2$ (positive), the displacement y(x, t) is given by the equation

$$\frac{\partial^2 y}{\partial t^2} = C^2\frac{\partial^2 y}{\partial x^2}$$

This partial differential equation is known as one dimensional wave equation.

Solution of the one dimensional wave Equation

The one dimensional wave equation is

$$\frac{\partial^2 y}{\partial t^2} = C^2\frac{\partial^2 y}{\partial x^2} \qquad \text{(i)}$$

Assume that a solution of (i) is of the form y = X(x) T(t). where X is a function of x alone and T is a function of t only.

Then $\frac{\partial^2 y}{\partial t^2} = X\frac{d^2T}{dt^2}$ and $\frac{\partial^2 y}{\partial x^2} = T\frac{d^2X}{dx^2}$

putting these values in (i), we get

$$\frac{1}{X}\frac{d^2X}{dx^2} = \frac{1}{C^2T}\frac{d^2T}{dt^2} = k(\text{say}) \qquad \text{(ii)}$$

The (ii) leads to the ordinary differential equations.

$$\frac{d^2X}{dx^2} - kX = 0 \quad \text{and} \quad \frac{d^2T}{dt^2} - kC^2T = 0 \qquad \text{(iii)}$$

solving (iii) we get

(1) when K is negative say $-p^2$, then

$$X = c_1 \cos px + c_2 \sin px$$

$$T = c_3 \cos cpt + c_4 \sin cpt \qquad \text{(iv)}$$

(2) when k =0, then

$$X = C_5 x + C_6$$

$$T = C_7 t + C_8 \qquad \text{(v)}$$

(3) when k is positive say p^2, then

$$\left.\begin{array}{l} X = C_9 e^{px} + C_{10} e^{-px} \\ T = C_{11} e^{cpt} + C_{12} e^{-cpt} \end{array}\right\} \qquad \text{(vi)}$$

of these three solutions we have to choose that solution which is consistent with the physical nature of the problem. As we are dealing with problems on vibration y must be a periodic function of x and t. Hence, the solution must involve trigonometric terms.

Accordingly the solution given by (iv) i.e. of the form

$y = (C_1 \cos px + C_2 \sin px)(C_3 \cos cpt + C_4 \sin cpt)$

is the only suitable solution of the wave equation.

Example 5. A string of length L is stretched and fastened to two fixed points. Find the solution of the wave equation $y_{tt} = a^2 y_{xx}$, when initial displacement is $y(x, 0) = f(x) = b \sin\left(\frac{\pi k}{L}\right)$ where symbols have usual meaning.

(U.P.T.U. 2009)

Solution. Consider an elastic string tightly stretched between two points O and A. Let O be the origin and OA as x-axis on giving a small transverse displacement i.e. the displacement perpendicular to its length.

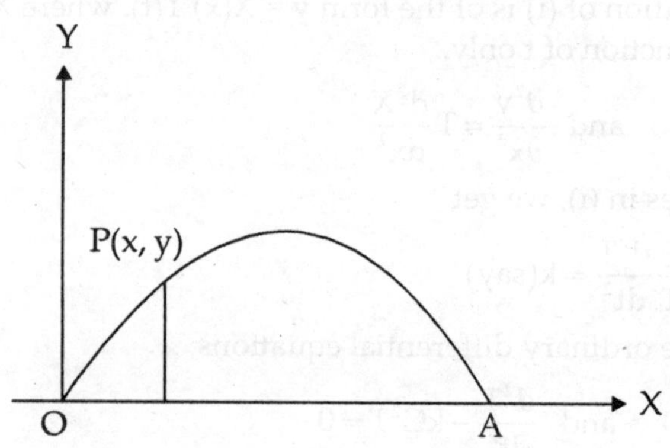

let y be the displacement at the point P (x, y) at any time, the wave equation

$y_{tt} = a^2 y_{xx}$

or $\frac{\partial^2 y}{\partial t^2} = a^2 \frac{\partial^2 y}{\partial x^2}$

As the end points of the string are fixed, for all time,

$y(0, t) = 0$ (ii)

$y(L, t) = 0$ (iii)

Since, the initial transverse velocity of any point of the string is zero, therefore

$\left(\frac{\partial y}{\partial t}\right)_{t=0} = 0$ (iv)

Also $y(x, 0) = b \sin \frac{\pi x}{L}$ (v)

The general solution of (i) is

$y = (C_1 \cos px + C_2 \sin px)(C_3 \cos apt + C_4 \sin apt)$ (vi)

Applying the boundary condition

$y = 0$ at $x = 0$

$0 = C_1 (C_3 \cos apt + C_4 \sin apt)$

$\therefore C_1 = 0$

Therefore,

$y = C_2 \sin px (C_3 \cos apt + C_4 \sin apt)$ (vii)

Again applying $\frac{\partial y}{\partial t} = 0$, at $t = 0$ on (vii)

$\frac{\partial y}{\partial t} = C_2 \sin px\ ap.(-C_3 \sin apt + C_4 \cos apt)$

$0 = C_2 \sin px.\ ap.\ C_4 \Rightarrow C_4 = 0$

Then (vii) becomes $y = C_2 C_3 \sin px \cos atp$ (viii)

Applying $y = 0$ at $x = L$

$0 = C_2 C_3 \sin pL \cos atp$

$\therefore \sin pL = 0 = \sin n\pi, n = 0,1,2,3.................$

$\therefore pL = n\pi$

or $p = \frac{n\pi}{L}$

putting $p = \frac{n\pi}{L}$ in (viii), we have

$$y = C_2 C_3 \sin\frac{n\pi x}{L}\cos\frac{n\pi at}{L} \quad \text{(ix)}$$

At $t = 0$, $y = b \sin \dfrac{\pi x}{L}$

$$b \sin \frac{\pi x}{L} = C_2 C_3 \sin \frac{n\pi x}{L}$$

$\therefore C_2 C_3 = b, n = 1$

putting $C_2 C_3 = b$, $n = 1$ in (ix) we get

$$y = b\sin\left(\frac{\pi x}{L}\right)\cos\left(\frac{\pi at}{L}\right)$$

which is our required solution.

Example 6. A string is stretched and fastened to two points l apart. Motion is started by displacing the string the form $y = a \sin \dfrac{\pi x}{l}$ from which it is released at a time $t = 0$. Show that the displacement of any point at a distance x from one end at time t is given by

$$y(x,t) = a\sin\left(\frac{\pi x}{l}\right)\cos\left(\frac{\pi ct}{l}\right)$$

(U.P.T.U. 2004, S.V.T.U. 2007)

Solution : Solving exactly just like as example 5.

Example 7 : A tightly stretched string with fixed end points $x = 0$ and $x = l$ is initially in a position given by $y = y_0 \sin^3 \dfrac{\pi x}{l}$.

If it is released from the rest from this position find the displacement y (x, t).

Solution The equation to the vibrating string be

$$\frac{\partial^2 y}{\partial t^2} = C^2 \frac{\partial^2 y}{\partial x^2} \quad \text{(i)}$$

Here the initial conditions are

$y(0, t) = 0$, $y(l, t) = 0$

$\dfrac{\partial y}{\partial t} = 0$ at $t = 0$, $y(x, 0) = y_0 \sin^3 \dfrac{\pi x}{l}$

The general solution of (i) is of the form

$$y = (C_1 \cos px + C_2 \sin px)(C_3 \cos cpt + C_4 \sin cpt) \quad \text{(ii)}$$

Now $y = 0$ at $x = 0$ gives $C_1 = 0$

$\therefore$ y = $C_2 \sin px (C_3 \cos cpt + C_4 \sin cpt)$ (iii)

Again $\frac{\partial y}{\partial t} = 0$ at t = 0 gives $C_4 = 0$

$y = c_2c_3 \sin px$ cost cpt (iv)

At x = l, y = 0

$0 = C_2 C_3 \sin pl \cos cpt$

$\Rightarrow \quad \sin pl = 0 = \sin n\pi, n = 0,1,2.........$

$$\therefore p = \frac{n\pi}{l}$$

$$\therefore y = C_2C_3 \sin\frac{n\pi x}{l}\cos\frac{n\pi ct}{l} \quad \text{(v)}$$

Let $C_2 C_3 = b_n$, As b_n is arbitrary constants

Therefore general solution is

$$y(x,t) = \sum_{n=1}^{\infty} b_n \sin\frac{n\pi x}{l}\cos\frac{n\pi ct}{l} \quad \text{(vi)}$$

At t = 0, y = $y_0 \sin^3\frac{\pi x}{l}$, so from equation (vi) we have

$$y_0 \sin^3\left(\frac{\pi x}{l}\right) = \sum_{n=1}^{\infty} b_n \sin\frac{n\pi x}{l}$$

$$\Rightarrow \frac{y_0}{4}\left(3\sin\frac{\pi x}{l} - \sin\frac{3\pi x}{l}\right) = b_1 \sin\frac{\pi x}{l} + b_2 \sin\frac{2\pi x}{l} + b_3 \sin\frac{3\pi x}{l} + ...$$

$$\therefore b_1 = \frac{3y_0}{4},, b_2 = 0,\ b_3 = -\frac{y_0}{4},\ b_4 = b_5 = b_6 = ...= 0$$

Hence (vi) becomes

$$y(x,t) = \frac{y_0}{4}\left(3\sin\frac{\pi x}{l}\cos\frac{\pi ct}{l} - \sin\frac{3\pi x}{l}\cos\frac{3\pi ct}{l}\right)$$

Example 8: A string is stretched and fastened to two points l apart. Motion is started by displacing the string into the form y =k (lx $-x^2$) from which it is released at time t =0. Find the displacement of any point on the string at a distance of x from one end at time t.

(U.P.T.U. 2002)

Solution

The vibration of the string is given by the equation

$$\frac{\partial^2 y}{\partial t^2} = C^2\frac{\partial^2 y}{\partial x^2} \quad \text{(i)}$$

As the end points of the string are fixed for all time,

$y(0, t) = 0$ (ii)

$y(l, t) = 0$ (iii)

Since the initial transverse velocity of any point of the string is zero, therefore

$$\left(\frac{\partial y}{\partial t}\right)_{t=0} = 0 \quad \text{(iv)}$$

and $y(x, 0) = k(lx - x^2)$ (v)

solution of (i) is

$y = (c_1 \cos px + c_2 \sin px)(c_3 \cos cpt + c_4 \sin cpt)$ (vi)

At $x = 0$, $y = 0$ gives $c_1 = 0$

$y = c_2 \sin px (c_3 \cos cpt + c_4 \sin cpt)$ (vii)

At $t = 0$, $\dfrac{\partial y}{\partial t} = 0$

$0 = c_2 \sin px.\ cp.\ c_4$

$\therefore c_4 = 0$

$y = c_2 c_3 \sin px \cos cpt$ (viii)

At $x = l$, $y = 0$

$0 = C_2 C_3 \sin pl \cos cpt$

$\therefore \sin pl = 0 = \sin n\pi$; $n = 0,1,2,3$............

$$\therefore p = \frac{n\pi}{l}$$

$$y = C_2 C_3 \sin\frac{n\pi x}{l}\cos\frac{n\pi ct}{l}$$

Let $C_2 C_3 = b_n$

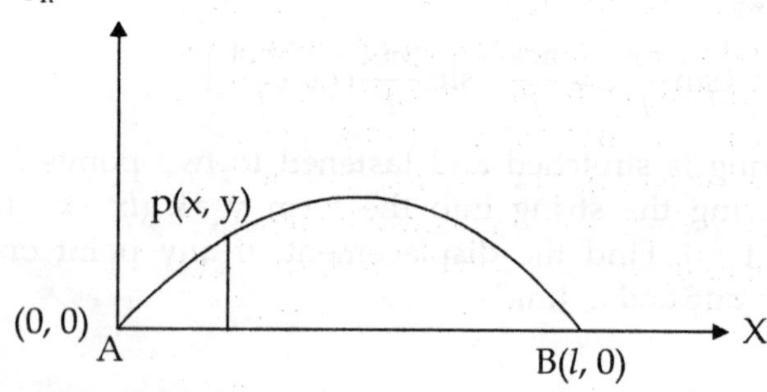

$$y = b_n \sin\frac{n\pi x}{l}\cos\frac{n\pi ct}{l}$$

As b_n is arbitrary constants and a differential equation satisfy solution for all constants. Then we can write

$$y=\sum_{n=1}^{\infty} b_n \sin\frac{n\pi t}{l}\cos\frac{n\pi ct}{l} \qquad \text{(ix)}$$

At t = 0, $y = k(lx - x^2)$

$$k\left(lx-x^2\right)=\sum_{n=1}^{\infty} b_n \sin\frac{n\pi x}{l}$$

Applying half range Fourier sine series

$$b_n=\frac{2}{l}\int_0^l f(x)\sin\frac{n\pi x}{l}dx$$

$$b_n=\frac{2k}{l}\int_0^l \left(lx-x^2\right)\sin\frac{n\pi x}{l}dx$$

$$b_n=\frac{2k}{l}\left[(-1)^{n+1}\frac{2l^3}{n^3\pi^3}+\frac{2l^3}{n^3\pi^3}\right]$$

$$b_n=\begin{cases}\dfrac{8kl^2}{n^3\pi^3}, & \text{when n is odd}\\ 0, & \text{when n is even}\end{cases}$$

$$y=\sum_{n=1}^{\infty}\frac{8kl^2}{n^3\pi^3}\sin\frac{n\pi x}{l}\cos\frac{n\pi ct}{l}, \text{ when n is odd}$$

$$\text{or } y=\sum_{n=1}^{\infty}\frac{8kl^2}{(2n-1)^3\pi^3}\sin\frac{(2n-1)\pi x}{l}\cos\frac{(2n-1)\pi ct}{l}$$

Which is required solution.

Example 9 Solve the boundary value problem

$$\frac{\partial^2 y}{\partial t^2}=4\frac{\partial^2 y}{\partial x^2}, \text{ given that } y(0,t)=0$$

$$y(5,t)=0,\ y(x,0)=0 \text{ and } \left(\frac{\partial y}{\partial t}\right)_{x=0}=5\sin\pi x$$

Solution: Applying the method of separation of variables to the wave equation

$$\frac{\partial^2 y}{\partial t^2}=2^2\frac{\partial^2 y}{\partial x^2}.$$

The suitable solution is

$$y = (C_1 \cos px + C_2 \sin px)(C_3 \cos 2pt + C_4 \sin 2pt)$$

Applying the initial condition

$y(x, 0) = 0$ we have

$0 = C_3 (C_1 \cos px + C_2 \sin px)$

$\Rightarrow \qquad C_3 = 0$

$\therefore\ y = C_4(C_1 \cos px + C_2 \sin px) \sin 2pt$

Now using $y(0, t) = 0$, we get

$0 = C_1 C_4 \sin 2pt \Rightarrow \qquad C_1 = 0$

$\therefore\ y = C_2 \sin 2pt \sin px$

Further $y(5, t) = 0$ we have $C_2 \sin 2pt \sin 5p = 0$

$\Rightarrow \qquad \sin 5p = 0 = \sin n\pi$

$\Rightarrow \qquad p = \frac{n\pi}{5}, n = 1, 2, 3, \ldots\ldots\ldots$

Therefore,

$$y = c_2 \sin\left(\frac{n\pi 2t}{5}\right) \sin\left(\frac{n\pi x}{5}\right)$$

Also the boundary condition $\left(\frac{\partial y}{\partial t}\right)_{x=0} = 5 \sin \pi x$

$$\therefore c_2 \frac{n\pi^2}{5} \cos\left(\frac{n\pi 2t}{5}\right) \sin\left(\frac{n\pi x}{5}\right) = 5 \sin \pi x$$

$\Rightarrow \qquad n = 5$ and $2\pi c_2 = 5$

Therefore, we have

$$y = \frac{5}{2\pi} \sin \pi x \sin 2\pi t$$

Example 10: A string of length *l* is fastened of both ends A and C. At a distance 'a' from the end A, the string is transversely displaced to a distance 'd' and is released from rest when it is in this position. find the equation of the subsequent motion.

OR

Find the half period sine series for f(x) given in the range (*l*, 0) by the graph ABC as shown in figure.

(U.P.T.U. 2009)

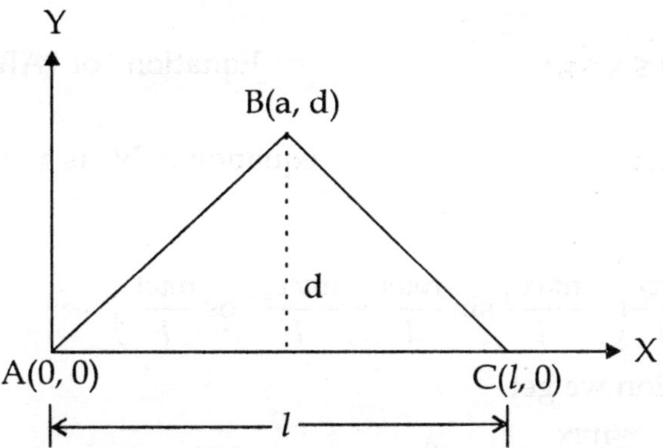

Solution let y(x, t) is the displacement of the string Now, by the one dimensional wave equation we have

$$\frac{\partial^2 y}{\partial t^2} = C^2 \frac{\partial^2 y}{\partial x^2} \qquad \text{(i)}$$

The solution of equation (i) is given by

$$y(x, t) = (C_1 \cos px + C_2 \sin px)(C_3 \cos cpt + C_4 \sin cpt) \qquad \text{(ii)}$$

Now using the boundary conditions as follows

The boundary conditions are

At x =0 (at A), y =0 $\Rightarrow$ y(0, t) =0

and At x =l (at C), y =0 $\Rightarrow$ y(l, t) =0

From (ii), we have

$$0 = C_1 (C_3 \cos cpt + C_4 \sin cpt) \Rightarrow C_1 = 0$$

using $C_1 = 0$ in equation (ii), we get

$$y(x,t) = C_2 \sin px\,(C_3 \cos cpt + C_4 \sin cpt) \qquad \text{(iii)}$$

using second boundary condition, from (iii), we have

$$0 = C_2 \sin pl\,(C_3 \cos cpt + C_4 \sin cpt)$$

$$\Rightarrow \quad \sin pl = 0 \quad \Rightarrow \quad \sin pl = \sin n\pi$$

$$\Rightarrow \quad p = \frac{n\pi}{l}$$

using the value of p in (iii) we have

$$y(x,t) = C_2 \sin\frac{n\pi x}{l}\left(C_3 \cos\frac{n\pi ct}{l} + C_4 \sin\frac{n\pi ct}{l}\right) \qquad \text{(iv)}$$

Next , the initial conditions are as follows:

velocity $\frac{\partial y}{\partial t} = 0$ at t =0

and displacement at t =0 is

$$y(x,0)=\begin{cases}\dfrac{d.x}{a}, & 0\le x\le a\\ \dfrac{d(x-l)}{a-l}, & a\le x\le l\end{cases}$$ $\because$ Equation of AB is $y=\dfrac{d.x}{a}$ and equation of BC is $y=\dfrac{d(x-l)}{a-l}$

From (iv)

$$\frac{\partial y}{\partial t}=c_2\sin\frac{n\pi x}{l}\left(-\frac{n\pi cc_3}{l}\sin\frac{n\pi ct}{l}+\frac{n\pi cc_4}{l}\cos\frac{n\pi ct}{l}\right)$$

using initial condition we get

$$0=c_2c_4\frac{n\pi c}{l}.\sin\frac{n\pi x}{l}\Rightarrow c_4=0$$

using $c_4=0$ in equation (iv), we get

$$y(x,t)=c_2c_3\sin\frac{n\pi x}{l}.\cos\frac{n\pi ct}{l}$$

$\therefore$ The general solution of the given problem is

$$y(x,t)=\sum_{n=1}^{\infty}b_n\sin\frac{n\pi x}{l}.\cos\frac{n\pi ct}{l} \qquad \text{(v)}$$

Using initial condition in equation (v), we get

$$y(x,0)=\sum_{n=1}^{\infty}b_n\sin\frac{n\pi x}{l}$$

which is half range Fourier sine series, so we have

$$b_n=\frac{2}{l}\int_0^l y(x,0).\sin\frac{n\pi x}{l}dx$$

$$=\frac{2}{l}\int_0^a\frac{d}{a}.x.\sin\left(\frac{n\pi x}{l}\right)dx+\frac{2}{l}\frac{d}{(a-l)}\int_a^l(x-l)\sin\frac{n\pi x}{l}dx$$

$$+\frac{2d}{l(a-l)}\left[(x-l)\left(\frac{-l}{n\pi}\right)\cos\frac{n\pi x}{l}-\left(\frac{-l^2}{n^2\pi^2}\right)\sin\frac{n\pi x}{l}\right]_a^l$$

$$\Rightarrow b_n=-\frac{2d}{n\pi}\cos\frac{n\pi a}{l}+\frac{2dl^2}{al\,n^2\pi^2}\sin\frac{n\pi a}{l}+\frac{2d}{n\pi}\cos\frac{n\pi a}{l}-\frac{2dl^2}{l(a-l)n^2\pi^2}\sin\frac{n\pi a}{l}$$

$$\Rightarrow b_n=\frac{2dl^2}{a(l-a)n^2\pi^2}\sin\frac{n\pi a}{l}$$

$\therefore$ From (v), we get

$$y(x,t)=\frac{2dl^2}{a(l-a)\pi^2}\sum_{n=1}^{\infty}\frac{1}{n^2}\sin\frac{n\pi a}{l}.\sin\frac{n\pi x}{l}.\cos\frac{n\pi ct}{l}$$

Example 11: A lightly stretched string of length l with fixed ends is initially in equilibrium position. it is set vibrating by giving each point a velocity $V_0 \sin^3 \frac{\pi x}{l}$. Find the displacement y(x, t).

(I.A.S. 2004, U.P.T.U. 2003)

Solution The equation of the vibrating string is $\frac{\partial^2 y}{\partial t^2} = C^2 \frac{\partial^2 y}{\partial x^2}$ (i)

The boundary condition are $y(0, t) = 0, y(l, t) = 0$ (ii)

Also the initial conditions are $y(x, 0) = 0$ (iii)

and $\left(\frac{\partial y}{\partial t}\right)_{t=0} = V_0 \sin^3 \frac{\pi x}{l}$ (iv)

Since the vibration of the string is periodic, therefore, the solution of (i) is of the form

$y(x, t) = (C_1 \cos px + C_2 \sin px)(C_3 \cos cpt + C_4 \sin cpt)$

by (ii) $y(0, t) = C_1(C_3 \cos cpt + C_4 \sin cpt) = 0$

For this to be true for all time $C_1 = 0$

$\therefore y(x, t) = C_2 \sin px (C_3 \cos cpt + C_4 \sin cpt)$

Also $y(l, t) = C_2 \sin pl (C_3 \cos cpt + C_4 \sin cpt) = 0$ for all t.

This gives $pl = n\pi$ or $p = \frac{n\pi}{l}$, n being an integer

Thus $y(x,t) = C_2 \frac{n\pi x}{l}\left(C_3 \cos \frac{cn\pi}{l} t + C_4 \sin \frac{cn\pi}{l} t\right)$

$\therefore y(x,t) = b_n \sin \frac{n\pi x}{l} \sin \frac{cn\pi t}{l}$ where $b_n = C_2 C_4$

Adding all such solutions the general solution of (i) is

$$y(x,t) = \sum b_n \sin \frac{n\pi x}{l} \sin \frac{cn\pi t}{l} \quad \text{(v)}$$

Now $\frac{\partial y}{\partial t} = \sum b_n \sin \frac{n\pi x}{l} . \frac{cn\pi}{l} \cos \frac{cn\pi t}{l}$

By (iv), $V_0 \sin^3 \frac{\pi x}{l} = \left(\frac{\partial y}{\partial t}\right)_{t=0} = \sum \frac{cn\pi}{l} b_n \sin \frac{n\pi x}{l}$

or $\frac{V}{4}\left(3 \sin \frac{\pi x}{l} - \sin \frac{3\pi x}{l}\right) = \sum \frac{cn\pi}{l} b_n \sin \frac{n\pi x}{l}$

$= \frac{c\pi}{l} b_1 \sin \frac{\pi x}{l} + \frac{2c\pi}{l} b_2 \sin \frac{2\pi x}{l} + \frac{3cx}{l} b_3 \sin \frac{3\pi x}{l} + \ldots\ldots\ldots\ldots\ldots$

Equating Coefficients from both sides, we get

$$\frac{3V_0}{4} = \frac{c\pi}{l} b_1, \quad 0 = \frac{2c\pi}{l} b_2, \quad -\frac{V_0}{4} = \frac{3c\pi}{l} b_3 \ldots\ldots\ldots\ldots$$

$$\therefore = \frac{3lV_0}{4c\pi}, \; b_3 = -\frac{lV_0}{12c\pi}, \; b_2 = b_4 = b_3 = \ldots\ldots\ldots\ldots = 0$$

Substituting in (v), the desired solution is

$$y = \frac{lV_0}{12c\pi}\left(9\sin\frac{\pi x}{l}\sin\frac{c\pi t}{l} - \sin\frac{3\pi x}{l}\sin\frac{3c\pi t}{l}\right)$$

Example 12: A lightly stretched string with fixed end points x =0 and x = l is initially at rest in its equilibrium position. If it is vibrating by giving to each of its points a velocity $\lambda x(l - x)$. find the displacement of the string at any distance x from one end at any time t.

(U.P.T.U. 2002)

Solution. The equation of the vibrating string is

$$\frac{\partial^2 y}{\partial t^2} = C^2 \frac{\partial^2 y}{\partial x^2} \quad \text{(i)}$$

The boundary condition are y(0, t) =0, y (l, t) =0 (ii)

Also the initial conditions are y (x, 0) =0 (iii)

$$\text{and} \left(\frac{\partial y}{\partial t}\right)_{t=0} = \lambda x(l - x) \quad \text{(iv)}$$

As in example 11, the general solution of (i) satisfying the conditions (ii) and (iii) is

$$y(x,t) = \sum_{n=1}^{\infty} b_n \sin\frac{n\pi x}{l} . \sin\frac{n\pi ct}{l} \quad \text{(v)}$$

$$\frac{\partial y}{\partial t} = \sum_{n=1}^{\infty} b_n \sin\frac{n\pi x}{l} . \cos\frac{n\pi ct}{l}\left(\frac{n\pi c}{l}\right)$$

$$\text{By (iv), } \lambda x(l - x) = \left(\frac{\partial y}{\partial t}\right)_{t=0} = \frac{\pi c}{l}\sum_{n=1}^{\infty} nb_n \sin\frac{n\pi x}{l}$$

$$\therefore \frac{\pi cn}{l} b_n = \frac{2}{l}\int_0^l lx(l - x)\sin\frac{n\pi x}{l} dx$$

$$= \frac{2\lambda}{l}\left[(lx - x^2)\left(-\frac{l}{n\pi}\cos\frac{n\pi x}{l}\right) - (l - 2x)\left(-\frac{l^2}{n^2\pi^2}\sin\frac{n\pi x}{l}\right) + (-2)\left(\frac{l^3}{n^3\pi^3}\cos\frac{n\pi x}{l}\right)\right]_0^l$$

$$= \frac{4\lambda l^2}{n^3\pi^3}(1 - \cos n\pi) = \frac{4\lambda l^2}{n^3\pi^3}\left[1 - (-1)^n\right]$$

or $b_n = \frac{4\lambda l^2}{c\pi^4 n^4}\left[1-(-1)^n\right]$

$$= \frac{8\lambda l^3}{c\pi^4 (2m-1)^4} \text{ taking } n = 2m-1$$

Hence, from (v) the desired solution is

$$y = \frac{8\lambda l^2}{c\pi^4}\sum_{m=1}^{\infty}\frac{1}{(2m-1)^4}\sin\frac{(2m-1)\pi x}{l}\sin\frac{(2m-1)\pi ct}{l}$$

Example 13. Solve completely the equation $\frac{\partial^2 y}{\partial t^2} = C^2\frac{\partial^2 y}{\partial x^2}$, representing the vibrations of a string of length l, fixed at both ends, given that $y(0, t) = 0$, $y(l, t) = 0$

$$y(x, 0) = f(x), \text{ and } \frac{\partial}{\partial t}y(x,0) = 0,\ 0<x<l$$

(U.P.T.U. 2005)

Solution Here the given equation is

$$\frac{\partial^2 y}{\partial t^2} = C^2\frac{\partial^2 y}{\partial x^2} \quad \text{(i)}$$

The solution of equation (i) is given by

$$y = (C_1 \cos px + C_2 \sin px)(C_3 \cos cpt + C_4 \sin cpt) \quad \text{(ii)}$$

Now, applying the boundary conditions $y = 0$ when $x = 0$, we get

$$0 = C_1(C_3 \cos cpt + C_4 \sin cpt)$$

$$\Rightarrow \quad C_1 = 0$$

Therefore, equation (ii) becomes

$$y = C_2 \sin px\,(C_3 \cos cpt + C_4 \sin cpt) \quad \text{(iii)}$$

Now putting $x = l$ and $y = 0$ in equation (iii), we get

$$0 = C_2 \sin pl\,(C_3 \cos cpt + C_4 \sin cpt)$$

$$\Rightarrow \quad \sin pl = 0 = \sin n\pi$$

$$\text{or } pl = n\pi \Rightarrow p = \frac{n\pi}{l}$$

Thus, equation (iii) becomes

$$y = C_2 \sin\frac{n\pi}{l}x\left(C_3 \cos\frac{n\pi ct}{l} + C_4 \sin\frac{n\pi ct}{l}\right) \quad \text{(iv)}$$

Differentiating equation (iv) with respect to t, we get

$$\frac{\partial y}{\partial t} = C_2 \sin\frac{n\pi x}{l}\left(-C_3\frac{n\pi c}{l}\sin\frac{n\pi ct}{l} + C_4\frac{n\pi c}{l}\cos\frac{n\pi ct}{l}\right)$$

Using given boundary condition $\frac{\partial y}{\partial t} = 0$, t =0 we get

$$0 = C_2 \sin\frac{n\pi c}{l}\left(0 + C_4\frac{n\pi c}{l}\right)$$

$\Rightarrow \qquad C_4 = 0$

Thus, equation (iv) becomes

$$y = C_2C_3 \sin\frac{n\pi x}{l}\cos\frac{n\pi ct}{l}$$

Now applying the last boundary condition given, we get

$$f(x) = b_n \sin\frac{n\pi x}{l},\ b_n = C_2\, C_3$$

Where

$$b_n = \frac{2}{l}\int_0^l f(x)\sin\frac{n\pi x}{l}dx$$

Thus, the required solution is

$$y = b_n \sin\frac{n\pi x}{l}\cos\frac{n\pi ct}{l}$$

Solution of wave Equation By D' Almbert's Method

Transform the equation $\frac{\partial^2 y}{\partial t^2} = C^2\frac{\partial^2 y}{\partial x^2}$ to its normal form using the transformation u = x + ct, v = x – ct and hence solve it. Show that the solution may be put in the form $y = \frac{1}{2}[f(x+ct) + f(x-ct)]$. Assume initial condition y = f(x) and $\frac{\partial y}{\partial t} = 0$ at t=0

(U.P.T.U. 2003)

Proof. Consider one dimensional wave equation

$$\frac{\partial^2 y}{\partial t^2} = C^2\frac{\partial^2 y}{\partial x^2} \qquad \text{(i)}$$

Let u =x + ct and v = x – ct, be a transformation of x and t into u and v.

then

$$\frac{\partial y}{\partial x} = \frac{\partial y}{\partial u}\frac{\partial u}{\partial x} + \frac{\partial y}{\partial v}\frac{\partial v}{\partial x} = \frac{\partial y}{\partial u} + \frac{\partial y}{\partial v} \qquad \because \frac{\partial u}{\partial x} = 1$$

$$\frac{\partial v}{\partial x} = 1$$

or $\quad \dfrac{\partial}{\partial x} = \dfrac{\partial}{\partial u} + \dfrac{\partial}{\partial v}$

$$\Rightarrow \frac{\partial^2 y}{\partial x^2} = \frac{\partial}{\partial x}\left(\frac{\partial y}{\partial x}\right) = \left(\frac{\partial}{\partial u} + \frac{\partial}{\partial v}\right)\left(\frac{\partial y}{\partial u} + \frac{\partial y}{\partial v}\right)$$

$$= \frac{\partial^2 y}{\partial u^2} + \frac{\partial^2 y}{\partial u \partial v} + \frac{\partial^2 y}{\partial v \partial u} + \frac{\partial^2 y}{\partial v^2}$$

$$\Rightarrow \frac{\partial^2 y}{\partial x^2} = \frac{\partial^2 y}{\partial u^2} + 2\frac{\partial^2 y}{\partial u \partial v} + \frac{\partial^2 y}{\partial v^2} \quad \text{(ii)}$$

and $\dfrac{\partial y}{\partial t} = \dfrac{\partial y}{\partial u}\dfrac{\partial u}{\partial t} + \dfrac{\partial y}{\partial v}\dfrac{\partial v}{\partial t} = c\dfrac{\partial y}{\partial u} - c\dfrac{\partial y}{\partial v} \qquad \therefore \dfrac{\partial u}{\partial t} = c, \dfrac{\partial v}{\partial t} = -c$

or $\dfrac{\partial}{\partial t} = c\left(\dfrac{\partial}{\partial u} - \dfrac{\partial}{\partial v}\right)$

$$\therefore \frac{\partial^2 y}{\partial t^2} = \frac{\partial}{\partial t}\left(\frac{\partial y}{\partial t}\right) = c\left(\frac{\partial}{\partial u} - \frac{\partial}{\partial v}\right)\left(c\frac{\partial y}{\partial u} - c\frac{\partial y}{\partial v}\right)$$

$$= c^2\left(\frac{\partial^2 y}{\partial u^2} - 2\frac{\partial^2 y}{\partial u \partial v} + \frac{\partial^2 y}{\partial v^2}\right)$$

$$\Rightarrow \frac{\partial^2 y}{\partial t^2} = c^2\left(\frac{\partial^2 y}{\partial u^2} - 2\frac{\partial^2 y}{\partial u \partial v} + \frac{\partial^2 y}{\partial v^2}\right) \quad \text{(iii)}$$

marking use of equation (ii) and (iii) in equation (i), we get

$$c^2\left(\frac{\partial^2 y}{\partial u^2} - 2\frac{\partial^2 y}{\partial u \partial v} + \frac{\partial^2 y}{\partial v^2}\right) = c^2\left(\frac{\partial^2 y}{\partial u^2} + 2\frac{\partial^2 y}{\partial u \partial v} + \frac{\partial^2 y}{\partial v^2}\right)$$

$$\Rightarrow 4c^2 \frac{\partial^2 y}{\partial u \partial v} = 0$$

$$\Rightarrow \frac{\partial^2 y}{\partial u \partial v} = 0 \quad \text{(iv)}$$

Integrating equation (iv) w.r.t. 'v' we get

$$\frac{\partial y}{\partial u} = \phi(u) \quad \text{(v)}$$

where $\phi(u)$ is a constant in respect to v

Again integrate equation (v), we get

$$y = \int \phi(u)du + \phi_2(v)$$

$$\Rightarrow \quad y = \phi_1(u) + \phi_2(v)$$

$$\Rightarrow \quad y(x, t) = \phi_1(x + ct) + \phi_2(x - ct) \quad \text{(vi)}$$

The solution (vi) is D' Alembert's solution of wave equation.

Now, we applying initial conditions y = f(t) and $\frac{\partial y}{\partial t} = 0$ at t =0

From (vi), we get at t =0

$$f(x) = \phi_1(x) + \phi_2(x) \qquad \text{(vii)}$$

and $\frac{\partial y}{\partial t} = c\phi'(x+ct) - c\phi_2^1(x-ct)$

$$\Rightarrow \left(\frac{\partial y}{\partial t}\right)_{t=0} = 0 = c\phi_1'(x+0) - c\phi_2'(x-0)$$

$$\Rightarrow \phi_1'(x) - \phi_2'(x) = 0$$

$$\Rightarrow \phi_1'(x) = \phi_2'(x)$$

on integrating, we get

$$\phi_1(x) = \phi_2(x) + c_1 \qquad \text{(viii)}$$

using equation (viii) in equation (vii), we get

$$f(x) = \phi_2(x) + c_1 + \phi_2(x) = 2\,\phi_2(x) + c_1$$

$$\Rightarrow \phi_2(x) = \frac{1}{2}[f(x) - c_1] \Rightarrow \phi_2(x-ct) = \frac{1}{2}[f(x-ct) - c_1]$$

and $\phi_1(x) = \frac{1}{2}[f(x) + c_1] \Rightarrow \phi_1(x+ct) = \frac{1}{2}[f(x+ct) + c_1]$

putting the values of $\phi_1(x + ct)$ and $\phi_2(x - ct)$ in eq^n(vi) we get

$$y(x, t) = \frac{1}{2}[f(x + ct) + f(x-ct)]$$

VIBRATING MEMBRANE - TWO DIMENSIONAL WVAE EQUATION

Consider a lightly stretched uniform membrane (such as the membrane of a drum) with tension T per unit length is the same in all directions at every point.

Consider the forces on an element $\delta x\ \delta y$ of the membranes. Due to its displacement u perpendicular to the xy plane, the forces $T\delta x$ and $T\delta y$ act on the edges along the tangent to the membrane

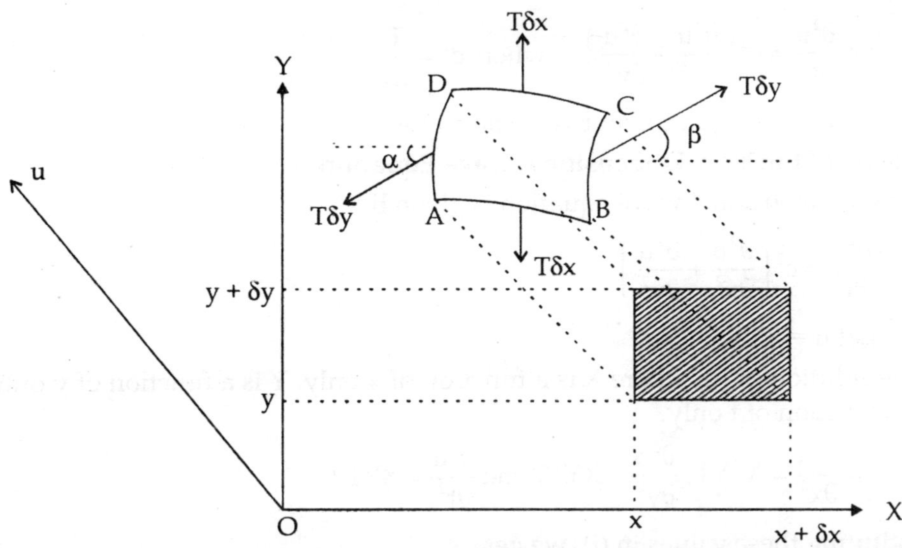

The forces Tδy (tangential to the membrane) on its opposite edges of length δy act at angles α and β to the horizontal. So, their vertical component

$= (T\delta y)\sin\beta - (T\delta y)\sin\alpha$

$= T\delta y(\tan\beta - \tan\alpha)$, $\quad \because$ α and β are very small i.e. $\sin\alpha \approx \tan\alpha$ etc.

$$= T\delta y\left\{\left(\frac{\partial u}{\partial x}\right)_{x+\delta x} - \left\{\frac{\partial u}{\partial x}\right\}_x\right\}$$

$$= T\delta y\,\delta x\frac{\left[\left(\frac{\partial u}{\partial x}\right)_{x+\delta x} - \left\{\frac{\partial u}{\partial x}\right\}_x\right]}{\delta x}$$

$= T\delta y\,\delta x\dfrac{\partial^2 u}{\partial x^2}$, upto a first order of approximation

similarly, the forces Tδx (the vertical component of the force) acting on the edges of length δx have the vertical component $= T\delta x\,\delta y\dfrac{\partial^2 u}{\partial y^2}$

If m be the mass per unit area of the membrane, then the equation of motion of the element ABCD is

$$m\delta x\,\delta y\frac{\partial^2 u}{\partial t^2} == T\left(\frac{\partial^2 u}{\partial x^2} + \frac{\partial^2 u}{\partial y^2}\right)\delta x\delta y$$

$$\text{or } \frac{\partial^2 u}{\partial t^2} = \frac{T}{m}\left(\frac{\partial^2 u}{\partial x^2} + \frac{\partial^2 u}{\partial y^2}\right)$$

$$\text{or } \frac{\partial^2 u}{\partial t^2} = c^2\left(\frac{\partial^2 u}{\partial x^2} + \frac{\partial^2 u}{\partial y^2}\right) \quad \text{where } c^2 = \frac{T}{m}$$

This in the wave equation in two dimensions.

Solution of the Two-Dimensional wave Equation

The two dimensional wave equation is given by

$$\frac{\partial^2 u}{\partial t^2} = c^2\left(\frac{\partial^2 u}{\partial x^2} + \frac{\partial^2 u}{\partial y^2}\right) \qquad \text{(i)}$$

Let u = XYT (ii)

be the solution of (i), where X is a function of x only, Y is a function of y only and T is a function of t only.

$$\Rightarrow \frac{\partial^2 u}{\partial x^2} = X''YT,\ \frac{\partial^2 u}{\partial y^2} = XY''T \text{ and } \frac{\partial^2 u}{\partial t^2} = XYT''$$

substituting these values in (i), we get

$$\frac{1}{c^2}XYT'' = X''YT + XY''T$$

Dividing by XYT throughout, we get

$$\frac{1}{c^2}\frac{T''}{T} = \frac{X''}{X} + \frac{Y''}{Y} \qquad \text{(iii)}$$

since each variable is independent, hence this will be true only when each member is a constant. Suitably choosing the constants, we have

$$\frac{d^2X}{dx^2} + k^2X = 0,\ \frac{d^2Y}{dy^2} + l^2y = 0$$

$$\text{and } \frac{d^2T}{dt^2} + \left(k^2 + l^2\right)c^2T = 0$$

Hence, the solution of these equations are given by

$$X = c_1 \cos kx + c_2 \sin kx$$

$$Y = c_3 \cos ly + c_4 \sin ly$$

$$\text{and } T = c_5 \cos\sqrt{(k^2 + l^2)}ct + c_6 \sin\sqrt{(k^2 + l^2)}ct$$

Hence form (ii), the solution of (i) is given by

$$u(x, y, t) = (c_1 \cos kx + c_2 \sin kx)(c_3 \cos ly + c_4 \sin ly)$$

$$\left[c_5 \cos\sqrt{(k^2 + l^2)}ct + c_6 \sin\sqrt{(k^2 + l^2)}ct\right] \qquad \text{(iv)}$$

Let is consider that the membrane is rectangular and stretched between the lines x =0, x =a, y =0, y = b

Now the boundary condition are

(1) u =0, when x =0, for all t

(2) u =0, when x =a, for all t

(3) u =0, when y =0, for all t

(4) u =0, when y =b, for all t

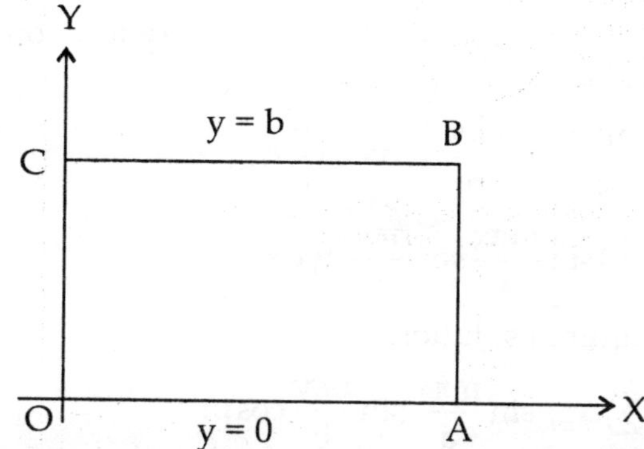

Now using condition (1) in (iv), we get

$$0 = c_1 (c_3 \cos ly + c_4 \sin ly)\left[c_5 \cos\sqrt{(k^2 + l^2)}ct + c_6 \sin\sqrt{(k^2 + l^2)}ct\right]$$

$\Rightarrow c_1 = 0$

substituting $c_1 = 0$ in (iv) and using condition (ii), we get

$$\sin ka = 0 \qquad \text{or } k = \frac{m\pi}{a}, \text{ where m is an integer}$$

Hence solution of (iii) becomes

$$u(x, y, t) = c_2 c_4 \sin\frac{m\pi x}{a} \sin\frac{n\pi y}{b}(c_5 \cos pt + c_6 \sin pt) \qquad \text{(iv)}$$

$$\text{where } p = \pi c\sqrt{\frac{m^2}{a^2} + \frac{n^2}{b^2}}$$

Now replacing the arbitrary constants, we can write the general solution as

$$u(x, y, t) = \sum_{m=1}^{\infty}\sum_{n=1}^{\infty} \sin\frac{m\pi x}{a} \sin\frac{n\pi y}{b}(A_{mn} \cos pt + B_{mn} \sin pt) \qquad \text{(v)}$$

Equation (v) is the solution of the wave equation (i) which is zero on the boundary of the rectangular membrane.

Suppose the membrane starts from rest from the initial position.

u = f(x, y) i.e. u(x, y, 0) = f(x, y)

Then using the condition $\frac{\partial u}{\partial t} = 0$, when t =0, we get $B_{mn} = 0$

Further using the condition u = f(x,y) when t =0, we get

$$f(x,y)=\sum_{m=1}^{\infty}\sum_{n=1}^{\infty}A_{mn}\sin\frac{m\pi x}{a}\sin\frac{n\pi y}{b} \quad \text{(vi)}$$

This is a double Fourier series. multiplying both sides by $\sin\frac{m\pi x}{a}\sin\frac{n\pi y}{b}$ and integrating from x =0 to x =a and y =0 to y =b, every term on the right except one become zero. Thus, we have

$$\int_0^a\int_0^b f(x,y)\sin\frac{m\pi x}{a}\sin\frac{n\pi y}{b}dydx=\frac{ab}{4}A_{mn}$$

$$\text{or } A_{mn}=\frac{4}{ab}\int_0^a\int_0^b f(x,y)\sin\frac{m\pi x}{a}\sin\frac{n\pi y}{b}dydx \quad \text{(vii)}$$

Therefore from (v) required solution is

$$u(x,y,t)=\sum_{m=1}^{\infty}\sum_{n=1}^{\infty}A_{mn}\sin\frac{m\pi x}{a}\sin\frac{n\pi y}{b}\cos pt$$

where A_{mn} is given by (vii) and $p=\pi c\sqrt{\left(\frac{m^2}{a^2}+\frac{n^2}{b^2}\right)}$

Example 14: Find the deflection u(x, y, t) of a square membrane with a = b = 1 and c =1, If the initial velocity is zero and the initial deflection is f(x,y) = A sin πx sin 2πy

Solution : The deflection of the square membrane is given by the two dimensional wave equation.

$$\frac{\partial^2 u}{\partial t^2}=c^2\left(\frac{\partial^2 u}{\partial x^2}+\frac{\partial^2 u}{\partial y^2}\right)$$

The boundary conditions are

u(x, 0, t) = 0 = u(x, 1, t) and u(0, y, t) =0 = u(1, y, t)

The initial conditions are u(x, y, 0) = f(x, y) = A sin πx sin 2xy, $\left(\frac{\partial u}{\partial t}\right)_{t=0}=0$

∴ Deflection

$$u(x,y,t)=\sum_{m=1}^{\infty}\sum_{n=1}^{\infty}A_{mn}\cos k_{mm}t\sin m\pi x.\sin n\pi y \quad \text{(i)}$$

where $A_{mn}=4\int_0^1\int_0^1 f(x,y)\sin m\pi x.\sin n\pi y\,dx\,dy \because a=b=1, c=1, k_{mn}^2=\pi^2(m^2+n^2)$

$$=4\int_0^1\int_0^1(\sin\pi x)(\sin 2\pi y)(\sin m\pi x)\sin n\pi y\,dx\,dy$$

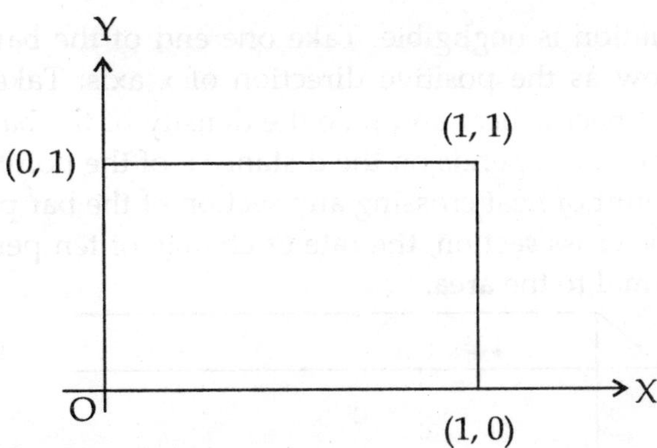

on integration, we find that $A_{m1} = A_{m3} = A_{m4} = \ldots\ldots\ldots\ldots = 0$

But

$$A_{m2} = 4A\int_0^1\int_0^1 (\sin \pi x)(\sin m\pi x)\sin^2 2\pi y\,dx\,dy$$

$$= 2A\int_0^1\int_0^1 (\sin \pi x)(\sin m\pi x)(1 - \cos 4\pi y)\,dx\,dy$$

$$= 2A\int_0^1 \sin \pi x \sin m\pi x\left(y - \frac{1}{4\pi}\sin 4\pi y\right)_0^1 dx$$

$$= 2A\int_0^1 \sin \pi x \sin m\pi x\,dx$$

on integration we find that $A_{22} = A_{32} = \ldots\ldots\ldots\ldots\ldots = 0$

Also we find $A_{12} = 2A\int_0^1 \sin \pi x \sin \pi x\,dx$

$$= A\int_0^1 2\sin^2 \pi x\,dx$$

$$= A\int_0^1 (1 - \cos 2\pi x)\,dx$$

$$= A\left(x - \frac{1}{2\pi}\sin 2\pi x\right)_0^1 = A$$

Thus, from (1), we have

$u(x, y, t) = A_{12}(\cos k_{12}t)(\sin \pi x)\sin 2\pi y$

$= A\cos \sqrt{5}\,\pi t \sin \pi x \sin 2\pi y$

$\because k_{12}^2 = \pi^2(1^2 + 2^2)$ i.e. $k_{12} = \pi\sqrt{5}$

since $c = 1$, $m = 1$, $n = 2$

ONE - DIMENSIONAL HEAT FLOW

Consider a homogeneous bar of uniform cross section A. Here we assume that the sides of the bar are insulated and the loss of heat from the sides by conduction or radiation is negligible. Take one end of the bar as the origin and the direction of flow as the positive direction of x axis. Take k be the thermal conductivity 's' the specific heat and ρ be the density of the bar. The temperature u at any point of the bar depends on the distance x of the point from one end and the time t. The amount of heat crossing any section of the bar per second depends on the area A of the cross section, the rate of change of temperature with respect to 'x' (distance) normal to the area.

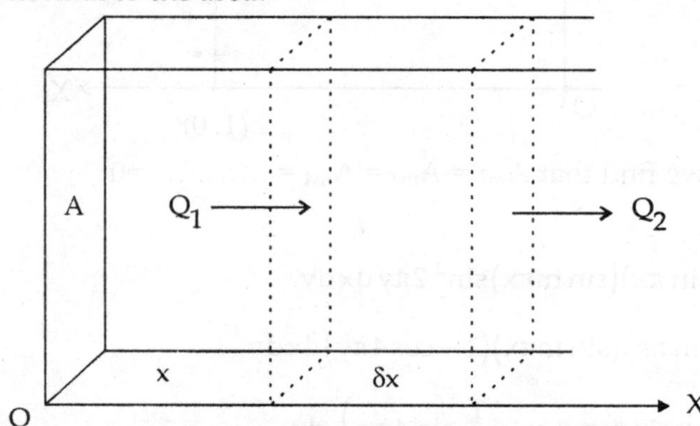

One dimensional heat flow

Therefore Q_1, the quantity of heat flowing into section at a distance $x = -kA\left(\frac{\partial u}{\partial x}\right)_x$ per second. (The negative sign indicates that as x increases, u decreases). Q_2, the quantity of heat flowing out of the section at a distance $x + \delta x = -kA\left(\frac{\partial u}{\partial x}\right)_{x+\delta x}$ per second.

Hence, the rate of increase of heat in the slab with thickness δx is

$$Q_1 - Q_2 = kA\left[\left(\frac{\partial u}{\partial x}\right)_{x+\delta x} - \left(\frac{\partial u}{\partial x}\right)_x\right] \text{ per second} \qquad \text{(i)}$$

But the rate of increase of heat of the slab = $s\rho A\delta x\frac{\partial u}{\partial t}$ (ii)

From (i) and (ii), we get

$$s\rho A\delta x\frac{\partial u}{\partial t} = kA\left[\left(\frac{\partial u}{\partial x}\right)_{x+\delta x} - \left(\frac{\partial u}{\partial x}\right)_x\right]$$

$$\text{or } s\rho\frac{\partial u}{\partial t} = k\left[\frac{\left(\frac{\partial u}{\partial x}\right)_{x+\delta x} - \left(\frac{\partial u}{\partial x}\right)_x}{\delta x}\right]$$

Taking limit as $\delta x \to 0$, we have

$$s\rho \frac{\partial u}{\partial t} = k \frac{\partial^2 u}{\partial x^2}$$

or $$\frac{\partial u}{\partial t} = \frac{k}{s\rho} \frac{\partial^2 u}{\partial x^2}$$

or $$\frac{\partial u}{\partial t} = c^2 \frac{\partial^2 u}{\partial x^2} \quad \text{(iii)}$$

where $c^2 = \dfrac{k}{s\rho}$ is called the diffusivity of the material.

Equation (iii) is called the one-dimensional heat flow equation.

Solution of one-dimensional heat Equation

The equation of one dimensional heat flow is $\dfrac{\partial u}{\partial t} = c^2 \dfrac{\partial^2 u}{\partial x^2}$ (i)

Assume that a solution of (i) is

$u = X(x). T(t)$

where X is a function of x alone and T is a function t alone.

Then $\dfrac{\partial u}{\partial t} = XT', \dfrac{\partial u}{\partial x} = X'T$ and $\dfrac{\partial^2 u}{\partial x^2} = X''T$

Putting these values in (i), we get

$$XT' = C^2X''T \Rightarrow \frac{X''}{X} = \frac{T'}{c^2T} \quad \text{(ii)}$$

Now L.H.S expression is a function of x only while R.H.S is a function of t only, so the two can be equal only when these are equal to a constant, say k.

$\Rightarrow \dfrac{X''}{X} = k$ i.e. $X'' = kX$

or $$\frac{d^2X}{dx^2} - kX = 0 \quad \text{(iii)}$$

and $$\frac{dT}{dt} - c^2kT = 0 \quad \text{(iv)}$$

There are arise following cases:

Case I. when $k > 0$, let $k=p^2$

Then $X = c_1 e^{px} + c_2 e^{-px}$ and $T= c_3 e^{c^2p^2t}$

Therefore $u(x, t) = (c_1e^{px} + c_2 e^{-px}) c_3 e^{c^2p^2t}$

Case II. when $k<0$, let $k= -p^2$

Then $X = c_4 \cos px + c_5 \sin px$ and $T = c_6 e^{-c^2p^2t}$

Therefore, $u(x, t) = (c_4 \cos px + c_5 \sin px)\, c_6 e^{-c^2p^2t}$

Case III when k =0, then

$X = c_7x + c_8$ and $T = c_9$

Therefore $u(x, t) = (c_7x + c_8)\, c_9$

Here we are dealing with the heat conduction problem so the temperature u decreases with the increase of time t and hence solution given by case II is appropriate.

i.e. $u(x, t) = (c_1\cos px + c_2 \sin px)\, c_3\, e^{-c^2p^2t}$ is the only suitable solution of the heat equation.

Fourier series solution of one-dimensional heat equation

Applying the boundary conditions and the initial condition, we get

$$u(x,t) = \sum_{n=1}^{\infty} b_n \sin\frac{n\pi x}{l} e^{\frac{-n^2\pi^2c^2t}{l}}$$

$$\text{where } b_n = \frac{2}{l}\int_0^l f(x)\sin\frac{n\pi x}{l}dx$$

Remark: In steady state $\frac{\partial u}{\partial t} = 0$, so $\frac{\partial u^2}{\partial x^2} = 0$

Example 15 An insulated rod of length l has its end A and B maintained 0ºC and 100ºC respectively until steady state condition Prevail. If B is suddenly reduced to 0ºC and maintained at 0ºC find the temperature at a distance x from A at time t.

(U.P.T.U.2004, 2005)

Solution: The equation of one dimensional heat flow be

$$\frac{\partial u}{\partial t} = c^2\frac{\partial^2 u}{\partial x^2} \qquad \text{(i)}$$

The boundary conditions are

(a) $u(0, t) = 0$ºC and (b) $u(l, t) = 100$ºC

In steady state condition $\frac{\partial u}{\partial t} = 0$, here from (i) we get

$$\frac{\partial^2 u}{\partial x^2} = 0$$

on integration, we get $u(x) = c_1x + c_2$ (ii)

where c_1 and c_2 are constants to be determined.

At x =0, from equation (ii), we have

$0 = C_2$

and at $x = l$, $100 = c_1 l + 0 \Rightarrow c_1 = \frac{100}{l}$

$\therefore$ form (ii)

$$u(x) = \frac{100}{l} x \quad \text{(iii)}$$

Now the temperature at B is suddenly changed we have again transient state. if u(x, t) is the subsequent temperature function, the boundary conditions are (a') $u(0, t) = 0^oC$ (b') $u(l, t) = 0^oC$ and the initial condition (c') $u(x,0) = \frac{100}{l} x$

since the subsequent steady state function $u_s(x)$ satisfies the equation

$$\frac{\partial^2 u_s}{\partial x^2} = 0$$

or $\frac{d^2 u_s}{dx^2} = 0 \Rightarrow u_s(x) = c_3 x + c_4$

at $x = 0$, wet get $0 = c_4$

and at $x = l$, we get

$0 = c_3 l + 0 \Rightarrow c_3 = 0$

Thus $u_s(x) = 0$ (iv)

If $u_T(x, t)$ is the temperature in transient state then the temperature distribution in the rod u(x, t) can be expressed in the form $u(x, t) = u_s(x) + u_T(x, t)$

$\Rightarrow u(x, t) = u_T(x, t)$ since $u_s(x) = 0$ (v)

Again from heat equation we have

$$\frac{\partial u_T}{\partial t} = c^2 \frac{\partial^2 u_T}{\partial x^2} \quad \text{(vi)}$$

The solution of equation (6) is

$u_T(x, t) = (c_4 \cos px + c_2 \sin px)\, c_3 e^{-c^2 p^2 t}$ (vii)

At x =0, u(0, t) =0

$\Rightarrow 0 = c_1 e^{-c^2 p^2 t} \quad \Rightarrow \quad c_1 = 0$

From (vii), we get

$u(x, t) = c_2 \sin px\, c_3 e^{-c^2 p^2 t}$ (viii)

Again at $x = l$, $u(l, t) = 0$

$\Rightarrow \quad 0 = c_1 c_2 \sin pl . e^{-c^2 p^2 t}$

$\Rightarrow \quad \sin pl = 0 = \sin n\pi$

$$\Rightarrow p = \frac{n\pi}{l}$$

From (viii), we get

$$u(x,t) = c_2 c_3 \sin\frac{n\pi x}{l}.e^{-\frac{n^2c^2\pi^2 t}{l^2}}$$

$$\Rightarrow u(x,t) = \sum_{n=1}^{\infty} b_n \sin\frac{n\pi x}{l}.e^{-\frac{n^2c^2\pi^2 t}{l^2}} \qquad \text{(ix)}$$

using initial condition i.e. at t = 0, $u = \frac{100}{l}x$, we get

$$u(x,0) = \frac{100}{l}x = \sum_{n=1}^{\infty} b_n \sin\frac{n\pi x}{l}$$

$$\therefore b_n = \frac{2}{l}\int_0^l \frac{100}{l} x \sin\frac{n\pi x}{l}\,dx$$

$$= \frac{200}{l^2}\int_0^l x \sin\frac{n\pi x}{l}\,dx = \frac{200}{l^2}\left[-\frac{xl}{n\pi}\cos\frac{n\pi x}{l} + \frac{l^2}{n^2\pi^2}\sin\frac{n\pi x}{l}\right]_0^l$$

$$\Rightarrow b_n = \frac{200}{l^2}\left[-\frac{l^2}{n\pi}\cos n\pi\right] = \frac{200}{n\pi}(-1)^{n+1}$$

Hence from equation (ix) we get

$$u(x,t) = \frac{200}{\pi}\sum_{n=1}^{\infty}\frac{(-1)^{n+1}}{n}\sin\frac{n\pi x}{l}.e^{-\frac{n^2c^2\pi^2 t}{l^2}}$$

Example 16: Determine the solution of one dimensional heat equation $\frac{\partial u}{\partial t} = c^2\frac{\partial^2 u}{\partial x^2}$ subjected to the boundary conditions u(0, t) =0, u(*l*, t) =0 (t >0) and the initial condition u(x, 0) =x, *l* being the length of the bar.

(I.A.S. 2007, U.P.T.U. 2006)

Solution: We have

$$\frac{\partial u}{\partial t} = c^2\frac{\partial^2 u}{\partial x^2} \qquad \text{(i)}$$

we know that the solution of equation (i) is given by

$$u(x, t) = (c_1\cos px + c_2 \sin px)\, c_3\, e^{-p^2c^2t} \qquad \text{(ii)}$$

At x =0, u =0

$\Rightarrow \quad 0 = c_1 c_3 e^{-p^2c^2t}$

$\Rightarrow \quad c_1 = 0$

From (ii) we get

$u(x, t) = c_2c_3 \sin px e^{-p^2c^2t}$ (iii)

Again at $x = l$, $u = 0$

$$\Rightarrow \quad 0 = c_2\, c_3 \sin pl.\, e^{-p^2c^2t}$$

$$\Rightarrow \quad \sin pl = 0 = \sin n\pi$$

$$\Rightarrow \quad p = \frac{n\pi}{l}$$

From (iii) the general solution of equation (i) is

$$u(x,t) = \sum_{n=1}^{\infty} b_n \sin\frac{n\pi x}{l}.e^{-\frac{n^2c^2\pi^2t}{l^2}} \quad \text{(iv)}$$

At $t = 0$, $u = x$

$$\Rightarrow x = \sum_{n=1}^{\infty} b_n \sin\frac{n\pi x}{l}$$

$$\therefore b_n = \frac{2}{l}\int_0^l x\sin\frac{n\pi x}{l}dx = \frac{2}{l}\left[x.\frac{l}{n\pi}\left(-\cos\frac{n\pi x}{l}\right) - \left(\frac{-l^2}{n^2\pi^2}\sin\frac{n\pi x}{l}\right)\right]_0^l$$

$$= \frac{2}{l}\left[\left(l.\frac{l}{n\pi}(-\cos n\pi) + \frac{l^2}{n^2\pi^2}\sin n\pi\right) - 0\right]_0^l$$

$$\Rightarrow b_n = \frac{2}{l}\left[-\frac{l^2}{n\pi}(-1)^n\right] = (-1)^{n+1}\frac{2l}{n\pi}$$

putting the value of b_n in equation (iv), we get

$$u(x,t) = \frac{2l}{\pi}\sum_{n=1}^{\infty}\frac{(-1)^{n+1}}{n}\sin\frac{n\pi x}{l}.e^{-\frac{n^2c^2\pi^2t}{l^2}}$$

Example 17: The ends A and B a rod 20 cm long have the temperature at 30°C and 80°C until steady state prevails. The temperature of the ends are changed to 40°C and 60°C respectively. Find the temperature distribution in the rod at time t. (I.A.S. 2005)

Solution : The heat equation in one dimensional is

$$\frac{\partial u}{\partial t} = c^2\frac{\partial^2 u}{\partial x^2} \quad \text{(i)}$$

The boundary conditions are

(a) $u(0, t) = 30°C$ (b) $u(20, t) = 80°C$

In steady condition $\frac{\partial u}{\partial t} = 0$

$\therefore$ From (i) we get $\frac{\partial^2 u}{\partial x^2} = 0$, on integration, we get

$u(x) = c_1 x + c_2$ (ii)

at $x = 0, u = 30$ so $30 = 0 + c_2 \Rightarrow c_2 = 30$

and at $x = 20, u = 80$ so $80 = c_1 20 + 30 \Rightarrow c_1 = \frac{5}{2}$

From equation (ii) we get

$$u(x) = \frac{5x}{2} + 30 \quad \text{(iii)}$$

Now the temperature at A and B are suddenly changed we have again gain transient state.

If $u_1(x, t)$ is subsequent temperature function then the boundary conditions are

$u_1(0, t) = 40^0C$ and $u_1(20, t) = 60^0C$

and the initial condition i.e. at $t = 0$, is given by (iii)

Since the subsequent steady state function $u_s(x)$ satisfies the equation

$$\frac{\partial^2 u_s}{\partial x^2} = 0 \quad \text{or} \quad \frac{d^2 u_s}{dx^2} = 0$$

The solution of above equation is

$u_s(x) = c_3 x + c_4$

At $x = 0, u_s = 40 \Rightarrow 40 = 0 + c_4 \Rightarrow c_4 = 40$ $\quad \because u_s(0) = 40^0C$

$u_s(20) = 60^0C$

and at $x = 20, u_s = 60 \Rightarrow 60 = 20\, c_3 + 40 \Rightarrow c_3 = 1$

$\therefore$ form (iv), we get

$u_s(x) = x + 40$ (v)

Thus the temperature distribution in the rod at time t is given by

$u(x, t) = u_s(x) + u_T(x,t)$

$\Rightarrow \quad u(x, t) = (x + 40) + u_T(x, t)$ (vi)

where $u_T(x, t)$ is the transient state function which satisfying the conditions

$u_T(0, t) = u_1(0, t) - u_s(0) = 40 - 40 = 0$

$u_T(20, t) = u_1(20, t) - u_s(20) = 60 - 60 = 0$

and $u_T(x, 0) = u_1(x, 0) - u_s(x)$

$$= \frac{5x}{2} + 30 - x - 40 = \frac{3x}{2} - 10$$

The general solution for $u_T(x, t)$ is given by

$$u_T(x,t) = \sum_{n=1}^{\infty} b_n \sin\frac{n\pi x}{20}.e^{-\frac{n^2c^2\pi^2 t}{l^2}} \qquad \text{(vii)}$$

At t=0, from (vii) we get

$$\frac{3x}{2} - 10 = \sum_{n=1}^{\infty} b_n \sin\frac{n\pi x}{20}$$

$$\therefore b_n = \frac{2}{20}\int_0^{20}\left(\frac{3x}{2} - 10\right)\sin\frac{n\pi x}{20}dx$$

$$= \frac{1}{10}\left[\left(\frac{3x}{2} - 10\right)\left(-\frac{20}{n\pi}\cos\frac{n\pi x}{20}\right) - \frac{3}{2}\left(-\frac{400}{n^2\pi^2}\cos\frac{n\pi x}{20}\right)\right]_0^{20}$$

$$= \frac{1}{10}\left[-20\left(\frac{20}{n\pi}\right)(-1)^n - (-10)\left(\frac{20}{n\pi}\right)\right]$$

$$= -\frac{20}{n\pi}\left[2(-1)^n + 1\right]$$

putting the value of b_n in equation (vii), we get

$$u_T(x,t) = -\frac{20}{\pi}\sum_{n=1}^{\infty}\frac{2(-1)^n + 1}{n}\sin\frac{n\pi x}{20}.e^{\frac{-n^2\pi^2c^2t}{400}} \qquad \text{(viii)}$$

From (vi) and (viii), we get

$$u_T(x,t) = (x + 40) - \frac{20}{\pi}\sum_{n=1}^{\infty}\frac{2(-1)^n + 1}{n}\sin\frac{n\pi x}{20}.e^{\frac{-n^2\pi^2c^2t}{400}}$$

TWO - DIMENSIONAL HEAT FLOW

Consider the flow of heat in a metal plate of inform thickness α (cm), density ρ (gr/cm[3]), specific heat s(cal/gr deg) and thermal conductivity k (cal/cm sec-deg) Let XOY plane be taken in one face of the plates as shown in figure. If the temperature at any point is independent of the z coordinate and depends only on x, y and time t, then the flow is said to be two-dimensional. In this case, the heat flow is in the XY plane only and is zero along the normal to the XY-plane.

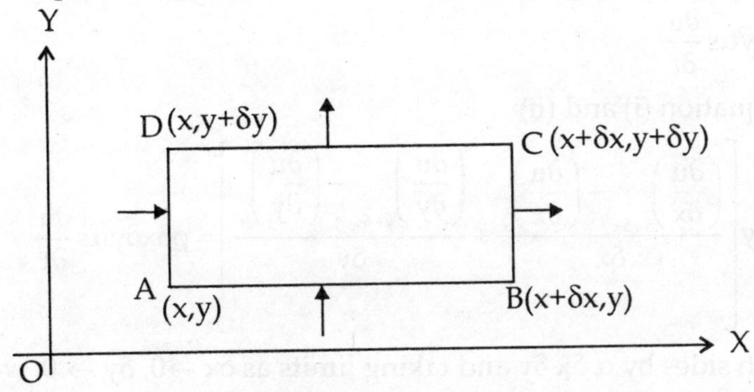

Consider a rectangular element ABCD of the plane with sides δx and δy. The amount of heat entering the element in 1 sec. from the side AB

$$= -k\alpha\delta x\left(\frac{\partial u}{\partial y}\right)_y \quad \text{(see one dimensional heat flow)}$$

and the amount of heat entering the element in 1 second from the side AD

$$= -k\alpha\delta y\left(\frac{\partial u}{\partial x}\right)_x$$

The quantity of heat flowing out through the side CD per sec

$$= -k\alpha\delta x\left(\frac{\partial u}{\partial y}\right)_{y+\delta y}$$

and the quantity of heat flowing out through the side BC per second

$$= -k\alpha\delta y\left(\frac{\partial u}{\partial x}\right)_{x+\delta x}$$

Hence the total gain of heat by the rectangular element ABCD per second

$$= -k\alpha\delta x\left(\frac{\partial u}{\partial y}\right)_y - k\alpha\delta y\left(\frac{\partial u}{\partial x}\right)_x + k\alpha\delta x\left(\frac{\partial u}{\partial y}\right)_{y+\delta y} + k\alpha\delta y\left(\frac{\partial u}{\partial x}\right)_{x+\delta x}$$

$$= k\alpha\delta x\left[\left(\frac{\partial u}{\partial y}\right)_{y+\delta y} - \left(\frac{\partial u}{\partial y}\right)_y\right] + k\alpha\delta y\left[\left(\frac{\partial u}{\partial x}\right)_{x+\delta x} - \left(\frac{\partial u}{\partial x}\right)_x\right]$$

$$= k\alpha\delta x\delta y\left[\frac{\left(\frac{\partial u}{\partial x}\right)_{x+\delta x} - \left(\frac{\partial u}{\partial x}\right)_x}{\delta x} + \frac{\left(\frac{\partial u}{\partial y}\right)_{y+\delta y} - \left(\frac{\partial u}{\partial y}\right)_y}{\delta y}\right] \qquad \text{(i)}$$

Also the rate of gain of heat by the element

$$= \rho\delta x\,\delta y\alpha s\frac{\partial u}{\partial t} \qquad \text{(ii)}$$

Thus from equation (i) and (ii)

$$k\alpha\delta x\delta y\left[\frac{\left(\frac{\partial u}{\partial x}\right)_{x+\delta x} - \left(\frac{\partial u}{\partial x}\right)_x}{\delta x} + \frac{\left(\frac{\partial u}{\partial y}\right)_{y+\delta y} - \left(\frac{\partial u}{\partial y}\right)_y}{\delta y}\right] = \rho\delta x\delta y\alpha s\frac{\partial u}{\partial t}$$

Dividing both sides by α δx δy and taking limits as δx →0, δy → 0, we get

$$k\left(\frac{\partial^2 u}{\partial x^2}+\frac{\partial^2 u}{\partial y^2}\right)=\rho s\frac{\partial u}{\partial t}$$

$$\text{i.e. } \frac{\partial u}{\partial t}=c^2\left(\frac{\partial^2 u}{\partial x^2}+\frac{\partial^2 u}{\partial y^2}\right) \qquad \text{(iii)}$$

where $c^2=\dfrac{k}{\rho s}$ is the diffusivity

Equation (iii) gives the temperature distribution of the plate in the transient state.

Cor. In the steady state, u is independent of t, so that $\dfrac{\partial u}{\partial t}=0$ and the above equation reduces to,

$$\frac{\partial^2 u}{\partial x^2}+\frac{\partial^2 u}{\partial y^2}=0$$

and this is called Laplace's equation in two dimensions.

Solution of Laplace Equation in Two Dimensions

Laplace equation in two dimensions is given by

$$\frac{\partial^2 u}{\partial x^2}+\frac{\partial^2 u}{\partial y^2}=0 \qquad \text{(i)}$$

Let u = XY be a solution of (i)

where X is a function of x alone and Y is the function of y alone.

Then $\dfrac{\partial^2 u}{\partial x^2}=X''Y$ and $\dfrac{\partial^2 u}{\partial y^2}=XY''$

Substituting these values in (i), we get

$$X''Y+XY''=0$$

$$\text{or } \frac{X''}{X}=-\frac{Y''}{Y} \qquad \text{(ii)}$$

Now in equation (ii) variables are separable since x and y are independent variable, this equation can hold only when both sides reduce to a constant, say k.

$$\text{i.e. } \frac{X''}{X}=-\frac{Y''}{Y}=k$$

$$\Rightarrow \frac{d^2X}{dx^2}-KX=0 \quad \text{and} \quad \frac{d^2Y}{dy^2}+KY=0 \qquad \text{(iii)}$$

Solving equation (iii), we get

(a) when k is positive say p^2, then we have

$X=c_1e^{px}+c_2e^{-px}$, $\quad Y=c_3\cos py+c_4\sin py$

(b) when k is negative say $-p^2$, then we have

$$X = c_5 \cos px + c_6 \sin px, \quad Y = c_7 e^{py} + c_8 e^{-py}$$

(c) when k = 0, then

$$X = c_9 x + C_{10}, \quad Y = C_{11} y + C_{12}$$

Thus the various possible solution of (i) are

$$u = (c_1 e^{px} + c_2 e^{-px})(c_3 \cos py + c_4 \sin py) \quad \text{(iv)}$$

$$u = (c_5 \cos py + c_6 \sin py)(c_7 e^{py} + c_8 e^{-py}) \quad \text{(v)}$$

$$u = (c_9 x + c_{10})(c_{11} y + c_{12}) \quad \text{(vi)}$$

of these we take that solution which is consistent with the given boundary conditions, i.e., physical nature of the problem.

Example 18: Solve the Laplace equation $\frac{\partial^2 u}{\partial x^2} + \frac{\partial^2 u}{\partial y^2} = 0$ subjected to the conditions

$$u(o, y) = u(l, y) = u(x, o) = 0 \text{ and } u(x, a) = \sin \frac{n\pi x}{l} \qquad \text{(U.P.T.U 2004)}$$

Solution. The three possible solution of

$$\frac{\partial^2 u}{\partial x^2} + \frac{\partial^2 u}{\partial y^2} = 0 \quad \text{(i)}$$

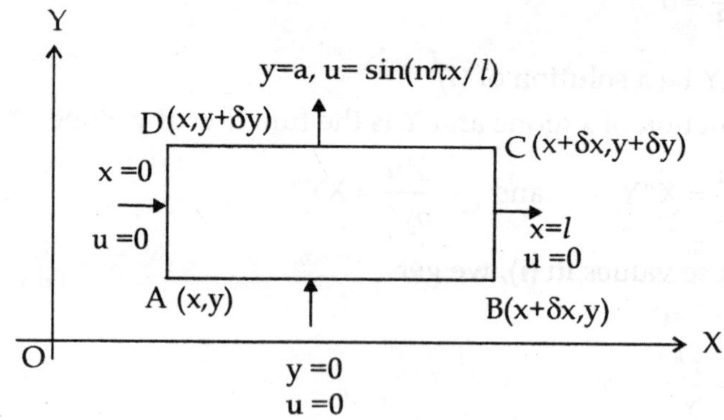

are

$$u = (c_1 e^{px} + c_2 e^{-px})(c_3 \cos py + c_4 \sin py) \quad \text{(ii)}$$

$$u = (c_5 \cos px + c_6 \sin px)(c_7 e^{py} + c_8 e^{-py}) \quad \text{(iii)}$$

$$u = (c_9 x + c_{10})(c_{11} y + c_{12}) \quad \text{(iv)}$$

we have to solve (i) satisfying the following boundary conditions

$$u(0, y) = 0 \quad \text{(v)}$$

$$u(l, y) = 0 \quad \text{(vi)}$$

$u(x, 0) = 0$ (vii)

$u(x, a) = \sin n\pi x/l$ (viii)

using (v) and (vi) in (ii), we get

$c_1 + c_2 = 0$, and $c_1 e^{pl} + c_2 e^{-pl} = 0$

Solving these equations, we get $c_1 = c_2 = 0$, which lead to trivial solution. similarly we get a trivial solution by using (v) and (vi) in (iv). Hence the suitable for the present problem is solution (iii), using (v) in (iii),

we have $c_5 (c_7 e^{py} + c_8 e^{-py}) = 0$ i.e. $c_5 = 0$

$\therefore$ (iii) becomes $u = c_6 \sin px\,(c_7 e^{py} + c_8 e^{-py})$ (ix)

using (vi), we have $c_6 \sin pl\,(c_7 e^{py} + c_8 e^{-py}) = 0$

$\therefore$ either $c_6 = 0$ or $\sin pl = 0$

If we take $c_6 = 0$, we get trivial solution

Thus $\sin pl = 0 \Rightarrow pl = n\pi$

$$\Rightarrow p = \frac{n\pi}{l}, \text{ where } n = 0,1,2\ldots\ldots\ldots\ldots$$

$$\therefore \text{(ix) becomes } u = c_6 \sin\left(\frac{n\pi x}{l}\right)\left(c_7 e^{n\pi y/l} + c_8 e^{-n\pi y/l}\right) \quad \text{(x)}$$

$$\text{Using (vii), we have } o = c_6 \sin\left(\frac{n\pi x}{1}\right).(c_7 + c_8) \text{ i.e. } c_8 = -c_7$$

Thus the solution suitable for this problem is

$$u(x,y) = b_n \sin\frac{n\pi x}{l}\left(e^{n\pi y/l} - e^{-n\pi y/l}\right) \text{ where } b_n = c_6 c_7$$

Now using the condition (viii), we have

$$u(x,a) = \sin\frac{n\pi x}{l} = b_n \sin\frac{n\pi x}{l}\left(e^{n\pi a/l} - e^{-n\pi a/l}\right)$$

$$\text{we get } b_n = \frac{1}{\left(e^{n\pi a/l} - e^{-n\pi a/l}\right)}$$

Hence the required solution is

$$u(x,y) = \frac{e^{n\pi y/l} - e^{-n\pi y/l}}{e^{n\pi a/l} - e^{-n\pi a/l}} \sin\frac{n\pi x}{l}$$

$$\Rightarrow u(x,y) = \frac{\sinh(n\pi y/l)}{\sinh(n\pi a/l)} \sin\frac{n\pi x}{l}$$

Example 19: A thin rectangular plate whose surface is impervious to heat flow, has at t =0 an arbitrary distribute of temperature f(x, y), if four edges x =0, x =a, y

=0, y =b are kept at zero temperature. determine the temperature at a point of the plate as t increases.

(U.P.T.U. 2002)

Solution The two dimensional heat equation is

$$\frac{\partial u}{\partial t} = c^2 \left(\frac{\partial^2 u}{\partial x^2} + \frac{\partial^2 u}{\partial y^2} \right) \qquad (1)$$

where $c^2 = \dfrac{k}{\sigma \rho}$

The initial temperature of the plate is f(x, y) and the temperature of the four edges of the plate are kept at zero temperature.

Now the function u(x, y, t) is required to satisfy (i) and the boundary and initial conditions given below:

The boundary conditions are

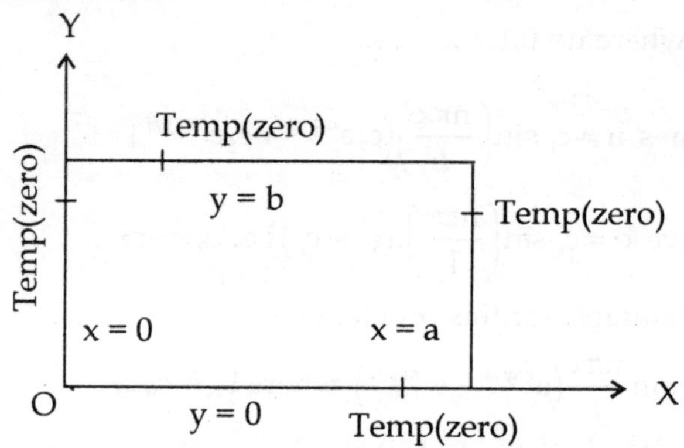

u(o, y, t) =0 (i)

u(a, y, t) =0 (ii)

u(x, 0, t) =0 (iii)

u(x, b, t) =0 (iv)

and the initial condition is

u(x, y, 0) = f(x, y) (2)

Let the solution of the heat equation (1) be of the form

u(x, y, t) = X(x) Y(y) T(t) = XYT(say) (3)

where X is a function of x only, Y is that of y only and T is that of t only.

using (3) in (1) we get

$$\frac{1}{c^2 T}\frac{dT}{dt} = \frac{1}{X}\frac{d^2X}{dx^2} + \frac{1}{Y}\frac{d^2Y}{dy^2}$$

Hence in order that (3) may satisfy (1) we have of these three possibilities :

(A) $\frac{1}{X}\frac{d^2X}{dx^2}=0,\quad \frac{1}{Y}\frac{d^2Y}{dy^2}=0,\quad \frac{1}{c^2T}\frac{dT}{dt}=0$

(B) $\frac{1}{X}\frac{d^2X}{dx^2}=p_1^2,\quad \frac{1}{Y}\frac{d^2Y}{dy^2}=p_2^2,\quad \frac{1}{c^2T}\frac{dT}{dt}=p^2$

(C) $\frac{1}{X}\frac{d^2X}{dx^2}=-p_1^2,\quad \frac{1}{Y}\frac{d^2Y}{dy^2}=-p_2^2,\quad \frac{1}{c^2T}\frac{dT}{dt}=-p^2$

where $p^2=p_1^2+p_2^2$

It can be easily observed that differential equation (C) only gives the solution for this situation, and the general of the differential equation in this case is

$$X = A_1\cos p_1x + B_1\sin p_1x$$

$$Y = A_2\cos p_2y + B_2\sin p_2y \text{ and } T = A_3e^{-c^2p^2t}$$

$$\Rightarrow \quad u(x,y,t) = (A_1\cos p_1x + B_1\sin p_1x)\left(A_2'\cos p_2y + B_2'\sin p_2y\right)e^{-c^2p^2t} \qquad (4)$$

where $A_2' = A_2A_3,\ B_2' = B_2B_3$

Under boundary condition (i) we get

$$u(o,y,t) = A_1\left(A_2'\cos p_2y + B_2'\sin p_2y\right)e^{-c^2p^2t} = 0$$

$$\Rightarrow \qquad A_1 = 0$$

Again using (ii)

$$u(a,y,t) = B_1\sin p_1a\left(A_2'\cos p_2y + B_2'\sin p_2y\right)e^{-c^2p^2t} = 0$$

$\sin p_1a = 0 \Rightarrow p_1a = m\pi \Rightarrow p_1 = \frac{m\pi}{a}$ where m = 1, 2, 3.........

similarly, making use of (iii) and (iv), we obtain

$$A_2' = 0 \quad \text{and} \quad p_2 = \frac{n\pi}{b} \quad (n = 1, 2, 3 \ldots\ldots\ldots\ldots)$$

Thus we have

$$u_{mn}(x,y,t) = A_{mn}e^{-c^2p^2_{mn}t}\sin\frac{m\pi x}{a}\sin\frac{n\pi y}{b}$$

$$\text{where } p^2 = p_{mn}^2 = \pi^2\left(\frac{m^2}{a^2}+\frac{n^2}{b^2}\right)$$

$$\Rightarrow u(x,y,t) = \sum_{m=1}^{\infty}\sum_{n=1}^{\infty}A_{mn}e^{-c^2p_{mn}^2t}\sin\frac{m\pi x}{a}\sin\frac{n\pi y}{b} \qquad (5)$$

This solution satisfies the boundary conditions.

Now to derive the solution which satisfies the initial condition also, we proceed as follows:

$$\Rightarrow u(x,y,0)=\sum_{m=1}^{\infty}\sum_{n=1}^{\infty}A_{mn}e^{-c^2p_{mn}^2t}\sin\frac{m\pi x}{a}\sin\frac{n\pi y}{b}=f(x,y)$$

Hence, the L.H.S is the double Fourier sine series of f(x,y)

$$\Rightarrow A_{mn}=\frac{2}{a}\cdot\frac{2}{b}\int_{x=0}^{a}\int_{y=0}^{b}f(x,y)\sin\frac{m\pi x}{a}\sin\frac{n\pi y}{b}dxdy \qquad (6)$$

Hence the required solution of heat equation (1) is given by (5) with coefficients given by (6)

Example 20 Solve the P.D.E by separation of variables method $u_{xx} = u_y + 2u$, $u(0, y) = 0$

$$\frac{\partial}{\partial x}u(0,y)=1+e^{-3y}$$

(U.P.T.U. 2009)

Solution Hint. $\frac{X''}{X}=\frac{Y'+2Y}{Y}=K(\text{say})$

$$\Rightarrow X=c_1e^{\sqrt{k}x}+c_2e^{-\sqrt{k}x},Y=c_3e^{(k-2)y}$$

$u = XY$

Therefore $u(x,y)=\frac{1}{\sqrt{2}}.\sinh\sqrt{2}x+\sin.e^{-3y}$

EXERCISE

Solve the following P.D.E. by the method of separation of variables.

1. $3\frac{\partial u}{\partial x}+2\frac{\partial u}{\partial y}=0$ given $u(x, 0) = 4e^{-x}$

Ans. $u(x,y)=4e^{\frac{1}{2}(3y-2x)}$

2. $\frac{\partial^2 z}{\partial x^2}+4\frac{\partial^2 z}{\partial y^2}=0$

Ans. $z=\left(c_1e^{\sqrt{k}x}+c_2e^{-\sqrt{k}x}\right)\left(c_3\sin\frac{1}{2}\sqrt{k}y+c_4\cos\frac{1}{2}\sqrt{k}y\right)$

3. $\frac{\partial u}{\partial x}+2\frac{\partial u}{\partial t}+u,$, $u(x, 0) = 6e^{-3x}$

(U.P.T.U. 2006)

Ans. $u(x, t) = 6e^{(-3x + 2t)}$

4. $2\frac{\partial u}{\partial x} - 3\frac{\partial u}{\partial y} = 0$, u(x, 0) = $5e^{3x}$

Ans. u(x, y) = $5e^{3x + 2y}$

5. $3\frac{\partial u}{\partial x} + 2\frac{\partial u}{\partial y} = 0$

(U.P.T.U. 2009)

Ans. $u(x,y) = ce^{\frac{1}{6}k(2x-3y)}$

6. The vibrations of an elastic string is governed by the P.D.E. $\frac{\partial^2 u}{\partial t^2} = \frac{\partial^2 u}{\partial x^2}$. The length of the string is π and the ends are fixed. The initial velocity is zero and the initial deflection is u(x, 0) = 2 (sin x + sin 3x). Find the deflection u(x, t) of the vibrating string for t > 0.

Ans. u(x, t) = 4 cos x cos 2t sin 2x

7. find the displacement of a string stretched between the fixed points (0, 0) and (1, 0) and released from rest from position a sin πx + cos 2πx.

Ans. u(x, t) = a sin πx cos πct + bsin 2πx cos 2πct

8. Solve the wave equation $\frac{\partial^2 u}{\partial t^2} = c^2\frac{\partial^2 u}{\partial x^2}$ subjected to the conditions

(i) $u(0, t) = u(l, t) = 0 \; t \geq 0$

(ii) $u(x,0) = \begin{cases} \frac{2x}{l}, & 0 \leq x \leq l/2 \\ \frac{2}{l}(l-x), & \frac{l}{2} \leq x \leq l \end{cases}$ (I.A.S 2006)

(iii) $\left(\frac{\partial u}{\partial t}\right)_{t=0} = 0, \quad 0 \leq x \leq l$

Ans. $u(x,t) = \frac{8}{\pi^2}\sum_{n=1}^{\infty}\frac{1}{n^2}\sin\frac{n\pi}{2}\sin\frac{n\pi x}{l}\cos\frac{n\pi ct}{l}$

9. A tant string of length 2*l* is fastended at both ends. The mid point of the string is taken to a height b and then released from the rest in that position. Find the displacement of the string.

Ans. $y(x,t) = \frac{8b}{\pi^2}\sum_{n=1}^{\infty}\frac{1}{(2n-1)^2}\sin(2n-1)\frac{\pi}{2}\sin\frac{(2n-1)\pi x}{2l}\cos\frac{(2n-1)\pi ct}{2l}$

10. find the temperature u(x, t) in a homogeneous bar of heat conduction material of length l cm with its ends kept at zero temperature and initial temperature is $dx(l-x)/l^2$

Ans. $u(x,t)=\frac{8d}{\pi^3}\sum_{n=1}^{\infty}\frac{1}{(2m-1)^3}\sin\frac{(2n-1)\pi x}{2l}e^{(2m-1)^2\pi^2c^2t/l^2}$

11. Solve the following Laplace equation

$$\frac{\partial^2 u}{\partial x^2}+\frac{\partial^2 u}{\partial y^2}=0$$

in a rectangle with u(0, y) =0, u(a, y) = 0, u(x, b) =0 and u(x, 0) = f(x) along x axis.

(U.P.T.U. 2008)

Ans. $u(x,y)=\sum_{n=1}^{\infty}b_n\sin\frac{n\pi x}{a}\left\{e^{\frac{n\pi y}{a}}-e^{-\frac{n\pi}{a}(y-2b)}\right\}$

where $b_n=\frac{2}{a\left(1-e^{\frac{2n\pi b}{a}}\right)}\int_0^a f(x)\sin\frac{n\pi x}{a}dx$

12. Find the temperature in a bar of length 2 whose end are kept at zero and lateral surface is insulated if the initial temperature is $\sin\frac{nx}{2}+3\sin\frac{5\pi x}{2}$

(U.P.T.U. 2009)

Solution

Hint. Heat equation $\frac{\partial u}{\partial t}=c^2\frac{\partial^2 u}{\partial x^2}$

$u(x,t) = (c_1\cos px + c_2\sin px)(c_3 e^{-c^2p^2t})$

$u(0, t) = 0 \Rightarrow c_1 = 0$

$u(l, t) = 0 \Rightarrow \quad p=\frac{n\pi}{l}$

$$u(x,t)=\sum_{n=1}^{\infty}b_n\sin\frac{n\pi x}{l}.e^{\frac{-c^2n^2\pi^2t}{l^2}}$$

where $b_n=\int_0^2\left(\sin\frac{n\pi}{2}+3\sin\frac{5\pi x}{2}\right)\sin\frac{n\pi x}{2}dx$

Choose the correct answer from the following:

1. One dimensional wave equation $\frac{\partial^2 z}{\partial t^2}=c^2\frac{\partial^2 z}{\partial x^2}$ is:

(a) elliptic (b) parabolic (c) hyperbolic (d) circular.

Ans. (c)

2. One dimensional heat flow equation $\frac{\partial u}{\partial t} = c^2 \frac{\partial^2 u}{\partial x^2}$ is:

(a) circular (b) hyperbolic (c) parabolic (d) elliptic.

Ans. (c)

3. The two dimensional heat flow equation in steady state $\frac{\partial^2 u}{\partial x^2} + \frac{\partial^2 u}{\partial y^2} = 0$ is:

(a) elliptic (b) circular (c) parabolic (d) hyperbolic.

Ans. (a)

4. The differential equation $4 \frac{\partial^2 u}{\partial x^2} + 4 \frac{\partial^2 u}{\partial x \partial y} + \frac{\partial^2 u}{\partial y^2} = 0$ is:

(a) parabolic (b) elliptic (c) hyperbolic (d) circular

Ans. (a)

5. The general solution of $\frac{\partial^2 u}{\partial x\, \partial y} = 0$ is:

(a) $u = f_1(x + y) + f_2(y)$ (b) $u = f_1(x) + f_2(y)$
(c) $u = f(xy)$ (d) $u = f_1(xy) + f_2(y)$.

Ans. (a)

6. The partial differential equation $\frac{\partial^2 u}{\partial x^2} = \frac{\partial^2 u}{\partial y^2}$ is:

(a) hyperbolic (b) parabolic (c) elliptic (d) circular.

Ans. (a)

7. The partial differential equation $9 \frac{\partial^2 u}{\partial x^2} - 6 \frac{\partial^2 u}{\partial x \partial y} + \frac{\partial^2 u}{\partial y^2} = 0$ is:

(a) hyperbolic (b) elliptic (c) circular (d) parabolic

Ans. (d)

8. The two dimensional heat equation in the transient state is:

(a) $\frac{\partial u}{\partial t} = c^2 \left(\frac{\partial^2 u}{\partial x^2} + \frac{\partial^2 u}{\partial y^2} \right)$ (b) $\frac{\partial^2 u}{\partial t^2} = c^2 \left(\frac{\partial^2 u}{\partial x^2} + \frac{\partial^2 u}{\partial y^2} \right)$

(c) $\frac{\partial^2 u}{\partial x^2} + \frac{\partial^2 u}{\partial y^2} = c$ (d) $\frac{\partial u}{\partial t} = c\left(\frac{\partial^2 u}{\partial x^2} + \frac{\partial u}{\partial y}\right)$

Ans. (a)

9. The two dimensional wave equation $\frac{\partial^2 u}{\partial t^2} = c^2\left(\frac{\partial^2 u}{\partial x^2} + \frac{\partial u}{\partial y^2}\right)$ is:

(a) circular (b) elliptic (c) parabolic (d) hyperbolic.

Ans. (d)

10. The Laplace's equation in polar coordinates is:

(a) $\frac{\partial^2 u}{\partial r^2} + \frac{\partial u}{\partial r} + \frac{1}{r^2}\frac{\partial^2 u}{\partial \theta^2} = 0$ (b) $\frac{\partial^2 u}{\partial r^2} - \frac{1}{r}\frac{\partial u}{\partial r} + \frac{1}{r^2}\frac{\partial^2 u}{\partial \theta^2} = 0$

(c) $\frac{\partial^2 u}{\partial r^2} + \frac{1}{r}\frac{\partial u}{\partial r} + \frac{1}{r^2}\frac{\partial^2 u}{\partial \theta^2} = 0$ (d) $\frac{\partial^2 u}{\partial r^2} - \frac{\partial u}{\partial r} + \frac{1}{r}\frac{\partial^2 u}{\partial \theta^2} = 0.$

Ans. (c)

11. The radio equations are:

(a) $\frac{\partial^2 v}{\partial x^2} = LC\frac{\partial^2 v}{\partial t^2}$ and $\frac{\partial^2 t}{\partial x^2} = LC\frac{\partial i}{\partial t}$ (b) $\frac{\partial^2 v}{\partial x^2} = LC\frac{\partial^2 v}{\partial t^2}$ and $\frac{\partial i}{\partial x} = LC\frac{\partial v}{\partial t}$

(c) $\frac{\partial^2 v}{\partial x^2} = LC\frac{\partial^2 v}{\partial t^2}$ and $\frac{\partial^2 i}{\partial x^2} = LC\frac{\partial^2 i}{\partial t^2}$ (d) none of these.

Ans. (c)

12. The telegraph equations are:

(a) $\frac{\partial^2 v}{\partial x^2} = RC\frac{\partial v}{\partial t}$ and $\frac{\partial^2 i}{\partial x^2} = RC\frac{\partial i}{\partial t}$ (b) $\frac{\partial v}{\partial x} = RC\frac{\partial v}{\partial t}$ and $\frac{\partial i}{\partial x} = RC\frac{\partial i}{\partial t}$

(c) $\frac{\partial^2 v}{\partial x^2} = RC\frac{\partial^2 v}{\partial t^2}$ and $\frac{\partial^2 i}{\partial x^2} = RC\frac{\partial^2 i}{\partial t^2}$ (d) none of these.

Ans. (a)

13. The equations for submarine cable are:

(a) $\frac{\partial v}{\partial x} = RGv$ and $\frac{\partial i}{\partial x} = RGi$ (b) $\frac{\partial^2 v}{\partial x^2} = RGv$ and $\frac{\partial^2 i}{\partial x^2} = RGi$

(c) $\frac{\partial^2 v}{\partial x^2} = RG$ and $\frac{\partial^2 i}{\partial x^2} = RG$ (d) $\frac{\partial^2 v}{\partial x^2} = Rv$ and $\frac{\partial^2 i}{\partial x^2} = Ri.$

Ans. (a)

Fill in the blanks in the following problems:

1. The partial differential equation $\frac{\partial u}{\partial t} = c^2 \frac{\partial^2 u}{\partial x^2}$ is of......type.

Ans. Parabolic

2. When a vibrating string has an initial velocity, its initial conditions are......

Ans. $\left(\frac{\partial y}{\partial t}\right)_{t=0} = v$

3. The solution of $\frac{\partial^2 u}{\partial x \partial y} = 0$ is of the form.

Ans. $u = f_1(y) + f_2(x)$

4. The equation $\frac{\partial^2 u}{\partial t^2} = c^2 \frac{\partial^2 u}{\partial x^2}$ is calssified as.......

Ans. Elliptic partial differential equation

5. In two dimensional heat flow, the temperature along he normal to the *xy*-plane is.....

Ans. Zero

6. D' Alembert's solution of the wave equation $\frac{\partial^2 y}{\partial t^2} = c^2 \frac{\partial^2 y}{\partial x^2}$ is......

Ans. $y(x, t) = \frac{1}{2}[f(x + ct) + f(x - ct)]$

Select True 'or' False answers in the following:

7. The solution of $\frac{\partial^2 z}{\partial x^2} = \sin(xy)$ is $-y^2 \sin(xy) + x f_1(x) + f_2(y)$.

(False)

8. The general solution of the equation $\frac{\partial^2 u}{\partial x^2} - \frac{\partial^2 u}{\partial t^2} = 0$ is ($C_1 \cos px + C_2 \sin px$) ($C_3 \cos pt + C_4 \sin pt$).

(True)

9. The two dimensional heat flow in transient state is $\frac{\partial u}{\partial t} = C^2 \left(\frac{\partial^2 u}{\partial x^2} + \frac{\partial^2 u}{\partial y^2}\right)$.

(True)

10. The telegraph equations are $\frac{\partial^2 v}{\partial x^2} = RC \frac{\partial v}{\partial t}$ and $\frac{\partial^2 i}{\partial x^2} = RC \frac{\partial i}{\partial t}$

(True)

11. The radio equations are $\frac{\partial v}{\partial x^2} = LC \frac{\partial^2 v}{\partial t^2}$ and $\frac{\partial i}{\partial x} = \text{LC} \frac{\partial^2 i}{\partial t^2}$

(False)

12. The Laplace's equation in three dimensions is $\frac{\partial^2 u}{\partial x^2} + \frac{\partial^2 u}{\partial y^2} + \frac{\partial u}{\partial z} = 0$

(False)

UNSOLVED QUESTION PAPERS

B. Tech.
(SEM. II) EXAMINATION. 2008-09
MATHEMATICS-II

Time : 3 Hours **Total Marks : 100**

SECTION – A **2 x 10 = 20**

Note: Attempt all questions. All questions carry equal marks

1. Fill up the appropriate answers in the space provided

(a) The differential equation

$$\left(\frac{d^3y}{dx^3}\right)^4 - 6x^2\left(\frac{dy}{dx}\right)^2 + e^x = \sin xy \text{ is of \_\_\_\_\_\_\_\_\_ order and \_\_\_\_\_\_\_\_\_}$$

degree.

(b) A particular solution of the differential equation $\frac{dy}{dt} = \frac{1}{2}(y^2 - 1)$ that satisfied the initial condition y(0) = 2 is __________

(c) The general solution of the differential equation $\frac{d^5y}{dx^5} - \frac{d^3y}{dx^3} = 0$ is given by __________.

Pick up the Correct Answer from the following

(d) The Rodrigues formula for Legendre Polynomial $P_n(x)$ is given by

(i) $P_n(x) = \frac{1}{|\underline{n}\, 2^n} \frac{d^n}{dx^n}(x^2 - 1)^n$

(ii) $P_n(x) = \frac{|\underline{n}}{2^n} \frac{dn}{dx^n}(x^2 - 1)^{n\ 1}$

(iii) $P_n(x) = \frac{|\underline{n}}{2^{n-1}} \frac{d^n}{dx^n}(x^2 - 1)^{n-1}$

(iv) $P_n(x) = \frac{1}{|\underline{n}\, 2^n}(x^2 - 1)^n$

(e) The Laplace transform of the function

$$f(t) = \begin{cases} 1, 0 \le t < 2 \\ -1, 2 \le t < 4, \end{cases} \quad f(t+4) = f(t) \text{ is given as}$$

(i) $\dfrac{1-e^{-2s}}{s(1+e^{-2s})}$

(ii) $\dfrac{1+e^{-2s}}{s(1+e^{-2s})}$

(iii) 0

(iv) $\dfrac{s+1}{s-1}$

(f) The inverse Laplace transform of $\log\left(\dfrac{s+1}{s-1}\right)$ is given by

(i) $\dfrac{2}{t}\cosh t$ (ii) $\dfrac{2}{t}\sin ht$

(iii) $2t\cos t$ (iv) $2t\sin t$

(g) Indicate True/False for the statements made therein-

(i) A function f(x) is even if $f(-x) = -f(x)$ (True or False)

(ii) A function f(x) is odd if $f(-x) = f(x)$ (True or False)

(iii) Most functions are neither even nor odd (True or False)

(iv) A function f(x) can always be expressed as an arithmetic mean of an even and odd function as $f(x) = \frac{1}{2}[f(x) + f(-x)] + \frac{1}{2}[f(x) - f(-x)]$

(True or False)

(h) (i) With usual symbols, the PDE $u_{xx} + u^2 u_{yy} = f(xy)$ is non-linear in 'u' and is of second order. (True or False)

(ii) The small transverse vibrations of a string are governed by one dimensional heat equation $y_t = a^2 y_{xx}$ (True or False)

(iii) Two dimensional steady state heat flow is given by Laplace's equation $u_t = a^2(u_{xx} + u_{tt})$.

(iv) The PDE of all sphere whose centre lie on z-axis and given by equations $x^2 + y^2 + (z-a)^2 = b^2$, a and b being constants, are governed by $x z_y - y z_x = 0$. (True or False)

(i) Applying the method of separation of variables techniques, the solution to the P.D.E. $3u_x + 2u_y = 0$ is ________ where $u_x = \dfrac{\partial u}{\partial x}$, $u_y = \dfrac{\partial u}{\partial y}$.

Match the column for the items of the Left side to that of right side:

(j) A second order P.D.E. in the function 'u' of two independent variables x, y given with usual symbols $Au_{xx} + Bu_{xy} + Cu_{yy} + F(u) = 0$, then

(i)	Hyperbolic	(a)	$B^2 - 4AC > 0$
(ii)	Parabolic	(b)	$B^2 - 4AC < 0$
(iii)	Elliptic	(c)	$B^2 - 4AC < 0$
(iv)	Not classified	(d)	$A = B = C = 0$

SECTION - B

Note: Attempt any three questions from this section. All questions carry equal marks.

3 x 10 =30

2. (a) Apply the method of variation of parameters to solve the ordinary differential equations

$$\frac{d^2y}{dx^2} + y = \tan x, 0 < x < \pi/2$$

(b) Show that Bessel's function $J_n(x)$ is an even function when n is even and is odd function when n is odd. Express $J_6(x)$ in terms of $J_0(x)$ and $J_1(x)$.

(c) Use convolution theorem to find the inverse of the function $\frac{1}{(s^2 + a^2)^2}$.

(d) Obtain the Fourier series of $f(x) = \left(\frac{\pi - x}{2}\right)$ in the interval $(0, 2\pi)$ and hence deduce $\frac{\pi}{4} = 1 - \frac{1}{3} + \frac{1}{5} - \frac{1}{7} + \ldots\ldots\ldots\ldots\ldots$

(e) A string of Length 'L' is stretched and fastened to two fixed points. Find the solution of the wave equation $y_{tt} = a^2 y_{xx}$ when initial displacement is $y(x, 0) = f(x) = b \sin\left(\frac{\pi x}{L}\right)$, where symbols have usual meaning.

SECTION - C

Note: Attempt all questions from this section, selecting any two parts from each question. All questions carry equal marks. [5 x 2] x 5 =50

3. (a) The equations of motion of a particle are given by $\frac{dx}{dt} + wy = 0$, $\frac{dy}{dt} - wx = 0$. Find the path of the particle and show that it is a circle.

(b) Integrate the differential equation $\frac{d^2x}{dt^2} + 2x\frac{dx}{dt} + w^2x = a \cos pt$ and give the physical interpretation of the complete solution. Also deduce that the solution takes the form

$$x = \frac{\alpha \cos(pt - \theta)}{\sqrt{(w^2 - p^2)^2 + 4k^2 p^2}} \left(\tan\theta = \frac{2kp}{w^2 - p^2}\right) \text{ As } t \to \infty$$

(c) Find the complete solution of the differential equation $\frac{d^2y}{dx^2} - 2\frac{dy}{dx} + y = x e^x \cos x.$

4. (a) For the Bessel's function, prove that $J_{\frac{1}{2}}(x) = \left(\sqrt{\frac{2}{\pi x}}\right) . \sin x.$

(b) Estimate the recurrence relation for Legendre's polynomials-

$x P_n'(x) = n P_n'(x) + P_{n-1}'(x)$

(c) Express the polynomial $f(x) = 4x^3 - 2x^2 - 3x + 8$ in terms of Legendre polynomials.

5. (a) Using Laplace transformation, show that

$$\int_0^{\infty} \frac{\sin t}{t} dt = \frac{\pi}{2}.$$

(b) A particle moves in a line so that its displacement x from a fixed point 0 at any time *t*, is given by $\frac{d^2x}{dt^2} + 4\frac{dx}{dt} + 5x = 80 \sin 5t$. Using Laplace transfrom, find its displacement at any time *t* if initially particle is at rest at $x = 0$.

(c) Find the Laplace transform of the function $f(t) = \begin{cases} t - 1, & 1 < t < 2 \\ 3 - t, & 2 < t < 3 \end{cases}$

6. (a) Find the half period sine series for f(x) given in the range (0, *l*) by the graph OPQ as shown in figure.

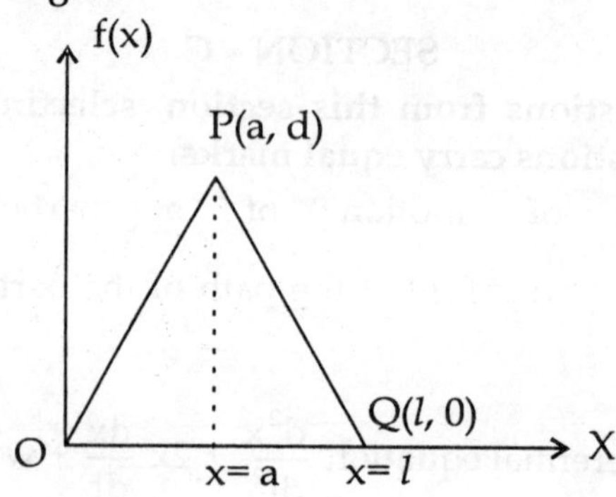

(b) Find the Fourier series expansion for $f(x) = x + \frac{x^2}{4}, -\pi \leq x \leq \pi.$

(c) Solve the P.D.E. $\frac{\partial^2 z}{\partial x^2} - 2\frac{\partial^2 z}{\partial x \, \partial y} + \frac{\partial^2 z}{\partial y^2} = \sin x.$

7. (a) Find the temperature in a bar of length 2 whose end are kept at zero and lateral surface is insulated if the initial temperature is $\sin\frac{nx}{2} + 3\sin 5\pi\frac{x}{2}.$

(b) Apply the method of separation of variables to solve $\frac{\partial^2 z}{\partial x^2} - 2\frac{\partial z}{\partial x} + \frac{\partial z}{\partial y} = 0.$

(c) Solve the P.D.E. by separation of variables method, $u_{xx} = u_y + 2u$, $u(0, y) = 0$, $\frac{\partial}{\partial x}u(0, y) = 1 + e^{-3y}$

UNSOLVED QUESTION PAPERS
B. Tech. Second Semester Examination
MATHEMATICS II
2007-08

Time : 3 Hours

Note : (1) Attempt all questions.
(2) All questions carry equal marks
(3) In case of numerical problems assume data wherever not provided.
(4) Be precise in your answer.

1. Attempt any four parts of the following: (5 x 4 = 20)

(a) Solve $\left(y + \sqrt{x^2 + y^2}\right) dx - xdy = 0, y(1) = 0.$

(b) Solve the following differential equation:
$(\cos x - x \cos y)\, dy - (\sin y + y \sin x)\, dx = 0$

(c) Find the solution of following differential equation:
$(D^2 - 4D - 5)y = e^{2x} + 3 \cos(4x + 3)$

where $D = \dfrac{d}{dx}$.

(d) Solve the following simultaneous differential equations:

$$\frac{dx}{dt} + 5x - 2y = t$$

$$\frac{dy}{dt} + 2x + y = 0$$

Also show that x = y = 0 when $t = 0$.

(e) Solve by method of variation of parameters

$$\frac{d^2y}{dx^2} + 2\frac{dy}{dx} + y = e^{-x} \log x.$$

(f) An inductance (L) of 2.0 H and a resistance (R) of 20 Ω are connected in series with an e.m.f. E volt. If the current (i) is zero, when $t = 0$, find the current (i) at the end of 0.01 second if $E = 100\ V$, using the following differential equation:

$$L\frac{di}{dt} + iR = E.$$

2. Attempt any four parts of the following: (5 x 4 = 20)

(a) Using Laplace transform, evaluate

$$\int_0^\infty e^{-t} \frac{\sin^2 t}{t} dt.$$

(b) State second shifting theorem for Laplace transform and hence find the Laplace transform of the following function

$$F(t) = \begin{cases} e^{t-a}, & t > a \\ 0, & t < a \end{cases}$$

(c) Using convolution theorem, find the inverse Laplace transform of the following $\dfrac{s}{(s^2+a^2)^3}$.

(d) Solve the following simultaneous differential equations by Laplace transform $\dfrac{dx}{dt} + 4\dfrac{dy}{dt} - y = 0$

$$\frac{dx}{dt} + 2y = e^{-t}$$

with condition $x(0) = y(0) = 0$.

(e) Solve the following differential equation using Laplace transform $\dfrac{d^2y}{dx^2} + 2\dfrac{dy}{dx} + 5y = e^{-x}\sin x$

where $y(0) = 0$. $\left(\dfrac{dy}{dx}\right)_{x=0} = 1$.

(f) Using unit step function, find the Laplace transform of:

(i) $(t-1)^2 . u(t-1)$

(ii) $\sin t.\ u(t-\pi)$.

3. Attempt any two parts of the following: (10 x 2 = 20)

(a) Solve the following differential equation in series:

$$2x^2\frac{d^2y}{dx^2} - x\frac{dy}{dx} + (x-5)y = 0.$$

(b) Show that

(1) $\displaystyle\int_{-1}^{+1} [P_n(x)]^2\, dx = \frac{2}{n+1}$

(2) $P_{2n}(0) = \dfrac{(-1)^n (2n)!}{2^{2n} (n!)^2}$

where $P_n(x)$ is the Legendre polynomial of degree *n*.

(c) Show that

(1) $J_{\frac{1}{2}}(x) = \left(1 - \dfrac{4}{x^2}\right)J_1(x) + \dfrac{2}{x}J_0(x)$

(2) $J_0^2 + 2\left(J_1^2 + J_2^2 + J_3^2 + ...\right) = 1$

where $J_n(x)$ is the Bessel's plynomial of degree n and dash denote the differentiation.

4. Attempt any two parts of the following: (10 x 2 = 20)

(a) Obtain a Fourier series to represent $f(x) = x^2$ in the interval $(0, 2\pi)$ and hence deduce that $\frac{1}{1^2} - \frac{1}{2^2} + \frac{1}{3^2} - \frac{1}{4^2} + ... = \frac{\pi^2}{12}$.

(b) Examine whether the function $f(x) = x \sin x$ is even or odd. Hence expand it in the form of Fourier series in the interval $(-\pi, \pi)$.

(c) Solve the following partial differential equations:

(1) $x(y^2 + z)\, p - y\,(x^2 + z)\, q = z\,(x^2 - y^2)$

(2) $\frac{\partial^2 z}{\partial x^2} - 2\frac{\partial^2 z}{\partial x \partial y} = \sin x. \cos 2y$

where $p = \frac{\partial z}{\partial x}$ and $q = \frac{\partial z}{\partial y}$.

5. Attem: any two parts of the following: (10 x 2 = 20)

(a) Solve the following equation by the method of separation of variables:

$\frac{\partial u}{\partial x} = 4\frac{\partial u}{\partial y}$ where $u(0, y) = 8e^{-3y}$.

(b) Solve the following Laplace equation:

$\frac{\partial^2 u}{\partial x^2} + \frac{\partial^2 u}{\partial y^2} = 0$

in a rectangle with $u(0, y) = 0$, $u(a, y) = 0$, $u(x, b) = 0$ and $u(x, 0) = f(x)$ along x-axis.

(c) Assuming the resistance of wire (R) and conductance to ground (C_r) are negligible, find the voltage $u(x, t)$ and current $i(x, t)$ in a transmission line of length l, t seconds after the ends are suddenly grounded. The initial conditions are $u(x, 0) = v_0 \sin\left(\frac{\pi x}{l}\right)$ and $i(x, 0) = i_0$.

2006-2007

Note: Attempt all the problems. The choice of problems is internal as indicated.

1. Attempt any four of the following: (5 x 4 = 20)

(a) A 4 kg object falls from rest at time t = 0 in a medium offering a resistance in kg numerically equal to twice its instantaneous velocity in m/sec. Find the velocity and distance travelled at any time t > 0, and also the limiting velocity.

(b) Solve $(3y - 2xy^3)\,dx + (4x - 3x^2y^2)\,dy = 0$

(c) Solve $2\frac{d^3y}{dx^3} - \frac{d^2y}{dx^2} - 4\frac{dy}{dx} - 2y = e^x$

(d) Solve $\frac{d^3y}{dx^3} + 4\frac{dy}{dx} = 4\cot 2x$.

(e) Solve $\frac{d^2x}{dt^2} + \frac{dy}{dt} + 3x = e^{-t}$.

$$\frac{d^2y}{dt} - 4\frac{dx}{dt} + 3y = \sin 2t$$

(f) Solve $x^2\frac{d^2y}{dx^2} + x\frac{dy}{dx} - \lambda^2 y = 0$

2. Attempt any four parts of the following: (5 x 4 = 20)

(a) Find the Laplace transform of

$$f(t) = \begin{cases} t^2 & 0 < t < 2 \\ t - 1 & 2 < t < 3 \\ 7 & t > 3 \end{cases}$$

(b) If L{f (t)} = F(s), show that

$L\left\{\frac{1}{t}f(t)\right\} = \int_s^{\infty} F(s)\,ds$. Hence, find the Laplace transform of the function

$$f(t) = \int_0^t \frac{\sin x}{x}\,dx$$

(c) Find the function whose Laplace transform is In $\left(1 + \frac{1}{s}\right)$

(d) State and prove convolution theorem for Laplace transform.

(e) Solve for y(t) the equation

$$y(t) = 1 + \int_0^t y(x)\cos(t - x)\,dx$$

(f) Solve, using Laplace transform method

$$y^n(t) + 4y'(t) + 4y(t) = 6e^{-t}$$
$$y(0) = -2,\ y'(0) = 8$$

3. Attmept any two parts of the following: (10 x 2 = 20)

(a) Expand $f(x) = K, 0 < x < 2$ in a half range

(1) Sine series

(2) Cosine series

(b) If

$$f(x) = \begin{cases} \pi x & ,0<x<1 \\ \pi(2-x), & 1<x<2 \end{cases}$$ using half range cosine series expansion, show that

$$\frac{1}{1} + \frac{1}{3^4} + \frac{1}{5^4} + \ldots\ldots\ldots\ldots = \frac{\pi^2}{96}$$ [out of syllabus]

(c) Solve $\frac{\partial^3 u}{\partial x^3} - \frac{\partial^3 u}{\partial x^2 \partial y} + 4\frac{\partial^3 u}{\partial y^3} = e^{x+2y}$

OR

(1) Find the Fourier transform of

$$f(x) = \begin{cases} x, & |x| < 1 \\ -x, & |x| > 1 \end{cases}$$

(2) Find z-transform of $\cos \alpha k$ where $k \geq 0$

Note: Following question number 4 and 5 are for New Syllabus only [TAS-204/MA-202(New)]

4. Attempt any two parts of the following: (10 x 2 = 20)

(a) Find the solution(s) using the power series method $x\frac{d^2y}{dx^2} + \frac{dy}{dx} + x^2 y = 0$

(b) Establish any two of the following recurrence formulae for the Legendre polynomials:

(1) $P_n(x) = \frac{1}{2^n \underline{|n}\, dx^n}(x^2 - 1)^n$

(2) $(x^2 - 1) P_n'(x) = n\{xP_n(x) - P_{n-1}(x)\}$

(3) $(n + 1) P_{n+1}(x) = (2n + 1) x P_n(x) - n P_{n-1}(x)$

(c) Prove any two of the following recurrence formulate for the Bessel's function $J_n(x)$:

(1) $2n J_n(x) = x (J_{n+1} + J_{n-1})$

(2) $J_{n+3}(x) + J_{n+5}(x) = \frac{2}{x}(n + 4) J_{n+4}(x)$

(3) $x^2 J_n''(x) = (n^3 - n - x^2) J_n(x) + x J(x)_{n+1}$

5. Solve any two of the following: (2 x 10 = 20)

(a) Derive the one dimensional wave equation for vibrating string under suitable conditions.

(b) Describe the method of separation of variables for solving a partial differential equation. Hence solve the one dimensional heat equation

$$\frac{\partial^2 u}{\partial x^2} = \frac{1}{c^2}\frac{\partial u}{\partial t}$$

under suitable initial and boundary conditions.

(c) Characterize the following partial differential equation into elliptic, parabolic and hyperbolic equations

$$A\frac{\partial^2 u}{\partial x^2} + 2B\frac{\partial^2 u}{\partial x \partial y} + C\frac{\partial^2 u}{\partial y^2} = F(x, y, u, u_x, u_y)$$

Here A, B, C may be functions of x and y.

. **Note:** Following question Numbers 4 and 5 are for Old Syllabus only (MA-202 (Old).

4. Attempt any two parts of the following:

(a) Evaluate the intergral by changing the order of integration

$$\int_0^{2a}\int_0^{\sqrt{2ay-y^2}} (x^2 + y^2)\, dx\, dy$$

(b) Evaluate $\iiint (x^2 + y^2 + z^2)\, dz\, dy\, dx$ over the volume enclosed by x = 0, y = 0, z = 0 and the plane x + y + z = p.

(c) Apply the Dirichlet's integral to find the mass of a sphere $x^2 + y^2 + z^2 = a^2$ where the density at any point being $\rho = kx^2y^2z^2$.

5. Attempt any two parts of the following: (10 x 2 – 20)

(a) Solve by Cardon's method $8x^3 - 9x^2 + 1 = 0$

(b) Fit a second degree parabola to the following data taking x as independent variable:

x	1	2	3	4	5	6	7	8	9
y	3	7	8	9	11	12	13	14	15

(c) Using the method of least square fit a linear relation of the form P = a + bw for the following data:

W(kg)	50	70	100	120
P(kg)	12	15	21	25

P being the pull required to lift a weight W by pulley block. Estimate P when W is 150 kg from this relation.

2005-2006

Note: (1) Attempt ALL questions.
(2) All questions carry equal marks.
(3) In case of numerical problems assume data wherever not provided.
(4) Be precise in your answer.

1. Attempt any four parts of the following: (5 x 4 = 20)

(a) Solve the following differential equation:

$$\frac{d^3y}{dx^3} - 3\frac{d^2y}{dx^2} + 4\frac{dy}{dx} - 2y = e^x + \cos x$$

(b) Solve $x^2\,dy + y(x + y)\,dx = 0$

(c) Solve the following simultaneous differential equations:

$$\frac{dx}{dt} + 4x + 3y = t$$

$$\frac{dx}{dt} + 2x + 5y = e^t$$

(d) Solve the following differential equation by changing the independent variable

$$x\frac{d^2y}{dx^2} + (4x^2 - 1)\frac{dy}{dx} + 4x^3y = 2x^3.$$

(e) Solve the following differential equation by method of variation of parameters

$$x^2\frac{d^2y}{dx^2} + x\frac{dy}{dx} - y = x^2 e^x.$$

(f) The equations of electromotive force in terms of current i for an electrical circuit having resistance R and a condenser of capacity C, in series, is

$$E = Ri + \int\frac{i}{C}\,dt.$$

Find the current i at any time t, when $E = E_0 \sin \omega t$.

2. Attempt any four parts of the following: (5 x 4 = 20)

(a) If $L\{F(t)\} = f(s)$ and $g(t) = \begin{cases} f(t-a), & t>a \\ 0, & 0<t<a \end{cases}$

then prove that $L\{g(t)\} = e^{-as} f(s)$

(b) If $L\{\cos^2 t\} = \dfrac{s^2+2}{s(s^2+4)}$ find $L\{\cos^2 at\}$

(c) State convolution theorem and hence evaluate

$$L^{-1}\left\{\frac{s^2}{(s^2+a^2)(s^2+b^2)}\right\}$$

(d) Using Laplace transform, find the solution of the initial value problem.

$$\frac{d^2y}{dt^2} + 9y = 6\cos 3t$$

where $y(0) = 2$, $y'(0) = 0$

(e) Solve the simultaneous differential equations by Laplace transform

$Dx - y = e'$

$Dy + x = \sin t$

(f) A function f(t) obeys the equation

$$f(t) + 2\int_0^t f(t) = \cos h\, 2t$$

Find the Laplace transform of f(t).

3. Attempt any two parts of the following: (10 x 2 = 20)

(a) Find the Fourier series of the function defined as

$$f(x) = \begin{bmatrix} x+\pi, & \text{for } 0 \le x \le \pi \\ -x-\pi, & \text{for } -\pi \le x < 0 \end{bmatrix}$$

and $f(x + 2\pi) = f(x)$

(b) Find the Fourier half range cosine series of the function

$$f(t) = \begin{cases} 2t, & 0<t<1 \\ 2(2-t), & 1<t<2 \end{cases}$$

(c) (i) Solve $(p^2 + q^2)\,y = qz$ where p and q are usual notations of P.D.E.

(ii) Solve the linear partial differential equation

$$\frac{\partial^2 z}{\partial x^2} + 2\frac{\partial^2 z}{\partial x \partial y} + \frac{\partial^2 z}{\partial y^2} = \sin(2x + 3y)$$

OR

(i) Find the Fourier transform of

$$f(x) = \begin{cases} 1, & |x| < a \\ 0, & |x| > a \end{cases}$$

(ii) Find z-transform of $\sin \alpha k$ where $k \ge 0$.

Note: Following Question Number 4 and 5 are for New Syllabus only

4. Attempt any two parts of the following: (10 x 2 = 20)

(a) Solve the following differential equation in series:

$$(1 - x^2)\frac{d^2y}{dx^2} - x\frac{dy}{dx} + 4y = 0$$

(b) Show that

(i) $nP_n = x\,P_n' - P_{n-1}'$

(ii) $\int_{-1}^{+1}(1 - x^2)\,P_m'\,P_n'\,dx = 0, m \neq n$

where dashes denote differentiation w.r.t. x,

(c) Prove that

(i) $xJ_n'(x) = -n\,J_n(x) + x\,J_{n-1}(x)$

(ii) $x^2\,J_n''(x) = (n^2 - n - x^2)\,J_n(x) + xJ_{n+1}(x), n = 0, 1, 2, \ldots\ldots\ldots$

5. Attempt any two parts of the following: (10 x 2 = 20)

(a) Using the method of separation of variable, solve

$$\frac{\partial u}{\partial x} = 2\frac{\partial u}{\partial t} + u$$

where $u(x, 0) = 6e^{-3x}$.

(b) Determine the solution of one dimensional heat equation

$$\frac{\partial u}{\partial x} = C^2\frac{\partial^2 u}{\partial x^2}$$

Subject to the boundary conditions $u(0, t) = 0$, $u(l, t) = 0$ $(t > 0)$ and the initial condition $u(x, 0) = x$, l being the length of the bar.

(c) If a string of length l is initially at rest in equilibrium position and each of its point given the velocity

$$\left(\frac{dy}{dt}\right)_{t=0} = b\sin^3\frac{\pi x}{l}, \text{ find the displacement } y(x, t).$$

Note: Following Question Number 4 and 5 are for Old Syllabus only (MA-202 Old)

4. Attempt any two parts of the following: (10 x 2 = 20)

(a) Evaluate the integral

$$\int_0^{2a}\int_0^{\sqrt{2ax - x^2}}(x^2 + y^2)\,dy\,dx$$

(b) Evaluate $\iiint xyz\,dz\,dy\,dx$ over the volume enclosed by the planes $x = 0$, $y = 0$, $z = 0$, $x + y + z = a$.

(c) Apply Dirichlet's integral to find the mass of an octant of the ellipsoid $\frac{x^2}{a^2} + \frac{y^2}{b^2} + \frac{z^2}{c^2} = 1$, the density at any point being $\rho = kxyz$

(c) Apply Dirichlet's integral to find the mass of an octant of the ellipsoid $\frac{x^2}{a^2} + \frac{y^2}{b^2} + \frac{z^2}{c^2} = 1$, the density at any point being $\rho = kxyz$

5. Attempt any two parts of the following: (10 x 2 = 20)

(a) Solve the following by Cardon's method

$x^3 - 3x^2 + 12x + 16 = 0$

(b) By the method of least squares, find the curve $y = ax + bx^2$ that best fits the following data:

x	1	2	3	4	5
y	1.8	5.1	8.9	14.1	19.8

(c) Solve $x^4 - 10x^3 + 35x^2 - 50x + 24 = 0$ by Ferrari's method.

2004-2005

Note: (1) Attempt All questions, all questions carry equal marks.
(2) The choice of questions is internal as indicated in each questions.
(3) In case of numerical problem assume data wherever not provided.

1. Attempt any four of the following: (5 x 4 = 20)

(a) Solve the equation

$$(2x^2 + 3y^2 - 7)\, x\, dx - (3x^2 + 2y^2 - 8)\, y dy = 0$$

(b) Solve $x dx + y\, dy = \dfrac{a^2\,(x dy - y dx)}{x^2 + y^2}$

(c) Solve $(D^2 - 2D + 1)\, y = xe^x \sin x$

(d) Solve the following simultaneous equation:

$$\frac{d^2x}{dt^2} - 3x - 4y = 0,\ \frac{d^2y}{dt^2} + x + y = 0$$

(e) Apply the method of variation of parameters to solve

$$\frac{d^2y}{dx^2} - 3\frac{dy}{dx} + 2y = \frac{e^x}{1 + e^x}$$

(f) The damped LCR circuit is governed by the equation $L\dfrac{d^2 Q}{dx^2} + R\dfrac{dQ}{dt} + \left(\dfrac{1}{C}\right) Q = 0$ where L, R, C are positive constants. Find the conditions under which the circuit is over-damped, underdamped and critically damped. Find also the critical resistance.

2. Attempt any two of the following: (10 x 2 = 20)

(a) Solve the following equation in power series about $x = 0$.

$$2x^2\frac{d^2y}{dx^2} + x\frac{dy}{dx} - (x + 1)\, y = 0$$

(b) Show that $(1 - 2xz + z^2)^{-1/2} = \sum_{n=0}^{\infty} z^n\, P_n\,(x)$

where $P_n\,(x)$ are Legendre polynomials of degree n.
Hence or otherwise prove that

(i) $P_{2m+1}\,(0) = 0.$

(ii) $P_{2m}\,(0) = (-1)^m \dfrac{(2m)!}{2^{2m}\,(m!)^2}$

(c) Prove that

(i) $\dfrac{d}{dx}\{x^n J_n\,(x)\} = x^n\, J_{n-1}(x)$

(ii) $\frac{d}{dx}\left(J_n^2(x) + J_{n+1}^2(x)\right) = 2\left(\frac{n}{x}J_n^2(x) - \frac{n+1}{x}J_{n+1}^2(x)\right)$

where $J_n(x)$ is a Bessel function of the first kind of order n.

3. Attempt any four of the following: (5 x 4 = 20)

(a) If L[F (t)] = f (s) then prove that $L\left[\frac{F(t)}{t}\right] = \int_s^\infty f(x)\,dx$

Provided the integral exists.

Using this, find the Laplace transform of $\frac{\sin \alpha t}{t}$.

(b) If L [F (t)] = f (s) then prove that

$L[t^n F(t)] = (-1)^n \frac{d^n}{ds^n} f(s)$, for n = 1, 2, 3,.........

Hence or otherwise show that $\frac{1}{s^{1/2}} = L\left[\frac{1}{\sqrt{\pi t}}\right]$.

(c) Find be inverse Laplace transform of

$\frac{5s+3}{(s-1)(s^2+2s+5)}$.

(d) Using convolution theorem, prove that

$L^{-1}\left[\frac{1}{s^3(s^2+1)}\right] = \frac{t^2}{2} + \cos t - 1.$

(e) Apply Laplace transform to solve the equation

$\frac{d^2y}{dt^2} + y = t\cos 2t$, t>0, given that

$y = \frac{dy}{dt} = 0$ for t = 0

(f) Solve the differential equation using Laplace transform method:

$\frac{\partial y}{\partial t} = 3\frac{\partial^2 y}{\partial t^2}$ where $y\left(\frac{\pi}{2}, t\right) = 0$

$\left(\frac{\partial y}{\partial x}\right)_{x=0} = 0$ and y(x, 0) = 30 cos 5x.

4. Attempt any two of the following: (10 x 2 = 20)

(a) Prove that

$x \sin x = 1 - \frac{1}{2}\cos x - \frac{2\cos 2x}{1.3} + \frac{2\cos 3x}{2.4} + \frac{2\cos 4x}{3.5} + \ldots\ldots$

$-\pi < x < \pi$.

Hence show that

$$\frac{\pi}{4} = \frac{1}{2} + \frac{1}{1.3} - \frac{1}{3.5} + \frac{1}{5.7} - \ldots\ldots.$$

(b) Find a Fourier series for the function defined by:

$$f(x) = \begin{cases} -1, & \text{for } -\pi<x<0 \\ 0, & \text{for } x = 0 \\ 1, & \text{for } 0 < x < \pi \end{cases}$$

Hence prove that

$$\frac{\pi}{4} = 1 - \frac{1}{3} + \frac{1}{5} - \frac{1}{7} + \frac{1}{9} \ldots..$$

(c) Solve the partial differential equation:

$$\frac{\partial^2 z}{\partial x^2} - 3\frac{\partial^2 z}{\partial x \partial y} + 2\frac{\partial^2 z}{\partial y^2} = e^{2x-y} + e^{x+y} + \cos(x + 2y).$$

5. Attempt any two of the following: (10 x 2 = 20)

(a) Solve the following equation $\frac{\partial^2 z}{\partial x^2} - 2\frac{\partial z}{\partial x} + \frac{\partial z}{\partial y} = 0$

by the method of separation of variables

(b) Solve completely the equation:

$$\frac{\partial^2 y}{\partial t^2} = C^2 \frac{\partial^2 y}{\partial x^2},$$

representing the vibrations of a string of length l, fixed at both ends, given that $y(0, t) = 0$, $y(l, t) = 0$;

$$y(x, 0) = f(x) \text{ and } \frac{\partial}{\partial t} y(x, 0) = 0,\ 0 < x < l.$$

(c) An insulated rod of length l has its ends A and B maintained at 0°C and 100°C respectively until steady state conditions prevail. If B is suddenly reduced to 0°C and maintained at 0°C, find the temperature at a distance x from A at time t.

2003-04

Time : 3 Hours

Note: (1) Attempt ALL questions.
(2) The choice of questions is internal as indicated in each questions.
(3) All questions carry equal marks.

1. Attempt any four of the following:

(a) Solve the equation:
$(y \log y)\, dx + (x - \log y)\, dy = 0$

(b) Solve the following differential equation:
$(D^2 - 4D + 4)\, y = 8x^2 e^{2x} \sin 2x$

(c) Using variation of parameters method, solve:

$$x^2 \frac{d^2y}{dx^2} + \frac{dy}{dx} - 12\, y = x^3 \log x$$

(d) Find the solution of the equation: $y'' - 4xy' + (4x^2 - 2)\, y = 0$,
given that $y = e^{x^2}$ is a solution.

(e) Solve the following simultaneous equations:

$$\frac{d^2x}{dt^2} + y = \sin t, \quad \frac{d^2y}{dt^2} + x = \cos t$$

(f) A body executes damped forced vibrations given by the equation:

$$\frac{d^2x}{dt^2} + 2k\frac{dx}{dt} + b^2x = e^{-kt} \sin nt$$

Solve the equation, when $n^2 = b^2 - k^2$.

2. Attempt any four of the following: (5 x 4 = 20)

(a) Solve the following equation in power series:

$$2x(1 - x)\frac{d^2y}{dx^2} + (1 - x)\frac{dy}{dx} + 3y = 0$$

(b) Prove the Rodrigues formula: $P_n(x) = \frac{1}{2^n \underline{|n}} \frac{d^n}{dx^n}(x^2 - 1)^n$.

where $P_n(x)$ is Legendre Polynomial.

(c) Prove $\frac{d}{dx}\left(x^n J_{n-1}(x)\right) = x^n J_{n-1}(x)$

where $J_n(x)$ is a Bessel function of the first kind of order n. Further, show that

$$\int x J_0^2(x)\, dx = \frac{1}{2} x^2 [J_0^2(x) + J_1(x)].$$

3. Attempt any four of the following: (5 x 4 = 20)

(a) Find the Laplace transform of:

$$\frac{\cos at - \cos bt}{t}$$

(b) Find the Laplace inverse transform of:

$$\frac{2s^2 - 6s + 5}{s^3 - 6s^2 + 11s - 6}$$

(c) Solve the differential equation using Laplace transform method:

$$\frac{d^2y}{dt^2} + 2\frac{dy}{dt} + 5y = e^t \sin t$$

where $y(0) = 0$, $y'(0) = 1$

(d) Use Laplace transform to solve: $\frac{dx}{dt} + y = \sin t$, $\frac{dy}{dt} + x = \cos t$ given that $x = 2$, $y = 0$ at $t = 0$

(e) Use Convolution theorem to evaluate:

$$L^{-1}\left[\frac{s^2}{(s^2 + a^2)(s^2 + b^2)}\right].$$

(f) Find the Laplace transform of the square wave function of period a defined as:

$$f(t) = \begin{cases} 1, & 0 \le t \le a/2 \\ -1, & a/2 < t < a \end{cases}$$

4. Attempt any two of the following: (10 x 2 = 20)

(a) Prove that $x^2 = \frac{\pi^2}{3} + 4\sum_{n=1}^{\infty}(-1)^n \frac{\cos nx}{n^2}$, $-\pi<x<\pi$

hence show that $\frac{\pi^2}{12} = \frac{1}{1^2} - \frac{1}{2^2} + \frac{1}{3^2} - \frac{1}{4^2} + \ldots\ldots$

(b) Express $f(x) = x$ as a half – range sine series in $0 < x < 2$.

(c) Solve the partial differential equation.

$$\frac{\partial^2 z}{\partial x^2} - 2\frac{\partial^2 z}{\partial x \partial y} + \frac{\partial^2 z}{\partial y^2} = \sin x$$

5. Attempt any two of the following: (10 x 2 = 20)

(a) A string is stretched and fastened to two points at a distance l apart. Motion is started by displacing the string in the form $y = a \sin\left(\frac{\pi x}{l}\right)$ from which it is released at time $t = 0$. Show that the displacement of any point at a distance x from one end at time t is given by:

$$y_{(x,\,t)} = a \sin \frac{\pi x}{l} \cos \frac{\pi ct}{l} t.$$

(b) Solve $\frac{\partial^2 u}{\partial x^2} + \frac{\partial^2 u}{\partial y^2} = 0$, subject to the conditions $u(0, y) = u(l, y) = u(x, 0) = 0$ and

$$u(x, a) = \sin \frac{n\pi}{l} x.$$

(c) An insulated red of length l has its ends A and B maintained at 0°C and 100°C respectively until steady state conditions prevail. If B is suddenly reduced to 0°C and maintained at 0°C, find the temperature at a distance x from A at time t.